# A GREAT ARIDNESS

# A GREAT ARIDNESS

CLIMATE CHANGE
AND THE FUTURE
OF THE AMERICAN
SOUTHWEST

WILLIAM
DEBUYS

**OXFORD**
UNIVERSITY PRESS

# OXFORD
UNIVERSITY PRESS

Oxford University Press is a department of the University of Oxford.
It furthers the University's objective of excellence in research,
scholarship, and education by publishing worldwide.

Oxford    New York

Auckland    Cape Town    Dar es Salaam    Hong Kong    Karachi
Kuala Lumpur    Madrid    Melbourne    Mexico City    Nairobi
New Delhi    Shanghai    Taipei    Toronto

With offices in

Argentina    Austria    Brazil    Chile    Czech Republic    France    Greece
Guatemala    Hungary    Italy    Japan    Poland    Portugal    Singapore
South Korea    Switzerland    Thailand    Turkey    Ukraine    Vietnam

Oxford is a registered trade mark of Oxford University Press
in the UK and certain other countries.

Published in the United States of America by
Oxford University Press
198 Madison Avenue, New York, NY 10016

© William deBuys 2011

First issued as an Oxford University Press paperback, 2013.

Library of Congress Cataloging-in-Publication Data
DeBuys, William Eno.
A great aridness : climate change and the future of the
American southwest / William deBuys.
p.   cm.
Includes bibliographical references and index.
ISBN 978-0-19-977892-8 (hardcover); 978-0-19-997467-2 (paperback)
1. Climatic changes—Southwest, New.    2. Water-supply—Effect of
global warming on—Southwest, New.    3. Droughts—Southwest, New.
4. Southwest, New—Environmental conditions.    I. Title.
QC903.2.U6D43 2011
551.6979—dc23        2011033298

*for Joanna*

# CONTENTS

# MAPS

# PREFACE

Westerners call what they have established out here a civilization, but it would be more accurate to call it a beachhead. And if history is any guide, the odds that we can sustain it would have to be regarded as low.

Marc Reisner, *Cadillac Desert*

THIS IS A book about the warming of the globe where it matters most to the people of the United States. Climate change is transforming the polar regions of North America fastest—melting ice, thawing permafrost, and drowning polar bears—but for most people who live in the Lower 48, those transformations are nearly as remote as a tsunami hitting Borneo. While we lament such calamities at a distance, the warming climate is working changes closer to home that promise sweeping transformations. The place where those changes might best be observed is a region already straining from rapid growth, whose water resources are stretched to the utmost—the aridlands of the North American Southwest. What happens under the turquoise skies of the continent's most celebrated landscapes will presage changes that people the world over can expect to experience.

In that land of exceptional beauty and complexity, the remotest arroyos and mountaintops bear witness to a changing climate. They give their testimony in the form of vanishing wildlife, insect outbreaks, tree-ring patterns, the dust-blown ruins of ancient villages, and the behavior of wildfires (which filled southwestern skies with smoke and set new records for destruction in the summer of 2011). The story they tell is already reshaping the politics of the Colorado River, on whose water nearly 30 million people depend. It also adds a subtext to the violence and human suffering along the U.S.–Mexico border. Adapting to the changes now under way will ultimately require the metamorphosis of such cities as Phoenix and Tucson, if those cities are to preserve a modicum of the quality of life they now possess.

Readers of the following narrative will visit quite a few of those arroyos and mountaintops, often in the company of the foresters, archaeologists, and landscape ecologists who have teased out their secrets. They will navigate the rapids of the Colorado River, hike migrant trails along the border, and glimpse rare wildlife in the battered desert grasslands of Chihuahua and atop Arizona's Mount Graham, perhaps the most "biopolitical" mountain on the continent.

The Southwest is hardly alone in its predicament. It stands as proxy for the large portion of the world that will experience similarly powerful impacts from climatic change. The die-off of forests in the Mediterranean Basin, uncontrollable fires in Australia and Russia, floods in Pakistan, and the drying-out of southern Africa are all part of the global dynamic. What sets the Southwest apart, endowed as it is with abundant financial, human, and technological resources, is that nowhere else will the drama will be acted out more elaborately or reported more thoroughly.

The story that follows is about *now* and about *soon* in a region that is already in the throes of rapid change. By its reactions and adaptations, this small corner of the world will establish patterns that other regions will soon be striving to copy—or avoid.

A GREAT ARIDNESS

# INTRODUCTION:

# THE TRACKS AT CEDAR SPRINGS

MARCH 1919. SOMEBODY killed the trader at Cedar Springs. The murderer also set fire to the trading post, and soon the dried-out floor planks and the roof beams and split-cedar ceiling erupted in smoke and flame. Except for its sandstone walls, the building would have burned to the ground. The next day a plume of smoke still hung in the Arizona sky, and people from miles around came to see what was the matter. Few automobiles had reached the windy expanses of the Navajo Reservation in those days, and men and women of all ages filed in by foot and horseback and in their buggies and buckboards. The first ones to arrive found the corpse of Charley Hubbell.

Cedar Springs was less than a crossroads, not really a hamlet. It stood within a circle of rough-sided buttes a few miles east of the main road between Hopi and Winslow. It consisted of a combined general store, trading post and post office, plus various pens and corrals. Navajos from a considerable distance came to exchange sheep and handcrafts for the things they needed. Charley Hubbell lived at the trading post and presided over what went on.

The people who found his body dispatched a messenger on a fast horse toward Ganado, fifty miles northeast, to find Charley's nephew Lorenzo, who was known throughout Navajo country. A man of giant girth and matching gravitas, Juan Lorenzo Hubbell had bought the Ganado trading post forty years earlier. With acumen and persistence he built a string of almost thirty such establishments from one end of Navajo country to the other. In many cases he appointed members of his large, mixed Anglo and Hispanic family to run them. The post near Ganado, to which the messenger was sent, served as his home and headquarters. Together with its broad sprawl of barns, corrals, and storage sheds, it is preserved today as Hubbell Trading Post National Historic Site, a unit of the National Park System.

The Hubbell operation was a mercantile enterprise, not a charitable one, and its trading posts were outfitted accordingly. If the main counter where business was transacted at Cedar Springs was anything like its equivalent at Ganado, it presented physical testimony that the relations between traders and their clients sometimes grew testy. The counter would likely have been uncomfortably high, at least breast high for a man of average height standing in front of it, and the outside face—before which stood the semi-supplicant Navajos who brought their wares for exchange—would have been canted well past the vertical, like an overhanging cliff. The architecture was purposeful. If you were pawning your grandmother's silver or trading a year's worth of weaving, and if you wanted to leap over the counter and grab the throat of the steely-eyed *bilagana* who you thought was cheating you, you'd have been discouraged—first by the height of the counter, and second by the overhang. The first might have been surmountable if the second had not been in place, but you couldn't plant your foot against a surface that pointed toward the floor and so you couldn't get the lift you needed to scale the counter and get your man before he could knock you down with the barrel of the shotgun he kept behind the counter.

Right there in the silent design of the trading room was a wealth of commentary on class, race, and economic relations in the reservation world, and none of it need cast a negative light on Lorenzo Hubbell and his associates, who—let it be said—were esteemed by many Navajos. Hubbell no doubt would have pointed out that the best way to avoid trouble was to provide the least chance for letting it develop. But despite his precautions, in March 1919, the worst of all possible troubles had arisen. Someone had killed his uncle and almost certainly robbed the trading post, which was now a smoking ruin. And the killers were loose. As soon as he heard the news, Hubbell ordered his wagon to be readied.

He finally pulled in at Cedar Springs a day and a half after his uncle's murder. By then, a hundred or more people, both Navajo and white and maybe a few Hopi, had milled around the scene of the crime for hour upon hour. Dogs ran back and forth, and untethered horses drifted through. People ambled, strode, drove their wagons, and shuffled across the dusty terrain that held the only clues that might unlock the mystery of the crime and bring the guilty to justice. Hubbell surveyed the situation and climbed to a low perch on the ruins of the trading post. He spread his arms to quiet the crowd and then he said, "They did it. We will find the ones who did it. I want you all to go home."[1]

The murmuring crowd soon dispersed, leaving behind two Navajo men, Bohokishi Begay and Quinani, whom Hubbell had asked to stay. At Hubbell's behest, they set to work, heads down, sometimes bent over, speaking little. Over the next several hours they practiced an art whose depiction would later become a staple of B-level westerns and Hopalong Cassidy episodes. But the art Bohokishi Begay and Quinani were practicing was the real thing, not the repetition of an oater cliché. They were trackers; they read the history that was written in the ground.

In spite of the crowd's innumerable footprints, in spite of the scuffing, trampling, lollygagging, and milling about, in spite of the trails people had made coming to the trading post and the trails they made leaving it, and in spite of the culprits' strenuous efforts to erase their own tracks by sweeping the ground with branches and driving loose horses this way and that—in spite of all these distractions, erasures, and false clues, Bohokishi Begay and Quinani teased out from an acre or two of disturbed ground the thread of the story they sought. Their clues might have been a bent twig by a corral post or some grass with the dust knocked off behind the spring; they might have included a moccasin's imprint, pressed deep as though with urgency or the outline of a hoof that was chipped just so. From a chaos of physical facts, they selected the relevant fragments of a tale, and its contours emerged. Here the thieves loaded their horses with heavy saddlebags, there they walked back to set the trading post afire; here they rode out sweeping their tracks, there they changed course for their true destination. Once the trackers had identified the hoofprints of the departing horses, the easy part was to follow their slight, windworn trail northward. They followed it over rock and sand, down washes and up canyons, for eighty miles, finally coming to a hogan, where jaded horses stood outside. They entered through the flap of hide that served as a door. Two men sat within, saddlebags crammed with silver pawn on the floor beside them. The trackers said, "It looks like you killed the Mexican trader."

The capture was peaceful. The judgment and punishment of the two men were carried out according to the ways of the tribe, which neither Lorenzo Hubbell nor anyone else appears to have recorded.

THE INTELLIGENCE USED to solve Charley Hubbell's murder depended on fluency in a language that is now largely lost and forgotten, like Hittite or

Catawba. It was a language born from the wastes of the Painted Desert and the piney slopes of the Chuska Mountains, a language that developed out of intimacy with the land and everything that moved within it—wind, water, animals, people. The language served the core business of existence, which was survival. It interpreted a universe of signs whose meaning was shrouded by noise, chaff, and dissonance. Fluency enabled Bohokishi Begay and Quinani to read the landscape as you or I might read the newspaper.

The problem the two Navajo trackers faced in teasing meaning from the scuffed grounds of the trading post was cousin to the essential problem of history or science. In each case one confronts a sea of data so profligate as to conceal the patterns, large and small, that lie within it. The trick is to pull signal out of all that noise, to convert some portion of the data into verifiable information, establishing facts that can be connected to other facts (footprints made by the same moccasin), and then to draw inferences (this footprint is deeper; he was carrying something); and ultimately to generate and

RUINS OF CEDAR SPRINGS TRADING POST, WHICH WAS DESTROYED BY FIRE IN 1919 WHEN ITS PROPRIETOR, CHARLEY HUBBELL, WAS MURDERED. THE TRADING POST WAS REBUILT, THEN SUBSEQUENTLY ABANDONED. *AUTHOR PHOTO.*

test hypotheses (the thing he carried is what he stole; there should be deep footprints where he loaded his horse). Each hypothesis is the nucleus of a story, and gradually the stories accumulate and link to each other, merging into a master narrative that explains what happened.

Smart people in every culture proceed in this manner to reveal what they can of the mysteries of their day, or at least the mysteries they feel motivated and permitted to pursue. Among the "trackers" who are unraveling the *whodunits* or, more important, the *who's-gonna-do-its* of the contemporary world, climatologists are at the front of the posse. The trampled ground they study is the chaos of the weather—years, decades, and reconstructed centuries of it. The bent twigs and scuffs they translate into tracks consist of enormous quantities of measurements—of the temperature of sea and air, of the location and direction of wind and ocean currents, of precipitation at both local and regional scales, of the paths and behavior of storms. The trails some of them follow bear shorthand names like ENSO (the El Niño Southern Oscillation), PDO (the Pacific Decadal Oscillation), and AMO (the Atlantic Multi-decadal Oscillation). Others try to decipher the dynamics of Hadley cells, giant atmospheric circuits transferring heat from the tropics toward the poles; still others trace the pattern and timing of monsoons. Nowadays the hypotheses they generate—the climatological equivalent of "they went thataway"—bear sufficient weight and portent to influence the affairs of nations.

Sometimes the tracks are grizzly-sized. Already the Arctic icepack has shrunk to a fraction of its former size, and the lands of the Far North are in the throes of rapid change. Vast flights of birds are nesting two weeks early. Brown bears, the grizzlies of the far north, now rove where they never went before, and prey on musk oxen. Lakes and wetlands are draining as the permafrost they perch on melts. At stake are not just the stranded polar bears, starved for seals, that drown trying to swim to pack ice, but an entire biome that is slowly warming like an unplugged refrigerator. The contents will fill no one with delight.

After the Arctic and apart from coastal disasters like Hurricane Katrina (the Ur-location for overheated refrigerators), the North American Southwest promises to be center stage for the continent's drama of climate change. Most models predict that the Southwest will outstrip other regions in both the rate and the amount of change, and already data from the field suggest that the models are correct—except in one important respect: the changes are occurring faster than expected.[2]

Already temperatures in many areas are averaging 1°C (1.7°F) higher than the norm of the twentieth century. This includes the headwaters of the Colorado River, lifeline of the region, where warming has caused earlier melt of the snowpack, resulting in lower flows in the river. The implications for water security, symbolized by broad chalky bathtub rings around Lake Mead, Lake Powell, and other diminished reservoirs, are plainly dire. Forests, too, are hard hit. Not only have insect outbreaks (a symptom of a warmer world) resulted in die-offs across millions of acres, but the background rate of tree mortality in ostensibly healthy forests has been increasing rapidly, along with the occurrence of large wildfires, which set one dismal new record after another during the brutal fire season of 2011. Notwithstanding the steady din of climate-denial propaganda, all of these changes can be confidently linked to human-caused climate change.[3] The most sobering news, however, is that the changes are just beginning. In the future even higher temperatures are a virtual certainty. A decline in precipitation, both rain and snow, is strongly probable. Singly or together these two changes portend that substantially less water will be available for people, crops, trees, grass, and critters of all kinds. Even the unlikely advent of semi-generous rains won't bail out the region, because a hotter environment will lose more of its water to evaporation—leaving the aridlands effectively more arid. We are on the verge of a new form of desertification. It won't be the result of overgrazing, failed agricultural schemes, or other familiar forms of land abuse. Instead, it will ensue from industrial society's abuse of the atmosphere.

HOTTER AND DRIER are not the only themes that promise to shape the path ahead.

There is a third variable that bears watching as much as temperature and precipitation, and this, somewhat confusingly, is variability itself. A stable, predictable environment is easier to inhabit than a highly variable one. Agriculture, for instance, can adjust to the timing of annual rains, even if they are usually scanty. Calamity arises when the expected rains fail to fall, or when, as has been the case in past centuries in the Southwest, they fail for years in a row— exceeding the capacity of people to store what they need and withstand hard times. More than a few ancient pueblos collapsed for this reason.[4]

The climate of the Southwest has long been distinguished by its variability, but an emerging theme in current climatological research is that this characteristic is likely to increase, with the highs of temperature and precipitation potentially becoming higher, the lows lower, and the transition between the states more rapid.[5] In such a scenario the region might find itself afflicted with cold waves or heat waves that are equally crippling. Or it might experience drought and flood in quick succession, bringing fires on the one hand and massive erosion on the other, a one-two punch leaving large areas bereft of vegetation, and so deeply cut by arroyos that the hydrology of the land is changed, making recovery to anything resembling the former landscape impossible.

But not only the climate is changing. For decades leading up to the so-called Great Recession of 2007–2009 (its exact dates vary depending on the metric employed), the Southwest led the nation with meteoric growth. Extrapolating from pre-recession trends, the Brookings Institution predicted that the population of the five states of Arizona, Colorado, Nevada, New Mexico, and Utah would nearly double between 2005 and 2040.[6] Although the recession and particularly the collapse of the real estate industry have slowed migration, the "dry Sunbelt" remains poised to resume its expansion, and a key question attending its future is whether the pause imposed by the recession will be followed by wiser, more sustainable patterns of growth, or will yield to the same feverish free-for-all of recent years. An essential problem of continuous growth is obvious: with population expanding, collective demand for water and other natural resources commensurately climbs. Even now, the watersheds of the Rio Grande and Colorado River, the two most important renewable sources of freshwater in the region, are overallocated and hard pressed to serve their users. The arrival of millions of residents more, together with what they bring by way of new industry, construction activities, and energy demands (traditional forms of energy generation consume large quantities of water), bodes well to place the region in a perpetual state of shortage, even if the rivers keep flowing at historically "normal" levels. And if the effective yield of water declines, whether because of higher temperatures or less precipitation, or a likely combination of the two, southwesterners and their oasis-based, aridland, hydraulic society will have to reinvent themselves in new and undoubtedly painful ways.

IN APOCALYPTIC VISIONS of global climate change, the North American Southwest makes an easy protagonist, the geographical equivalent of a stalled car on railroad tracks with a speeding train approaching. Its aridity exposes it to shocks of all kinds. It has big cities dependent upon water infrastructure as colossal and centralized as any the world has produced. It has ecosystems susceptible to multiple afflictions that include catastrophic fire, insect infestation, broadscale die-off, plant invasion, and multiple combinations of the foregoing. Split by an international border separating rich from poor, mostly white from entirely brown, it also has a powder-keg potential for violence and jingoistic overreactions of all kinds.

Metaphorically speaking, the silhouetted riders just now appearing over the ridge are not the Seventh Cavalry bringing aid and deliverance, but a rougher crew that includes Drought, Fire, Pestilence, Discord, and other apocalyptic horsemen. They pose a threat to every community in their path. The sooner the people of those communities come to terms with them, the better their chances of continuing to live in the aridlands with a measure of grace.

Jonathan Overpeck, a climate scientist who codirects the Institute of the Environment at the University of Arizona, put matters simply. Climate change, he said, will produce winners and losers, and "in the Southwest we're going to be losers. There's no doubt."[7]

THAT NOT VERY cheerful thought is the germ of this book, which relies on Overpeck and several dozen other thoughtful and experienced guides to track, as it were, the climate-driven challenges that confront the Southwest.

The path ahead will not be straight. The long-term trajectory of warming and drying will be interrupted by wet spells and cold snaps, some of them spectacular. The basic physics of climate change work like this: greenhouse gases trap more of the heat that Earth would otherwise radiate back into space. The retained heat charges the atmosphere and oceans—the main drivers of the planetary climate system—with more energy, loading them with more oomph to do the things they already do, but more powerfully than before. In general, wet places will become wetter; dry places, like the Southwest, drier; and extraordinary events even more extraordinary—and more common. It is not for nothing that some say "global warming" is a misnomer; "global weirding" may get closer to the truth.

THIS IS A thematic book, not a comprehensive one. My hope is that it has acquired a form that fits the busy lives of its readers. To be sure, the stories that it tells are a small subset of the stories that might be told—a condition true of all books, even long ones. The accounts contained herein are those that have seemed to me, first, to represent fairly the breadth of the region's natural history and, second, to be most responsive to the influence of a changing climate.

What I have tried to do is to paint a living portrait of a region that, for my money, has no rival in beauty, richness, or fascination. This is a love story, of sorts. I came to the Southwest as a young man in 1972, and, although I spent a few years elsewhere in the early 1980s, I never really left, and certainly never wanted to. For virtually all of my adult life, the Southwest has been my home, and I have felt blessed to be its student. Part of that blessing has been the opportunity to roam a few (but never enough) of its wild places and to meet and sometimes work with a few (but never enough) of its talented and unusual people. They hail from a score or more of native tribes, multiple branches of Hispanic heritage, and so many other backgrounds that it is common to lump the non-Indian and non-Hispanic folks together in the catch-all term *Anglo*, which encompasses a United Nations of both genetics and irony. Missing from these three categories, broad as they must be, however, are the children of multiple ethnicities, such as the half-Anglo, half-Mexican, and the half-Apache, quarter-Hispano, quarter-Hopi, and all their cousin permutations, who in their multiculturalism have always been a core component of the population of the Southwest, and who are a burgeoning portion today. Part of what is so fascinating about the region is the culture and identity that are forever being forged here.

The physical landscape of the Southwest is as rich as its culture. One way of visualizing the region is to imagine an ecological core defined by two sets of "four corners." The better known is the northern Four Corners, where Utah, Colorado, Arizona, and New Mexico meet. A tiny unit of the Navajo Nation Tribal Park system celebrates the exact point of intersection, which is monumented by a concrete plinth where tourists contort themselves to place a part of their body simultaneously into each of the four states, while relatives or friends on an adjacent platform shout advice and snap pictures.

The lines that define the northern Four Corners are abstractions of the grid survey of the United States. While their point of intersection is arbitrary, it nevertheless marks the approximate center of the vast, canyon-furrowed upland called the Colorado Plateau, a land fabled for its beauty, severity, and geological nakedness. The Colorado Plateau is also an area of convergence where the Great Basin Desert of Utah touches the Rocky Mountains of Colorado. From there, both systems grade southward through the cold high steppes and scattered mountains of Navajo country, or Dinetah, toward the deserts of the Borderlands.

The second Four Corners lies in the heart of those Borderlands. It is less precise because the respective surveyors of Mexico and the United States had no need or desire to coordinate their efforts. But it is not less interesting. The juncture of the adjacent northern corners of the Mexican states of Sonora and Chihuahua lies about 10 miles east of the mutual southern corners of Arizona and New Mexico. The land where these two junctures narrowly miss each other is harsh and rugged, broken by canyons and sudden mountain fronts that hide the infrequent lakebed plain. In broad terms this little-visited convergence of political jurisdictions is also the meeting place of four of the great biotic systems of North America: the Sonoran Desert, the Chihuahuan Desert, the Rocky Mountains, and the Sierra Madre. If you had to pick one place in the entire continent where the greatest number of surprising plants and animals mingle in proximity to each other, you would do well to stick your pin in this part of the map. Or, better, you would draw a modest circle around it. Hardly moving the tip of your pen, you would envelop within your circle the iconic spiny marvels of saguaro cactus, ocotillo, and sotol, plus multiple species of agave and yucca; you would include the majestic ponderosa and its skinny southern cousin the Chihuahua pine; you would capture high-country Engelmann spruce and desert canyon Arizona sycamore. You would encircle squadrons of frantic hummingbirds drawn from a dozen species, raucous parrots, itinerant jaguars, a strangely isolated subspecies of pronghorn antelope, Mexican gray wolves (a particular subspecies, not just wolves from Mexico), Rocky Mountain elk, a small herd of wild bison of indeterminate origin, and at least three species of deer (white tail, Coues, and mule). You would also enclose whole regiments of rooting and grunting javelinas, which are a kind of wild pig; squads of coatimundis, which are like raccoons on amphetamines; herds of desert bighorn sheep; a liberal sprinkling of mountain lions; innumerable choruses of weird, throaty frogs; the rare ridgenosed rattlesnake; and the ghost of Cochise.

If we match our circle over the southern Four Corners with a similar one in the north, and then join the two circles in a plump ellipse, we will have a fair start toward a definition of the region. It happens that the result will fairly neatly contain the homeland of the settled people of the medieval Southwest—the Hohokam in the area of present-day Phoenix and Tucson and the ancestral Puebloans from Kayenta and Mesa Verde in the north down to Casas Grandes in Chihuahua. Our ellipse is the heartland of the Southwest. From within it we can range outward, to California, to Texas, or deep into Mexico, along any trail we care to follow.

THE TRAILS WILL lead to places high and low, to the conifer forests of the Rocky Mountains and to the dried-out delta of the Colorado River. They will lead to the sidewalk-and-gutter, two-car-garage subdivisions of Las Vegas and to the raw cinderblock and shipping-crate *colonias* of Ciudad Juárez. They will lead to backcountry wilderness as well as ordinary backyards, to mile-deep canyons and to canals hundreds of miles long. They will also lead to ideas and themes that are fairly consistent from one instance to another. One of these is the notion that *no big thing happens for just one reason.* This may sound like a trivial matter, but in fact it is quite important. Notwithstanding the attraction of tidy, sound-bite-ready, just-so stories, most things happen not for one reason, but for many. It takes a lot to destabilize, let alone dismantle, a civilization—like that which once dwelled at Chaco Canyon or Mesa Verde—and it also takes a lot to cause an otherwise stable and self-replicating forest community to begin behaving in self-extinguishing ways in which it rarely or never behaved before. The more one looks at any situation, the more one marvels at the interlocking gyre of causes and effects. Part of the beauty of the world, even including its disasters, is its complexity.

Almost inevitably, wherever multiple causes are gathered, one or more will prove to have been produced by humans, a fact which leads to another theme: *the human contribution to change in the natural world more often catalyzes than dictates the outcome.* It energizes and speeds the interaction of forces that are already present and at work, rather than deciding the result outright and single-handedly. We see this in the "flip" of desert grassland to mesquite and other shrubs, where overgrazing helped bring about vast landscape changes in the space of a few generations that otherwise might have taken centuries. The idea that *it would have happened anyway,* however, should console no one. There is always chance. No set of potentials has just one outcome. If humans had not influenced the desert grasslands as they did, they

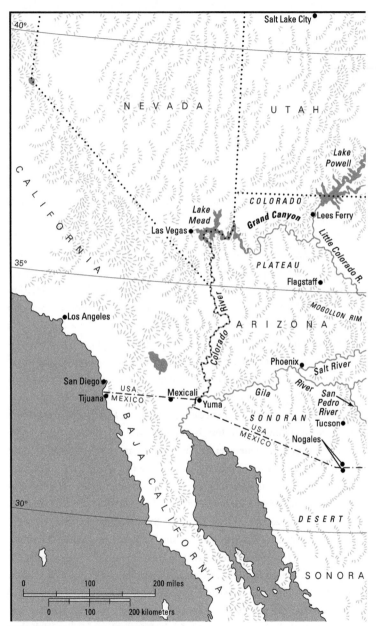

**MAP 1: THE NORTH AMERICAN SOUTHWEST.**

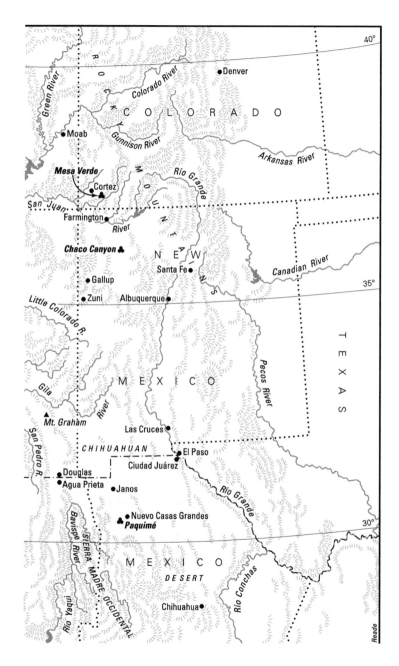

40°

● Denver

Green River

R O C K Y

Colorado River

C O L O R A D O

● Moab

Gunnison River

Arkansas River

M O U N T A I N S

**Mesa Verde**
▲ Cortez

Río Grande

San Juan

Farmington ●
River

**Chaco Canyon ♣**

N E W

Santa Fe ●

Canadian River

● Gallup

35°

● Zuni    Albuquerque ●

Little Colorado R.

Gila

M E X I C O

T E X A S

River

Pecos River

▲ Mt. Graham

● Las Cruces

San Pedro R.

*CHIHUAHUAN*

● El Paso

● Douglas
● Agua Prieta    ● Janos

Ciudad Juárez

Río Grande

Bavispe River

SIERRA

● Nuevo Casas Grandes
**Paquimé ♣**

30°

M E X I C O

MADRE

*D E S E R T*

Río Conchas

Río Yaqui

OCCIDENTAL

Chihuahua ●

Reade

15

would surely have influenced them in a different way. Indeed, they did so for eons before Europeans arrived in North America, by collecting mesquite beans for food, possibly on a very large scale. Long ago, in such a case, with a different catalyzing influence, yet another result ensued. The forces latent in nature have the potential to move ecological systems toward multiple future states; human activities help select the direction of the move.

A third theme to be found in the pages that follow concerns the *enormity of human capacity for adaptation*. At various times and under various circumstances—from the great droughts of the medieval era to the cardboard shanties of today's Mexicali—simply surviving in the Southwest has been a signal achievement. But the people of the Southwest have done far more than merely survive. They have erected a great hydraulic civilization in one of the most intimidating environments on the continent. Vast quantities of ingenuity, wealth, and cooperation have made this achievement possible. It is a heritage fit to inspire both pride and emulation. Today the robustness of the Southwest's past achievements is perhaps matched by its vulnerability, and the people of the region will need to call heavily on the same qualities that allowed their past successes if they are to meet the challenges of the decades to come.

The answers to these challenges will be plural, not singular, but no answer will count for much—will, in fact, be an answer—if it is not backed by strong social will and collective commitment. A professional class of policy wonks, land managers, and water buffaloes—the usual suspects to whom the public defers on environmental matters—cannot handle this set of problems on their own. There needs to be broad debate and broader understanding. There needs to be a citizenry aware of the stakes. This book is written for them. It is an attempt to peer into the future of the Southwest, to embark on a quest, in a way, for some hint of the character of a new and still-forming land. It is a land under threat, but it is also a land blessed with abundant human and natural capital, a land that can provide a hopeful example for other similarly threatened regions around the globe.

# 1

# HIGH BLUE: THE GREAT DOWNSHIFT OF DRYNESS

MAPMAKERS TYPICALLY DEPICT the aridlands of the world in colors like buff and buckskin, in contrast to the greens of wetter regions. Their choice is true to reality, for dry places usually produce scant vegetation, and the bare ground, baked by unobstructed sun, tends to wear a washed-out shade of dun, or one of its cousins. In the North American Southwest, you might add a touch of rust to reflect the widespread iron-rich geology. In many areas, oxides of iron produce the pinkish flesh tones that make it easy to think the landscape is alive. If you also brush in some piney greens and spruce black for upland woods and forests, and dab smaller areas white to represent high-country snowcaps, you have a fair start toward capturing the palette of the region.

But you would still be missing the most definitive color of the Southwest, which is found not beneath the feet, but overhead. You can look up, straight up, almost any day of the year, and there it is: an intense, infinite blue, miles deep and beyond reach. It is not merely bluish, not the watery blue of Scandinavian eyes, not the black-mixed blue of dark seas or bachelor buttons, not the hazy blue of glacier ice or distant mountains, but an Ur-blue, an *über*-blue, a defining quintessence. It is to other blues as brandy is to wine: a distillation, pure and heady.

It can be a little deflating to reflect that the ethereal blue of southwestern skies results from mundane forces, that it is the product of solar radiation and atmospheric gases interacting in an environment shaped by climate. If the air held more water vapor, the sky would whiten overhead, as it does at the horizons, where the light that reaches our eyes has more atmosphere and diffusing vapor to travel through. Overhead the light we see is blue because the gases of the atmosphere allow reds and yellows to pass unobstructed, but absorb the shorter-wave blue portion of the sun's spectrum and bounce it around before reflecting it down to us. That's why the otherwise white sun appears yellow in a blue sky: blues and violets have been subtracted from its light.

It is an oversimplification to say that the blue of southwestern skies is an artifact of aridity, but aridity, to be sure, plays a major role. So does dust—or, rather, the absence of dust, most of the time—and so does altitude: the higher you go, the bluer the sky. The thread we need to follow here is aridity, and it traces back to the core dynamics that produce the planet's climate. It leads to laboratories where scientists have undertaken to model global climate and predict its future behavior. What they are learning does not bode well for the North American Southwest, but an understanding of what they have to say may prove vital to the region's survival. Their insights inform the pages that follow. But first, a few basics.

THE ARIDITY OF the Southwest is a function of one of the primary features of world climate, the so-called Hadley cell. Earth is unevenly heated; it receives the greatest part of its solar energy near the Equator. Tropical seas, which absorb much of the heat, discharge part of it by evaporating water into the hot air above them. The hot, moist air cools as it rises, condensing its water into clouds and then into vast amounts of rain, which fall on sea and land alike. The upwelling air and associated raininess is especially pronounced at the Intertropical Convergence Zone (ITCZ), where the trade winds of the two hemispheres blow into each other. On some satellite photographs, it looks like a belt of clouds around the world. The location of the ITCZ varies seasonally north and south of the Equator, drifting in the direction of the greatest solar radiation—which is to say, toward hemispheric summer. Because the movement of air in the ITCZ is more vertical than horizontal, mariners in the Age of Sail dreaded encountering it; they called it the Doldrums.

The wringing out of water from the rising air of the tropics produces the well-drenched jungles of low latitudes, places like Amazonia, the Congo Basin, and the Indonesian archipelago. Shedding their moisture, the air masses of the tropics rise until they can rise no more (they bump against the bottom of the stratosphere, but that's another story), then take their heat poleward: north in the Northern Hemisphere, south in the Southern.

Ocean currents also carry tropical heat to higher latitudes. Roughly half of the tropics' export of heat is transmitted via the oceans, the other half by air.

The air that rises from the equatorial regions eventually cools enough to descend, and its descent defines the lands on the poleward side of the tropics in both hemispheres. The descending air is dry, its moisture squeezed out during its earlier ascent, and it warms with compression as it falls. This great downwelling of dry air gives birth to many of the aridlands of our planet. South of the Equator, the Atacama Desert in Chile, the Kalahari in southern Africa, and the arid outback of Australia are among its children, all roughly sharing the same band of latitude. In the Northern Hemisphere, the Sahara and the deserts of Arabia have a similar kinship, as do the deserts of North America—the Sonoran, Chihuahuan, Mohave, and Great Basin. Along with their dry air and blue skies, they are the necessary inverse of the moist tropics, where their fate is largely cast. (To be sure, local factors, like the rain shadows of tall mountains, also play important roles.)

Meteorologists refer to the circulation of air masses rising from the tropics and descending in the subtropics as Hadley cells, after George Hadley, an eighteenth-century British lawyer whose ambition was to explain the existence of the trade winds. (He disagreed with the then prevailing theory of Edmond Halley, namesake of the famous comet.) The ideas Hadley proffered have not escaped revision, but his name remains fixed to the phenomenon of tropical air circulation.[1] The pattern Hadley described can be imagined as a giant air pump powered by heat, lifting air in the tropics, moving it poleward, and depositing it in the subtropics, whence it returns toward the Equator at low elevation. While the Hadley circulation endows the subtropics, in general, with aridity, there are also humid subtropical areas, like Florida and Southeast Asia, that have oceans to their east. Because of Earth's rotation, the low-altitude return of Hadley air to the tropics moves diagonally, east to west (hence the trade winds), picking up moisture from the ocean and bringing it downwind.[2] The Hadley circulation helps account for the blueness of southwestern skies: not only is the air dry, but the relative high pressure of its downwelling mass discourages the formation of clouds.

THE FATE OF the Southwest, then, is tied to the Northern Hemisphere Hadley cell, and the fate of the Hadley cell is tied to the heat budget of the planet, which is shaped by the chemistry of the atmosphere. Scientific understanding of that chemistry took a big step forward in 1958 when David Keeling, a thirty-year-old postdoc at the Scripps Institute of Oceanography, began measuring atmospheric $CO_2$ from a station atop Mauna Loa in Hawaii. Fairly quickly, Keeling made two important discoveries. He found that $CO_2$ levels varied with the seasons. There was measurably less in the air during the Northern Hemisphere summer and more during the winter. It was a surprising observation, but one that was soon explained. Since the Northern Hemisphere holds most of Earth's land mass and therefore most of its plant life, and since growing plants extract $CO_2$ from the air, it made sense that when the world was most leafed out, $CO_2$ should drop. No one had previously suspected this, let alone observed it; Keeling had not been looking for it. Yet his discovery was profound. Keeling had documented the annual breathing in and breathing out of the planet.

Keeling soon noticed something else: atmospheric values for $CO_2$ rose year by year. When he began his measurements, he detected $CO_2$ at an average value of 315 parts per million (ppm). Within a few years it was nudging 320. For the rest of his working life Keeling continued the measurements, and his successors carried on after he retired, and then after his death in 2005. In 2009 and 2010, readings from Mauna Loa, which by then were being corroborated by many similar stations around the world, crept above 390 ppm—a 24 percent increase in fifty-two years and a significant change in the chemistry of the atmosphere. Plotting these values produces what is called the Keeling Curve—an arc of gradual, relentless, and somewhat accelerating increase. Keeling's findings led the National Science Foundation to issue its first warnings on global warming, which were delivered to President Lyndon Johnson in 1963. Increased $CO_2$ in the atmosphere, the NSF said, would reduce the amount of solar energy Earth re-radiates back into space. Unable to give off as much heat as it used to, Earth would gradually warm.

Among the processes presumed to arise from the steady heating of Earth, one of the most widely accepted holds that the Hadley cells of both hemispheres will expand. With more heat—which is to say, more thermodynamic power—the engine of the Hadley cell will run faster and stronger, and the cell will grow larger. If today the downwelling of dry Hadley air masses centers along, say, latitude 30° N in northern Mexico, and if indeed

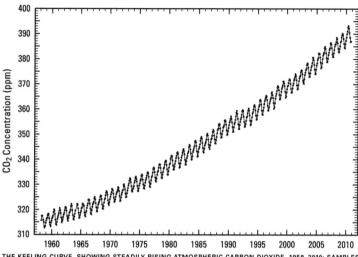

THE KEELING CURVE, SHOWING STEADILY RISING ATMOSPHERIC CARBON DIOXIDE, 1958–2010; SAMPLES OBTAINED AT MAUNA LOA, HAWAII. *COURTESY OF SCRIPPS $CO_2$ PROGRAM, LA JOLLA, CALIFORNIA.*

David Keeling's work inspired others to reconstruct the history of Earth's atmosphere in the distant past. Researchers analyzing samples from air bubbles within the Antarctic and Greenland ice sheets have learned that over the past 8,000 years atmospheric carbon crept slowly upward from 260 ppm to 275–280 ppm at the outset of the Industrial Revolution. The Holocene increase was likely due to gradual changes in both marine and terrestrial systems. The important thing to note is that "the rate of change of atmospheric $CO_2$ over the Holocene is two orders of magnitude smaller than the anthropogenic $CO_2$ increase since industrialization" (Indermühle, et al., 1999). The increase in atmospheric carbon from that point forward (275 to 390 ppm) now approaches 42 percent. This is the chemistry we know.

The chemistry we don't know—not with exactitude—is how much carbon and other greenhouse gases (methane and nitrous oxide) the atmosphere and oceans can absorb before their heat dynamics shift into radically different and essentially permanent patterns. One estimate

*Continued*

---

*Continued*

of what the atmosphere can tolerate belongs to James Hansen and nine coauthors, who presented their thinking in a 2008 paper entitled "Target Atmospheric $CO_2$: Where Should Humanity Aim?" Hansen, a scientist at NASA's Goddard Institute for Space Studies, has long been in the public eye because of his vigorous warnings about the perils of climate change and, more particularly, because of government efforts during the presidency of George W. Bush to silence him (see Mark Bowen, *Censoring Science*). The main point of the paper is made in language unusually direct for a scientific publication:

"If humanity wishes to preserve a planet similar to that on which civilization developed and to which life on Earth is adapted, paleoclimate evidence and ongoing climate change suggest that $CO_2$ will need to be reduced from its current 385 ppm [this was 2008] to at most 350 ppm, but likely less than that.... If the present overshoot of this target $CO_2$ is not brief, there is a possibility of seeding irreversible catastrophic effects" (James Hansen, et al., "Target Atmospheric $CO_2$: Where Should Humanity Aim?" *Open Atmospheric Science Journal* 2 [2008]: 217–231).

---

the Hadley cell expands, then years or decades from now the downwelling may center on the latitude of Tucson, Arizona, near 32° N, or, after that, Socorro, New Mexico, at 34° N. Some scientists say they have already detected evidence of such movement.[3] Indeed, some say that the Hadley cells of both hemispheres are expanding poleward faster than prevailing global change models predict. It is not a happy prospect. In any case, southwesterners who are curious about their climate can glimpse its future in the simplest of ways: they can look south.

THE GOOD NEWS, if you ask Richard Seager, an oceanographer at the Lamont-Doherty Earth Observatory in Palisades, New York, is that the Northern Hemisphere Hadley cell is not expanding as fast as some recent analyses have warned. Those projections, he says, depend upon comprehensive satellite data, which have been available only since the late 1970s, a brief slice of time. He thinks the observed Hadley cell expansion is better attributed to a naturally occurring decadal shift in Pacific Ocean currents.

The bad news, however, is that, because of Hadley cell expansion and other factors, the Southwest is moving—and to some extent has already moved—into a new set of climate conditions "unlike any climate state we have seen in the instrumental record." Seager thinks that the droughts of the past, like the Dust Bowl drought of the 1930s and the 2000–2004 drought that killed millions of piñons and ponderosas in New Mexico and Arizona, will be the norm of the future. He is saying, in effect, that what we think of as drought today will cease to be drought in the light of some not-too-distant tomorrow.

Drought is relative—and exceptional. One does not say that the Sahara is experiencing drought. The Sahara is dry by nature, not by exception. According to Seager, the future Southwest will be much drier than the conditions its present inhabitants consider "normal"; it will likely be as dry as its driest recent decades, and people will have to recalibrate their expectations to a new idea of what "normal" stands for. Moreover, he says that the droughts of the future will be superimposed on that dryness and that they will be unlike anything people in the region have known since late medieval times.[4]

Seager holds an endowed position at Lamont-Doherty, a research arm of Columbia University. He's a Brit who came to the United States for education and stayed. I visited him at his Morningside Heights apartment on a day when the lab was closed and his son Angus was home from school, contending with homework at the dining table. The humid smell of laundry was in the air. We had plenty of time to talk, but not an unlimited amount— an important football game (real football, played with the feet) would be televised from England that afternoon. Seager lives close to Columbia, only a block from the diner made famous by the sitcom "Seinfeld." He takes an amused pride in this link to American pop culture, marveling at the tourists posing for snapshots and the occasional celebrity guiding a tour. Seager is a bit of a cultural sponge. As we talked, he alluded frequently to the intricacies of Colorado River politics and to Dust Bowl histories and books on western water. Except for his accent, he could have been a good-old-boy insider.

Seager calls himself an oceanographer and meteorologist, avoiding the term *climatologist*, which suggests, to him, a duffer "with a cloth cap and a rain gauge in his garden." He came to the study of the Southwest "by pure chance, which is the way science goes, you know?" He and his team had been looking at Asian monsoon variability in the nineteenth century, going back to 1856, the year the British and American navies began recording air and sea surface temperatures and winds from all their ships. Other navies and merchant marines eventually followed suit, making it possible to compile a nearly global century-and-a-half record of oceanic weather. Seager and his team drew upon that data to model relationships between sea surface temperature and changes in atmospheric circulation. In 2004, Siegfried Schubert, a scientist at the nearby NASA Goddard Institute for Space Studies, a leading center for climate science, wrote a paper that linked the Dust Bowl drought to small temperature changes in surface waters of the tropical Pacific.[5] Seager had access to an advance copy. "When Schubert published that paper, and actually before it came out," says Seager, "I just looked in the model that day and said, well, does our model produce a Dust Bowl drought? And sure enough, it produced a Dust Bowl drought."

Further investigation showed it "made a nice big drought in the Southwest in the 1950s," which Schubert's simulation had missed, and it also showed the Southwest turning dry in 1998, the approximate onset of the drought that produced widespread tree die-offs in 2002–2003. Things were getting interesting. Seager checked the model against a new drought atlas that Lamont-Doherty's tree-ring researchers had just produced and found that they jibed: the model tracked the wet and dry periods recorded in the tree rings for the mid- to late nineteenth century with high fidelity. "So we just sort of dropped stuff, whatever we had been doing, and misused government funds to fund this research instead of what the government had actually given us funds to do." His program managers didn't complain, because they understood the importance and opportunities of the new work. "So we pushed it forward."

The forwardness was in two dimensions: forward in terms of refining an understanding of historical drought records and forward also in time, projecting climate for the Southwest into the twenty-first century. The Intergovernmental Panel on Climate Change (IPCC), which the United Nations established in 1988, had also done this. Modelers contributing to the IPCC's "Fourth Assessment Report," which appeared in 2007, concluded that the North American Southwest would grow appreciably drier in the twenty-first century, but their predictions focused far in the future, past 2080. They also concerned themselves solely with precipitation, not the finer-grained interaction of precipitation less evaporation, which better describes the moisture available to nourish soil and discharge into streams and rivers. "So there's a bit more to the story than the precipitation"—and Seager and his team set out to tell it.

Their simulations showed that the Southwest would not have to wait until 2080, or even 2050, to experience drier times; greater aridity was already on its doorstep—hence the considered use of the word *imminent* in the title of their paper, "Imminent transition to a more arid climate in southwestern North America," which appeared in the journal *Science* in 2007, attracting considerable attention. Their research also told them that the increase in aridity would not result from the same forces that triggered the droughts of the past. Those droughts, possibly including the extended droughts of medieval times, appear to have occurred when the currents of the tropical Pacific, for whatever reasons, carried relatively cool surface waters far to the east, toward the Americas. This pattern, commonly called La Niña, seems to give a northward push to the winter storm track that brings moisture to western North America.

The opposite pattern, El Niño, pulls the storm track southward, bringing winter rain to U.S. deserts and snow to the southern Rockies. (Mexico is not so lucky; it depends principally on summer monsoons.) Together El Niño, named for the Christmastime arrival of warm waters off the coast of Peru, and La Niña, its contrary sister, are known as the El Niño Southern Oscillation (ENSO). The mnemonic for this relationship, from a southwestern point of view, is sexist: the boy child brings welcome winter precipitation, filling the reservoirs crucial to both cities and agriculture, but the girl brings drought.

Or so it has been in the past. The model simulations analyzed by Seager and his team showed the Southwest to be drying out "for a different reason, and that different reason is just a simple consequence of warming up." Seager explains that when Earth's surface and its atmosphere warm up, two things happen. One is the intensification of the hydrologic cycle: warmer air holds more water vapor; more vapor converges in wet regions, and it rains more there. "The pattern of water vapor transport intensifies. That makes dry regions drier and wet regions wetter. So that's part of why the Southwest dries, because it's already in one of these places where important parts of the atmosphere circulation take moisture away." The other reason Seager cites is the expansion of the subtropical subsidence zone, where the dry, downwelling air of the Hadley cell returns to the surface of the land.

"And that's not the same as what causes the natural droughts," he concludes. "The cycle of natural dry periods and wet periods will continue, but they continue around a mean that gets drier. So the depths—the dry parts of the naturally occurring droughts—will be drier than we're used to, and the wet parts won't be as wet as we're used to, because they're both happening around a mean state that gets drier and drier." That new "mean state" is what he calls the new climatology of the Southwest, something similar to the Dust Bowl or the 1950s—a far cry from the anomalously wet period that ran from the late '70s into the '90s, which forms many people's notion of what is "normal" for the region.

Not everyone agrees with Seager's message, and certainly not with the blunt, alarming language in which he delivers it. There are always cries for more data and improved models, for further analysis at a finer spatial scale, and for the consideration of other possible explanations. There are also critics who say that even if he is right, he should be more careful: predictions as strident as his risk undermining the public's faith in the scientific community. What if the Southwest enjoys a string of generously wet years? Every system exhibits variation, and all long droughts include wet episodes. What if the new climatology arrives late, or not at all? Perhaps then the urgency he has tried to engender will turn to cynicism, commitment turn to inaction, and the public's grudging acceptance of the reality of climate change degrade into disbelief. Perhaps, in the end, his startling warning might have an effect opposite to the one he intended.

Seager shrugs: "It's because of things like that that the Bureau of Reclamation and their like can sort of throw their hands in the air and say, 'Well, what do we do with these climate model projections? They don't have information at the spatial scale we need to predict what the Colorado River is going to do.' And they're right, strictly speaking—that's correct, but that's not a good excuse for not doing anything, because the chances are, once you get a model with all the right spatial scales to have all the topography, it's probably not going to change the sign of the result." The sign, plus or minus, would still be a minus in terms of water availability—things will be getting drier. "So if you just do nothing and wait for that, you're just wasting time.... It could be ten, fifteen years before you get the models at the resolution you need, by which time you're probably deep into this drying trend."

JONATHAN OVERPECK SEES merit in Seager's analysis. Overpeck is professor of geosciences at the University of Arizona, where he also codirects the Institute of the Environment. He is articulate, broad-browed, and quite frequently harried, the last because he is much in demand as an interpreter of climate change for government groups and water managers, planners, and academics. He also submits to a steady stream of interviews with members of the media and the odd book-writer, like me. He says, "I think the safe bet now, based on what I've seen, is that even if Seager and his team didn't get it exactly right, in the end I think they got the picture. The main point for society is there's a good chance things are going to dry out here.... There's nothing that looks good for the Southwest, I'll tell you that."

If there is such a thing as a climate change establishment, Overpeck is part of it. He has been deeply involved in the labyrinthine world of the IPCC for many years. A child of the United Nations, the IPCC is charged with issuing periodic assessments of the status of climate change research and the predicted effects of climate change on global society and the environment. Overpeck has contributed to multiple assessments, and when the time came to produce the "Fourth Assessment Report," which appeared in 2007, he was one of twenty-two coordinating lead authors responsible for presenting the physical science of climate change. It was a monumental job of information integration and editorial cat-herding, and Overpeck's gentle, calming manner must have stood him in good stead. Thousands of scientists from around the world, as well as representatives from 180 participating governments, took part in the process at some level. Overpeck also helped prepare the report's technical summary, which demanded a broad view of the range of opinion within climate change science. A few months after the "Fourth Assessment Report" became public, the IPCC, along with former U.S. vice president Al Gore, was awarded the Nobel Prize for Peace. It was nice, says Overpeck, to feel appreciated.

In his office at the university, Overpeck offers a quick primer on the climate of the Southwest and the forces changing it: "The Southwest is going to dry out on average. We're going to have more drought, more frequent drought, and longer drought, probably, if you were a betting person. And when it rains, it's going to rain more intensely on average, meaning more floods." The predicted increase in floods arises because of the intensification of the hydrologic cycle, of which Seager spoke: warmer air, holding more moisture, leads not just to wet places getting wetter, but to wet *events* getting wetter: storms will carry more moisture, and when they let it go, the impacts will be greater. This is also the theoretical basis for the idea that a warmer world will see a greater number of hurricanes, and that more of them will be extremely powerful.

It stands to reason that global change modelers worked first to capture the most salient relationships of atmospheric and oceanic circulation in their simulations. Subtler feedbacks naturally took longer to describe. The trouble is, as these secondary relationships come into focus, the outlook for the Southwest grows worse, not better. An example of this involves mountain snowpack and albedo.

Albedo is a measure of reflectance. A snowfield has high albedo and reflects back into the atmosphere most of the light and heat energy that falls upon it as sunshine. Dark, bare ground, with low albedo, has the opposite effect; it warms up and stores heat comparatively well. Overpeck explains that the global change models the IPCC used "don't get the Rockies." They smooth them out, reducing their choppy, steep topography to rounder, more evenly shaped lumps, which means that they underestimate the surface area of the mountains—and by extension the albedo-feedback effect.

With fewer winter storms and less snowpack—in a word, less albedo—the mountains will be warmer in winter, which will cause more snow to melt sooner, which will further diminish albedo and accelerate warming. It's a vicious circle of robust proportions. Overpeck hunts on his desk for a paper, recently arrived, that attempts to quantify the process. He shoves it my way. "It means we can get about 50 percent more warming than the IPCC thought in this part of the world, and that means the snow and runoff impacts will be substantially larger than anyone else is suggesting. So that's pretty depressing, but it is a very physically consistent idea."

There's more. Forest fires and vegetation die-offs like those described elsewhere in this book can leave large areas barren and exposed to the wind. The resulting windborne dust and ash, or soot from tailpipes and power plants, falling on mountain snowpack will reduce albedo and speed warming, too.[6] It is a case of positive feedbacks increasing the effects nobody wanted in the first place: more warming leads to more warming.

And still more. Another group of researchers has demonstrated that, even apart from albedo dynamics, the high altitudes of the world are warming faster, and to a greater degree, than lower elevations. This is a result of the differing capacities of discrete portions of the atmosphere to store heat, so that if lands at sea level warm by 3–4° C, high altitude lands like the altiplano of the Andes might warm by 5–6° C, a phenomenon that bodes ill for lingering snowpacks and mountain glaciers and the farms and cities that rely on them for water.[7]

Overpeck is quick to point out that nearly all the major predictions about climate change in the Southwest concern winter weather. They include earlier onset of spring conditions coupled with less late-winter precipitation.[8] They also include reduced albedo and augmented local warming, and northward migration of the winter storm track. The underlying cause may be not one thing, but several. "Is it the jet moving, or is it just [that] the Hadley cell is expanding northward and we're getting all this subsidence here?" asks Overpeck. "Well, it's probably both. And yeah, you see the same thing going on symmetrically around the world. But each region has its own little nuances."

Echoing Richard Seager, he adds that if the Hadley cells are expanding and "if the jets push farther north, it means that even if there is a good moderate El Niño, maybe even a strong El Niño, it won't be able to get the moisture into this part of the world as easily as it would prior to global warming."[9]

And if winter moisture fails, what of the summer, the season of the monsoon, when air masses charged by warm waters of the Pacific and the Gulf of Mexico make their way inland and release their moisture in violent thunderstorms? "We don't know what the monsoon is going to do. The jet stream and the Hadley cell—that's all winter. Seager's work: that's all winter, too, and spring actually—early spring is where we're going to see the big

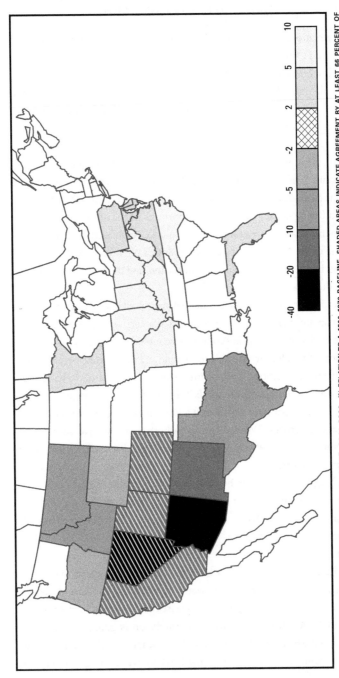

MAP 2: PROJECTED SURFACE WATER RUNOFF BY STATE, 2041–2060  IN RELATION TO A 1900–1970 BASELINE. SHADED AREAS INDICATE AGREEMENT BY AT LEAST 66 PERCENT OF MODELS ON DIRECTION OF CHANGE. DIAGONAL HATCHING INDICATES MORE THAN 90 PERCENT AGREEMENT. *ADAPTED FROM P. C. D. MILLY, ET AL., GLOBAL PATTERN OF TRENDS IN STREAM-FLOW AND WATER AVAILABILITY IN A CHANGING CLIMATE, NATURE 438 (NOVEMBER 17, 2005): 347–350.*

decrease. And wouldn't it be nice if the monsoon could intensify and make up for some of that?"

But there are problems with that scenario. "Even if the monsoon does intensify," Overpeck adds, "it might not be all that great because the rain will be falling harder, and you'll have more flood." And probably more erosion and arroyo cutting. "And you won't be able to use that water; it'll run off"—which is to say, it won't be stored and metered out in the convenient manner of a snowpack.

In the summer, according to Seager, the Southwest's high rate of evaporation makes it a net exporter of water. In the winter, it is an importer; that's when the region can save up. If winter moisture declines as much as predicted, then short of repealing the law of evaporation, the monsoons won't be able to bail out the region, even if the summer stormwaters are efficiently captured. The fundamental structure of the Southwest's water budget points toward sustained and severe shortage.

WHEN YOU BUILD a model of the world's climate, you begin by chopping the atmosphere, the oceans, and the land surfaces of the planet into units. A perfect model would describe every point within the biosphere, but of course the number of possible points is infinite, and you would need an infinite amount of computer processing power to describe them. So you compromise. You divide the atmosphere and the oceans and the land surface into pieces as small as you can afford to get onto a computer. Where the atmosphere is concerned, this involves sectioning it vertically as well as horizontally: you have a column of air miles high, doing different things at different altitudes. So you chop it all up into volumes that would look tiny from space but that are gigantic from an earthly point of view, and then you begin to assign processes to each of the volumes: what happens thermodynamically with different levels of $CO_2$ and other gases, and how that relates to temperature, wind, moisture transport, and other variables. You factor in the interrelationships among these variables over time and space, ramifying onward, potentially forever, and you express all this in terms of equations that are interdependent throughout the model. You also account for certain inputs: the amount of sunlight hitting Earth, which varies in luminosity, or volcanic eruptions, which change the concentrations of certain aerosols in the atmosphere. So then you have a whole mass of equations and you crank them forward. Basically you are simulating things at the timescale of weather—usually in half-hour steps.

I have paraphrased here what Chris Milly tried to teach me about the modeling work he pursues at the Geophysical Fluid Dynamics Laboratory (GFDL) outside Princeton, New Jersey. I don't think I was a very good student. Milly could not fail to notice the intermittent incomprehension in my eyes, but politely, generously, he pressed on: "So there's thousands of equations for each thirty-minute period that have to be solved. They're all coupled to each other, these big simultaneous equations. Yeah, you need big computers. So you put in all these equations and you solve them."

We are in a small sunlit room with a child's paintings on one wall and a greaseboard crowded with the hieroglyphics of higher mathematics on another. Outside a stiff fall wind is stripping the last leaves from the trees. The GFDL houses the kinds of supercomputers Milly is describing: powerful, state-of-the-art machines worth uncounted millions of dollars, and loaded with worlds of data. The computers themselves, let alone the fact that the facility, set amid groomed lawns in a handsome research park, is a federal installation, prompt tight security for the lab. The facility belongs to the National Oceanic and Atmospheric Administration (NOAA). It has keycoded doorlocks and a secure, glass-walled room next to the guard station where outsiders sign in. The architecture is corporate and the atmosphere almost martial, yet Milly epitomizes casualness. He's wearing jeans, running shoes, and a navy T-shirt with a breast pocket. The round lenses of his eyeglasses emphasize the arch of his eyebrows and lend him a quizzical air. He bears an expression of perpetual inquiry.

He has been describing a "eureka moment," a realization that came almost without warning and that fundamentally changed his skepticism about climate change, but the story of how it came about has required a number of digressions. The indecipherable scrawling on the greaseboard is a good reminder why that might be: the surprises in Chris Milly's professional life spring from a pretty rarefied context. For his story to make sense to a stranger, he has to reconstruct the context.

Milly is an exception at the GFDL: he doesn't work for NOAA, but is a senior researcher with the U.S. Geological Survey, posted at the NOAA site to create crossover connections. NOAA does air and the oceans; USGS does land. It makes sense to pull the elements together. Milly explained that he is neither an atmospheric scientist nor a climatologist, as though these were lamentable limitations. He was almost apologetic when he said, "I can't show you the equations for potential vorticity." Rather, he is a hydrologist. His forte is describing processes that take place on *terra firma*, and what he particularly brings to the research group he leads is an instinct for connecting the hydrological dynamics of the land to the dynamics of sea and air.

Milly and his team decided to test the hypothesis, offered a few pages back, that the water cycle intensifies as Earth warms: "Because everything's warmer, the atmosphere can hold more vapor, and therefore there's a lot of water just moving around and therefore all the water fluxes in the world get bigger: precipitation, evaporation, runoff." He pauses to let this sink in.

"That's not a bad zero-order characterization, but it is overly simplistic. And there is no reason to go by that anyway because we do have these models that have much more structure, and much more detail spatially, too. So we can look at what they say, and that's what we did."

In the "Fourth Assessment Report," IPCC scientists ran their models chiefly to generate outputs of temperature and precipitation. Although the models simulated runoff, which Milly describes as "the water that the land has left over after it's gotten its precipitation and kind of wrestled with the atmosphere over evaporation," nobody had taken a serious look at it.

"So we took those outputs from all these models"—there were twenty-one of them—"and we did some ensembling, which is just trying to combine the results of many different models to get one best estimate. Then we looked at what this ensemble model said about what one might have expected to see happen globally during the twentieth century."

To verify the model, he and his group also set about obtaining actual streamflow measurements for major rivers around the globe. Fortunately there is a central repository for such information. "You get all these records—the Mississippi at Vicksburg and so forth"—actual historical observations for the Ganges, the Euphrates, the Nile, the Zambezi, the Orinoco, and scores more, and then you "go into the models and make the observations at the same times and places in the models as where you have observations."

Milly is gesturing as though the models are in one hand and the firsthand, twentieth-century river measurements in the other. He brings his hands together, touching fingertips. "It's just a one-to-one correspondence between nature and the model, so you're not comparing apples and oranges, you're comparing simulated apples to observed apples." The team mapped both sets of outputs and colored them to show where things got wetter and where they got drier. Finally, Milly and the others had a look at them, side by side.

"The first time I saw those two together, that was that eureka moment. There was this feeling inside: 'You're validated.' Oh, my gosh! The correspondence between these maps was just too striking, at least subjectively at first. I mean, you have to do rigorous statistics to say whether, well, that could have just happened by chance. But the first impression upon seeing that was 'Wow,' because there just appeared to be too much correspondence to be explained by random variability."

The implication of the correspondence between the actual observed data and the output from the models was that the major twentieth-century regional patterns—things like the ups and downs of precipitation, the drying in the Sahel, and the increasing wetness in the eastern United States—were not random. Rather, they were largely the result of forces that drive global climate—varying levels of solar radiation, changes in greenhouse gases, and other climate-altering factors.

Randomness is the default assumption that hydrologists like Milly are trained to make—that "it's just natural variability. It's just bouncing around." If they want to be taken seriously as scientists, they have to be skeptics about causes of a higher order.

"So my biggest surprise was seeing that and seeing the success of the models that my colleagues had built and improved over the years, seeing a success in quantifying a variable that they had never looked at. They never tried to tune [the models] to get the hydrology right per se other than having precipitation in the tropics and no precipitation in the deserts."

Milly had been a self-described doubter on matters pertaining to climate change. Not a disbeliever, but cautious, wondering whether more was claimed for the phenomenon than facts would support. "Appropriately so, I think—that's sort of the background that I come from, just highly skeptical of any violation of the null hypothesis: there's nothing there; don't get too excited." But then came the two maps and the unignorable correspondence between real-world data and model simulations.

"The modelers were looking at other things, but this fell out of it, and it looked like the pattern of what happened in the world. We subsequently did all kinds of statistical analyses: run the models with no forcing at all for a stable climate. It's easy to get constant climate in a model. You just turn off all those exogenous things that make climate change. You look at many, many different climate simulations. Do any of them develop patterns, just by chance, that look as similar to the observations as did the forced example? And none of them did. They just couldn't generate that. So that makes a pretty strong case for external forcing causing major variations in the hydrology."

If the models were good at describing the past, it stood to reason that their description of the future deserved attention. "This doesn't suddenly say, 'Hey, the models are right,' but it lends credence. It enhances their believability. It makes you take them a bit more seriously." And so Milly and his group ran the model simulations of global hydrology forward to see what might be likely to happen, assuming no unforeseen developments (asteroid impacts, major volcanic eruptions, or a global economic melt-down that turns off the engines of industry). They published the results in *Nature* in 2005.[10]

They considered two factors in particular, the first being the degree of agreement among twenty-one models, running some of them multiple times (because different initial conditions yield different results). They wanted to see how well the models agreed on the direction of change for a given region—whether it would get wetter or drier. They figured that the more the models agreed, the more their predictions deserved credence. The second factor was the amount of change the models predicted—how much the streamflow in a given region would increase or decrease.

The Southwest comes out as one of the big losers. More than two-thirds of the model runs agreed that the region would become substantially drier, and for some subregions within the Southwest, over 90 percent of the models agreed. The amount of drying predicted for the mid-twenty-first century was generally in the range of 10 to 30 percent, relative to a baseline calculated from the period 1900–1970. Put simply, the models were suggesting that the region would have available to it about a fifth less water than had been the case during most of the twentieth century, even including its droughts. And as the century progressed, things would keep getting worse.

This was not good news for an already overtapped region that suffered, as one wag has put it, not so much from a shortage of water as from a *longage* of people. And more people are arriving all the time.

I REMEMBER THE first time I saw the maps from Chris Milly's article. It was at a conference in January 2006, not long after the article appeared. Jonathan Overpeck was speaking. I am an indifferent conference attender: short on attention, quick to daydream. I had been in the hallway chatting with friends when Overpeck began. But then the hallway emptied. Overpeck was well into his presentation when I finally slipped into a chair in the back of the room and thumbed through my program to learn who he was. I expected to settle into a restful thirty minutes of dozing. Then Overpeck put one of Milly's maps on the screen. The Southwest was as red as an open sore. So was a band that stretched across northern Africa, the Mediterranean basin, the Middle East and deep into central Asia. High agreement among the models, Overpeck said. And the models agreed that surface water availability—the blood of the oasis civilization of the Southwest and all those other arid and semi-arid regions—would substantially, perhaps precipitously, decline.

In a sense, this book was born at that moment, although I did not know it at the time. The room was still. No shuffling. No coughing. I remember the colors of the room, the design of the chair in front of me, the texture of the cushions. I felt cold. I tried to remember a statistic I had cited in something I had written—how many people relied on the Colorado River for all or part of their water more or less currently, and how many would do so in the near future. I have since looked the numbers up: 23 million people dependent on the Colorado in 1996, 38 million projected for 2020.[11] It is widely known that the Colorado is already dangerously overallocated. If its water yield were to decline, let's say, 20 percent, or even just 10 percent, what would be the effect on the lives and livelihoods of 38 million Mexicans and Americans (or Norteamericanos, as the Mexicans prefer to say)? And if you made a similar calculation for the overtaxed Rio Grande, beside which live several million more Americans and Mexicans in uneasy codependence, and if you factored in thousands of other communities scattered throughout the region that rely on sources of water less secure than either of those two big rivers, well then, the trouble we were in was of a scale to match the giant, sprawling, brawny Southwest itself.

"THERE'S ANOTHER VERY striking pair of maps that I could show you," Chris Milly said. Again he gestured with both hands: one hand as though holding the first map. "When you look at the reds and blues of the 2005 map, where the big drying patterns and the big wetting patterns are projected, that's one map." Then the other hand, outstretched: "I can show you another map, which is a map of the water-stressed regions of the world." Again, the hands come together, fingertips to fingertips. "This map of water-stressed regions, it's like one-to-one compared to the map of projected drying."

He hypothesized that in the dry regions of the globe, human population and activity inexorably expand until they approach the maximum the system can accommodate. "So human water scarcity naturally correlates with climatic water scarcity."

"It sounds like the Law of Closets," I offered. "That however much space you have, you fill it up."

"Yeah, that's true. Or the Peter Principle, that people rise to their level of incompetence, kind of the same idea."

But it was a grim idea: that not just in the Southwest alone, but in regions throughout the world, the areas most likely to experience a decline in available water are those least able to withstand it. Milly explained that it was not merely a question of drought, which semantically suggests an impermanent condition. He was saying what Richard Seager and Jonathan Overpeck, each in his way, had also said: that for the water-stressed regions of the world, including the North American Southwest, "normal" was downshifting. And it was downshifting a lot. Then he was silent, waiting for me to ask another question. He had made his point. World society was going to weaken at one of its weakest points.

Hundreds of millions of people live in the water-stressed regions of the world, their situations endlessly variable but their dependence on a scarce resource uniting them in unwanted vulnerability. Unfortunately, their unity of condition breeds a division of interests, as communities and nations compete for precious water, sometimes violently. One of the reddest regions on Milly's map, which is to say one of the portions of the planet most threatened by a decline in water supply, stretches from Lebanon and Israel through Iraq and Iran to Afghanistan, lands beset by generations of intense conflict, where the stress of water shortage can inflame old grudges. The downshifting lands of North America have seen their share of conflict, too, and more seems on the way, presaged by the hundreds of miles of border wall the United States has built and the tens of thousands of border police it has employed to enforce the separation of its aridlands from the aridlands of Mexico. The maps we stared at said that the screws of want and thirst in these regions would only tighten. Milly and I sat in silence for an awkwardly long time. There was nothing more to say, without saying too little or too much.

Finally I broke the silence and asked another question. It was not an important question. It was just another on my list.

# 2

## ORACLE: GLOBAL-CHANGE-TYPE DROUGHT

IN THE SPRAWLING, climate-controlled, glass- and steel-ribbed tent of Biosphere II, Dave Breshears is killing trees. Together with Henry Adams, his graduate student accomplice, he arranged to have piñon pines dug up and hauled in from northern New Mexico. Next he saw to their replanting and watering, got them rooted, growing, and happy, and then—for some, but not for all—shut off the water. Breshears wants to find out what makes them die.

Biosphere II is an unlikely setting for controlled experiments in tree murder—or for anything else. Rising from the Arizona desert like the main terminal of a misplaced airport, its design sexy and futuristic, it is full of the pride of technology and the promise of discovery. In its integration of multiple volumetric forms and vast banks of windows, it might be the architectural love child of the Bilbao Guggenheim Museum and a backyard greenhouse. A scanty procession of tourists detours from the beaten path to pay twenty dollars a head to enter the structure, which was meant to be an indoor replication of everything outdoors. It is a proto–space station: Biosphere II was conceived as a living and breathing microcosm of Biosphere I, the planet Earth.

Bankrolled by the Texas oilman and investor Edward P. Bass, the creators of Biosphere II hoped that it would serve as a laboratory where earthlings might learn to package (and eventually export) their bubble of life to distant planets and solar systems. Toward that end, in September 1991, with much fanfare and not a little criticism, eight so-called biospherians, four men and four women, were sealed inside the three-acre complex for a two-year stay. Socially and biologically, the expectations were utopian: they were to grow their own food, maintain a life-supporting atmosphere, and get along as a team. Functionally the results were quite different. After thirteen months, said one, "We were starving, suffocating and going quite mad."[1] The experimental team, both inside and outside the enclosure, fractured irreparably, and the artificial ecosystem inside the glass walls, even with large inputs of outside oxygen that violated the original experimental plan, barely kept the biospherians alive, never mind healthy. But they lasted two years, hanging on as grimly as prisoners of war, finally exiting in a condition wan, wobbly, and far thinner than they had gone in. Victory was immediately declared, no matter how yellow they looked.

In the end, the lessons of Biosphere II said more about the difficulty of recreating the earthbound support system outside its windows than about the practicality of colonizing distant worlds. The chemical balance of the indoor atmosphere, its exchange of gases with bodies of water and masses of soil, vegetation, and masonry, the reciprocal relationships between plants, parasites, and parasite predators, the astounding difficulty of accounting for every important variable in a massively intricate, self-regulating system—the sum of these challenges exceeded the problem-solving capacity of the architects and implementers of Biosphere II, who in the end proved to be iconically hubristic and human. Perhaps the greatest lesson of Biosphere II arose from the psychology of the experiment's actors, not their science. It concerned the frailty of social relations and the high cost of flawed leadership. It illuminated the difficulty of maintaining solidarity in the face of adversity. In this respect, Biosphere II proved a true microcosm of Biosphere I, where venality, ideology, self-interest, and other elements of the globe's political ecology, much more than the workings of the nonhuman world, have generated the greatest obstacles to solving environmental problems, climate change foremost among them.

CLIMATE CHANGE WAS one of the reasons that spurred Breshears and Adams to kill trees in Biosphere II. The facility itself was another. In 2007, after a succession of faltering administrative arrangements, the great white elephant of Biosphere II finally landed in the care of the University of Arizona, where Breshears was a professor in the School of Natural Resources. When the call went out for "visible science"—experiments that the public would be able to see and appreciate—Breshears recognized that the facility's controlled environment provided an opportunity to simulate different climates in which to observe tree mortality. He had been seeking an appropriate experimental site for a while. Surprisingly, the actual mechanics of how trees die—specifically, how they die from drought—were not well understood, and a few years before coming to the University of Arizona, Breshears had witnessed a die-off of trees on a scale that impressed him deeply and left him wondering more than ever about the implications of a warming climate for the forests and woodlands of the Southwest.

A string of wet years had blessed the northern Southwest for roughly two decades (depending on who's counting), but abruptly ceased between late 1995 and early 1996. Winter snows failed to materialize and spring came hot and early. Dry years alternated with wet ones until 2000, which was exceptionally dry. That spring the Cerro Grande Fire partly engulfed the nuclear research city of Los Alamos in New Mexico's Jemez Mountains, leaving 43,000 acres charred. Breshears, then a researcher at Los Alamos National Laboratory, experienced the community's trauma first-hand as the fire consumed trees and houses near his own in the pine forest on the west side of town. The following year was also dry, and 2002, the nadir of the drought, was brutally so.

In June of '02, on the Mogollon Plateau of central Arizona, the Rodeo and Chediski Fires broke out separately, two days apart. The first was ignited by an idle firefighter hoping to generate work, the second by an out-of-gas motorist trying to attract a helicopter's attention. Both radically overshot their goals. The brittle land was as dry as a match head, almost panting for a spark, and gusty winds soon drove flames 400 feet into the sky, quintuple the length of the trees that literally exploded within them. Even in their early stages the fires spread at rates of thousands of acres an hour. Soon they merged into a single conflagration that continued raging for another two weeks, ultimately consuming 467,066 acres, and setting—by a large margin—a new record for the largest fire recorded in the Southwest in the modern era. The Rodeo-Chediski Fire redefined most people's idea of "big" for a southwestern forest fire.[2] (That idea was redefined again in 2011 when the Wallow Fire burned almost 540,000 acres in Arizona and New Mexico.)

But elsewhere in the woods in 2002, something still bigger was at work. The Rodeo-Chediski Fire was small compared to the combined effects of drought and bark beetles attacking pines across Arizona and New Mexico. In 2003, die-offs in the two states totaled 2,609,475 acres—an area more than twice the size of Delaware. Piñon pines accounted for almost three-quarters of the mortality, ponderosas for the rest.[3] The resulting tableaus were unsettling: mile upon mile of sideslope, plain, and mesa top bristling with red-needled trees, a panorama of death relieved only by the brooding dark green of junipers, which seemed strangely unaffected. It was as though a defoliant had been sprayed across the landscape, selectively killing the pines. Over a span of months the red needles fell, and whole forests of black skeletons stood sentinel across the land.

Neither extensive fires nor beetle kills, nor certainly drought, the enabler of both, is new to the Southwest. Long before he was a best-selling author of mystery novels, Tony Hillerman worked as a reporter and editor for the Santa Fe *New Mexican*. In 1957, at the height of the Southwest's worst drought of the twentieth century, he filed a story that could easily have been written during the die-off of 2003: "Two species of bark beetles, working as a deadly team, are stripping a vast area of Northern New Mexico of its piñon and ponderosa pine....Approximately a million acres of trees are already dead or currently being killed."[4]

The drought of the 1950s is a dim memory in the Southwest today, but it provides an instructive contrast to the dry years of the early 2000s. Craig Allen, a close colleague of Breshears, heads the Jemez Mountains Field Station of the U.S. Geological Survey, based at Bandelier National Monument near Los Alamos. It was he who surfaced the Hillerman quote. Boyish-looking and compactly built, he has earned a local reputation as a "field beast" who easily hustles up hills and ridges, leaving his companions gasping for breath. (I count myself among that clan of under-oxygenated clamberers, having survived multiple field excursions with Dr. Allen.) One of the qualities that sets Allen apart from many other field ecologists is his attachment to a single regional landscape. For thirty years he has been a student of northern New Mexico, and more specifically the Jemez Mountains, whose ecological history he reconstructed in his dissertation. Along the way he has developed a fine eye for reading the landscape of his adopted home.

One of the things that initially puzzled him was the amount of dead wood he was seeing, especially at the lower margin of the ponderosa pine zone. Trying to place the evidence of the land in context, he rummaged through old Park Service files and yellowed newspaper clippings, gradually assembling a snapshot of the 1950s drought, which accounted for the mortality he had observed. Comparison of aerial photographs from 1954 and 1958 revealed more: along the mesas at the base of the Jemez range, the lower edge of the pine forest had retreated upslope as much as two kilometers, as ponderosas died and surrendered their territory to piñon and juniper.

The drought of the 1950s was longer and drier than that of the early 2000s, and it killed vast numbers of trees of multiple species, not only pines. But the later drought appears to have induced mortality at an even more spectacular scale. The key difference seems to have been temperature. Measured "by several metrics including mean, maximum, minimum, and summer (June–July) mean temperature," the more recent event was hotter by 1–1.5°C.[5] This increment is no more than the difference between the high 80s on the Fahrenheit scale and 90 or a little above—not the kind of thing a well-watered human might worry about. But sustained for weeks, if not months, over the course of a drought, it was evidently a matter of grave import for thirsty piñons and ponderosas.

Working together, Craig Allen, Dave Breshears, and others concluded that the higher temperature of the drought of the early 2000s was the critical factor in provoking widespread tree mortality, and they thought it should earn a special kind of recognition. The prevailing models of global climate change predict higher temperatures for the Southwest by approximately 4°C (7.2°F) over the course of the current century, and the upsurge of 1–1.5°C in 2000–2004 looked for all the world like a down payment on that account. Nevertheless, as scientists they were careful to avoid saying that the recent drought was *caused* by climate change, or that it represented climate change. Instead they pointed out that in comparison to its cooler 1950s predecessor, the drought of 2000–2004 seemed to look like what climate change would bring, when and if the models were proved right. They named it a "global-change-type drought"—not a phrase that sings but one that lingers forebodingly in the mind.

If higher temperatures will indeed characterize the future climate of the Southwest—and at this point few serious scientists, if any, believe that they will not—the drought, fires, and die-offs of the early 2000s may be a preview of what is to come.

HIGHER TEMPERATURES CONTRIBUTED to the potency of the early 2000s drought in at least two ways.

The first involved the *Ips* bark beetle, different species of which prey on piñon and ponderosa, as well as on other species of pine—but the procedure in every case is the same. The beetles bore into a tree, mate, and lay their eggs in the inner bark, or phloem. The larvae, once hatched, feed by tunneling their way through the phloem's living tissues. The tree's strongest defense is to flush out the insects with pitch, but if drought has dehydrated the tree, it may fail to summon an adequate flow—or, alternatively, the insects may be so numerous that they overwhelm what flow there is. In either case the beetles can eventually girdle the tree with their tunnels, killing it.

The onset of higher temperatures in this drama favors the beetles: under warm conditions, they start their reproductive cycle earlier in the spring and continue later in the fall. With extra weeks to reproduce, they may spawn an additional generation or two over the course of a hot summer, boosting the number of their kin that make it to winter. Then if the winter is warm and fails to knock them back with arctic freezes, and if spring comes early again, the swarms that initiate reproduction that summer will be much greater than the year before. For as long as the winters remain mild, the summers long, and the trees vulnerable, the beetles' numbers can continue to build, resulting in swarms large enough to overwhelm the defenses of even the healthiest trees.

At my home in a rural mountain valley in northern New Mexico, the winter of 2001–2002 was so mild that a local bear came out of hibernation months early and ransacked my yard. It tossed chairs and tools and firewood about, grouchily searching for food. That winter was a bad one for the bear, but a good one for bark beetles, and once spring came, bringing weather so hot and dry the pastures never greened, you could walk among the piñons and hear a faint mechanical drone, as of a thousand tiny chisels rhythmically chipping away. It was hordes of beetles, tunneling and feeding.

The Southwest is not alone in its vulnerability to beetle-driven die-offs. In the northern Rockies of Colorado, Wyoming, and Montana, millions of acres of lodgepole pine forests have suffered fatal infestations of mountain pine beetles, belonging to the genus *Dendroctonus*, which affect their host trees in much the same way as *Ips*. Colorado foresters expect that virtually all of their state's mature lodgepole forests will eventually succumb—about 5 million acres, including some of the state's most famously picturesque and tourist-pleasing vistas. They place the blame on drought and warm winters, along with a century of fire suppression that produced unbroken expanses of mature forest ideal for the spread of beetles.[6]

The situation is even worse in British Columbia, where mountain pine beetles have laid waste to 33 million acres of lodgepole forest, killing a volume of trees three times greater than Canada's annual harvest of timber. Foresters there call the beetle epidemic "the largest known insect infestation in North American history," and they point to even more chilling possibilities. Until recent years the frigid climate of the Canadian Rockies had kept the beetles on the milder Pacific side of the Continental Divide, but warming temperatures, assisted by the occasional windstorm, have now allowed them to top the passes of the Peace River country and penetrate northern Alberta. Should the insects adapt to the otherwise beetleless jack pines of Canada's boreal forests, their prospects for continued spread become almost limitless—they might chew their way across the continent to the Atlantic and then turn south for New England, generating transformative ecological effects on an immense scale.[7]

It is only natural for us to think that the greatest threats to our world will wreak their havoc in stupendously dramatic ways. The calamities we are used to seeing are those that populate disaster films or play well on the evening news: floods, fires, volcanic eruptions, hurricanes, earthquakes. Few of us expect that tiny creatures in exoskeletons that gnaw or sip or drill their way through the world will be powerful agents of change, but they can be. And as our climate changes, they will be. They have the power of explosive, exponential reproduction, and when conditions are favorable and they swarm into plagues of biblical proportions, they show the rest of us what being an opportunist really means.

THE SECOND REASON why marginally higher temperatures pose a threat to drought-stressed trees accounts for Dave Breshears's decision to kill trees in Biosphere II.

Breshears is by nature a quiet man, comfortably self-possessed and reserved, but like most people who love what they do, he gets a little excited when he talks about his work: his words come fast, and enthusiasm spills out. In his campus office in Tucson, he peers at his computer screen, reviewing the slides of a presentation he will make the next day. Breshears describes the two ways a tree can die from dehydration. The first and simplest is that it dies directly from thirst—its intake of environmental moisture fails to replace the water evaporating from its surfaces. At the extreme of this process, in Breshears's words, the tree's water column "cavitates"—the network of dispersed channels in its xylem (through which water moves up the stem) becomes blocked as cells rupture and fill with air, producing embolisms not unlike the air bubbles that can dangerously develop in a human blood vessel.

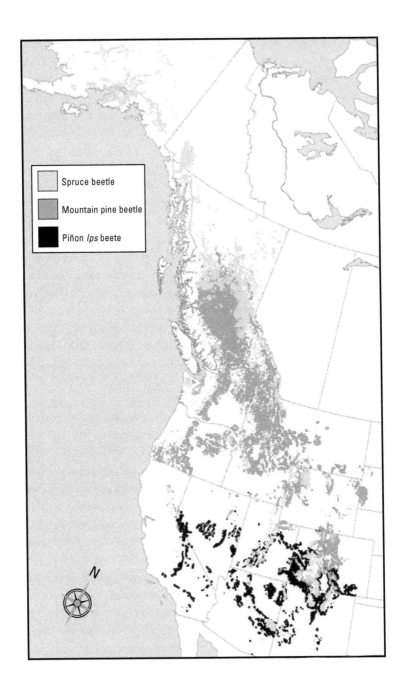

| | Spruce beetle |
| | Mountain pine beetle |
| | Piñon *Ips* beete |

N

A tree has no heart, no pump to clear away a major blockage—water simply wicks its way upward, rising against gravity because of the pull of evaporation and the adhesive attraction of water for itself. Low levels of cavitation are reparable and probably occur regularly. But when cavitation occurs broadly, the continuity of water in the living tissue of the tree breaks and cannot be repaired, rendering the tree incapable of supplying water to its twigs and leaves. If the break is only in a branch, the tree may die back but maintain a living core (this is a common response of junipers to drought). Yet if the trunk, afflicted with cavitation, ceases to move water upward, the tree will be unable to photosynthesize and feed itself, and it will die.

The second mechanism of tree death is more complex. In order to stave off death by cavitation, the tree can close down its stomata—the pores in its leaves through which it photosynthesizes and "breathes." This reduces the rate of the tree's water loss, but at the cost of fasting. With its stomata closed, the tree no longer absorbs its main food—the carbon it takes from carbon dioxide in the atmosphere and synthesizes into sugar. Thus sealed against loss and hoarding its water, the tree lives on borrowed time. The question controlling its fate becomes: will the rains return before fasting becomes starvation, and starvation death?[8]

"When plants get to a certain stress level, they shut down," Breshears explains. The point at which they do so varies from species to species. He points to a graph on the computer screen. "And for piñon this red line indicates where this value is."

"Where they shut down?"

"Where they shut down: they basically stop feeding themselves."

Piñon-juniper woodlands cover vast portions of the Southwest, mediating between the true forests of the high plateaus and mountains and the grass and shrublands of lower elevations. The piñon pine's chief partner—and competitor—in this enterprise is the one-seed juniper (*Juniperus monosperma*), which, says Breshears, has a different physiology: "It

doesn't shut down. It's kind of sloppy. It'll get stressed; it'll kind of keep trying to photosynthesize. It's not going to produce very much, but it's going to
keep chugging." He indicates another red line on the graph, far from the
piñon's. "It doesn't really shut down 'til here—a much lower value."

Breshears points out that piñon and juniper evolved amid drought, and
developed different strategies for handling it. Juniper gambled that it would
not dry out entirely, and so kept respiring and feeding itself, even at the cost
of continued water loss. Piñon took the opposite bet, hoarding water by clos-
ing stomata and living off reserves of stored carbon, like a hibernating animal.
Both piñon and one-seed juniper are well adapted to the failure of summer
rains. They can easily handle one season of dehydration or carbon fasting.
And if the following winter is also dry, which is not unusual in the Southwest,
they have their coping strategies; they place their bets. But if the drought continues for a second summer or a second winter, if the strain continues for
eighteen months or more, they approach their limits of adaptation. Even in
the absence of bark beetles or other pests to tip the odds, a tree of either
species cannot hold out forever. Without water, it will eventually lose its bet,
and its life.

Moisture is the dominant variable in the equation of dryland survival, but
temperature matters, too. Just how much it matters is what Breshears and
Adams were trying to get at in Biosphere II. The die-offs in the hot drought of
the early 2000s had suggested to Breshears that an improved understanding
of the role of temperature in tree mortality would be essential to modeling the
die-offs of the future. Drawing on the research and maintenance endowment
Ed Bass provided for Biosphere II, he and Adams hired a contractor to dig up
assorted young piñons near Las Vegas, New Mexico, and truck them down to
Biosphere II. They replanted half the trees in the Biosphere's desert environment and half in its savannah, allowing the temperature in both areas to fluc-
tuate with the seasons but keeping the desert consistently 4°C (7.2°F) hotter
than the savannah (and simulating the temperature increase projected for the
end of the century). They gave the trees time to adjust to their new environment and to begin growing again. Then they stopped watering half the trees
in each area and recorded what happened.

The water-deprived piñons in the comparatively cool savannah died, on
average, in twenty-five weeks—much faster than would have been the case in
the field, but not surprising considering the effects of transplantation. The

data seemed to show that the mechanism was carbon starvation, not cavitation. To no one's surprise, the trees in the desert environment died even faster—in just under nineteen weeks. As Breshears and (mainly) Adams, who would use this experiment in his dissertation, set about the task of analyzing a mass of data on foliar water content, diurnal stomatal conductance, and other esoterica, they also laid plans for a larger outdoor experiment to be conducted along similar lines near Flagstaff.

At first blush their work might seem to be a tedious elaboration of the obvious: of course trees die when you deprive them of water, and of course they die faster when you boost their water needs by raising the temperature. But with the right experimental data, you might be able to quantify the relationships linking tree survival, water stress, and temperature and thereby model, with some predictive accuracy, how whole landscapes will fare when drought—especially hot, "global-change-type drought"—next embraces the region. Possibly this information will help land managers respond to the challenges ahead. Of course, the word *respond* presumes that there will be something meaningful and effective to do, just as the word *manager* implies a measure of control. Full-out, implacable drought, however, has a way of proving both presumptions false. Even if it turns out that not much can be done to mitigate the die-offs of the future, the work of scientists like Breshears and Adams will at least help us know what we're in for.

Unsurprisingly, the news is not good. The team working in Biosphere II ultimately concluded that if mean temperature increased by 4°C in piñon habitat and nothing else changed, broadscale die-offs like the one that accompanied the drought of 2000–2004 would occur five times as often as they have in the past.[9] While scientific debate about experimental design and the precise physiology of tree death continue, the underlying dynamic is fairly simple. With higher temperatures, a short drought will produce the tree mortality that formerly only a long drought could generate, and short droughts are common in the Southwest. More troubling, however, is the potential for interactions that amplify the frequency and extent of die-offs. The prediction of a fivefold increase, dire as it is, does not take into account the influence of beetle outbreaks or other compounding factors. In particular, it does not contemplate a decline in precipitation, but if the rains fail and the snows fail, or if rain falls in place of snow and runs off too soon—if the droughts are long and cruel—then the odds for ever more spectacular and extensive die-offs, to say nothing of fires, come nearer to a sure bet.

"THERE IS SOMETHING else we should talk about," says Breshears. A one-word subject: "Dust."

Because aridlands tend to be underdressed in terms of vegetation, they are naturally dusty. Humans make them dustier. We construct and maintain roadways and pipelines; we raise cattle that trample the soil crust; we embark on agricultural misadventures that produce sodbusting on the Great Plains (a cause of the 1930s Dust Bowl) or result in abandoned cotton farms in Arizona; while we do these things for work, the motorheads among us roar off for fun on four-wheelers across the desert; and the list goes on. Humans are dust-making creatures.

Climate change is a pretty good dust-maker, too. If dust is characteristic of aridlands and if a warming climate will not only add to the inventory of aridlands but also thin the vegetation on the existing inventory, dust will be a big part of the dryland future. Add to that a likely epidemic of forest fire (with a lot of short-term ash and longer-term bare ground), extensive die-offs that also open large swaths of ground to the wind, and an increasing pace of human scratching and buzzing around, and pretty soon you start thinking that a breathing mask might be your next fashion statement.

Dust is a big deal. Breshears and another of his graduate students, Jason Field, determined that wind erosion was three times greater than water erosion on an experimental aridland site in a "normal" year. In a drought year—"a global-change-type year"—the disparity increased dramatically, with wind transporting roughly forty times more sediment than water.[10] That's only one site and one soil type, to be sure, and it would be wrong to extrapolate too much from the data, but the point is that wind-borne dust is a powerful force. Although there is already a lot of it, there will likely be considerably more in the years to come.

The impacts could be profound, but we don't yet know a lot about them. Tom Painter, a scientist at the National Snow and Ice Data Center in Boulder, Colorado, has done as much as anyone to quantify them. His research focuses on the effect of dust on the snowpack of the Colorado Rockies, a primary source of river flow sustaining billions of dollars of agriculture and millions of people. Sometimes he can tell that the dust settling on high-altitude snowfields originated half a world away in China or Mongolia—proof that dust storms are a worldwide phenomenon. Usually, however, the dust darkening the snow of the Colorado Rockies comes from aridlands immediately to the west—the Colorado Plateau and the Mohave Desert, which abound in dusty places. Inevitably Painter finds that dust accelerates the melting of snow. It darkens, sometimes with the merest blush of pink, the perfect white of the snow crystals, increasing the snow's absorption of light, and therefore heat. As a result, the snow melts faster.

Anybody with access to a snowfield can run a simple experiment to confirm this effect. Find (or create) a patch of darkened, dusty snow. Clear away the topmost inch in a small area to reveal the bright unblemished snow beneath. If your snow patch is out in the open and the day is sunny and moderately warm, you can come back in a few hours to find that the cleared area is now a pedestal, and the faster-melting, darker snow has subsided all around it. Magnify this phenomenon across hundreds of thousands of square miles of white-blanketed high country, and you get earlier spring runoff. You also get increased evapotranspiration (the sum of evaporation and plant transpiration) resulting from earlier exposure of vegetation and soils. Painter and his colleagues have estimated that the human-caused increment in dust creation over the past century may have shifted the timing of peak runoff of the Colorado River at Lees Ferry three weeks earlier. It may also have reduced net river flow by more than 800,000 acre-feet per year—a volume of water that might otherwise sustain a half dozen cities of moderate size and that an overallocated system can ill afford to lose.[11]

Dust can exacerbate drought in other ways, too. Too much dust in the atmosphere inhibits the formation of clouds and precipitation, and dust aerosols can accomplish this on massive scales. A strong case can be made, in fact, that the dust rising over the southern Great Plains in the 1930s converted what would have been a fairly typical La Niña drought into something far worse. The added dryness caused by the dust, coupled with other factors, may even have caused the "center" of the drought to shift from the Southwest to the area that would soon be known as the Dust Bowl, summoning that environmental disaster into being.[12]

A sliver of good news is embedded in these findings: dust is susceptible of management. If people stir up less of it, perhaps water yield in watersheds like the Colorado's might be nudged upward—or at least some of its decline from other causes might be offset. But the prospects aren't good that management will successfully constrain the growing numbers of people bent on pursuing either fun or profit where dust is born. Moreover, the potential for dust to magnify the impact of warmer temperatures has not been factored into estimates of future runoff and river flow. The feedback loops tend not to be favorable. Lucidly, Breshears outlines the prospects: "What's going to happen to the West can be summed up in three things: fire, dead trees, and dust."

To which I might add a fourth: thirst.

ONE OF THE things that Darwin's critics disliked most about *On the Origin of Species* was that the process of evolution, driven by natural selection, had no destination. It was purposeless. It simply happened. As Darwin wrote in the immortal final sentence of his treatise, "Whilst this planet has gone cycling on according to the fixed law of gravity, from so simple a beginning endless forms most beautiful and most wonderful have been, and are being, evolved." The march of species would continue forever, but it would never *arrive* anywhere in particular.

In some ways Darwin himself never fully acceded to his theory's implication of cosmic randomness. He was, after all, a man of his Victorian time, and he may have harbored a suspicion that natural selection was contained within a larger, possibly purposeful architecture of creation. Nevertheless, the exceedingly inconvenient truth that emerged from his theory was not simply that humans were descended from apes, but that evolution operated perfectly well without divine assistance, let alone direction. This, in turn, implied that the end result of evolution might not be divinely ordained, and indeed that no end result existed—there was no end.

Most people, knowing they are on a journey, hunger to have a map, and the more authoritative the mapmaker, the better. When the journey is spiritual or philosophical, the hunger is for teleology—the idea that things are directed toward a final result, that a sense of purpose guides the march of time. A number of thinkers sympathetic to Darwin strove to excise the inconvenience from his truth and graft onto it the desired sense of destiny. The most famous of these was the English philosopher Herbert Spencer (1820–1903), who sought to apply Darwin's theory to the development of human society. Spencer glossed over the disturbing indefiniteness of natural selection and produced a set of beliefs that came to be known as "Social Darwinism." He argued that if the "survival of the fittest" (his phrase, not Darwin's) were allowed to run its course without the useless coddling of the unfit, the result would be the perfection of civilization, with the fittest in charge and everyone else in their appropriate places. Spencer replaced Darwin's infuriatingly open-ended selective process with a utopian vision. He gave the future a locatable address and postulated this destination to be a kind of heaven on Earth, notwithstanding the arduousness and suffering of the journey. Spencer prophesized that civilized people possessed the means to make the journey long or short, and that to make it short they need only allow the law of selection to operate without hindrance, even if the result appeared cruel in its treatment of the poor, infirm, and luckless. Happily for Spencer, his vision was well fitted to the interests of the moneyed classes and earned him great celebrity at the apogee of the Industrial Revolution.

Ecology, as it developed, produced its own brand of teleology. Popular concepts like "the balance of nature" promised the existence of a set of conditions embodying how things were "supposed to be."[13] The theories of Frederic Clements concerning climax states delivered a similar kind of comfort, if in more sophisticated terms. The climax ecology of a given site represented the site's fully realized potential. It was a destination. It did not just happen—it was nature's plan. If by chance the climax community was disturbed or even destroyed, nature would reassert its intentions by deploying one or more seral communities—intermediate, preparatory stages—that would lead back to the reestablishment of the climax community.

Clements's formulation borrowed notably from Spencer. He postulated that a climax plant community was not merely a convergence of individual species, but a "complex organism." This mirrored Spencer's assertion that civilization was a superorganism embarked on its own evolutionary trajectory. Likewise, Spencer's belief that the evolution of society should not be hindered by reformist intervention found reflection in Clements's dictum that a climax organism optimized Nature's energies. He said that one altered its structure or dynamics at one's peril. Just as Spencer's ideas echo today in the pronouncements of free-market advocates, so also do Frederic Clements's theories linger in prescriptions for ecological management and preservation.[14]

Experience, however, has shown that actual ecological systems are more variable than Clements's schema allow. Chance, for instance, plays a big role in determining what species, whether native or introduced, become established after an ecological community is disturbed, and the effects of particular events in a site's natural history can ramify onward indefinitely. Moreover, seral and climax communities frequently overlap, sharing species and responding to successive disturbances according to the composition they happen to possess at a given time. Other systems can "flip," as in the case of desert grassland converting to mesquite shrubland, demonstrating the potential for multiple "climax" states.[15] The orderly world that Frederic Clements described, when examined in the field, has turned out to be surprisingly rowdy. It is more haphazard, Darwinian, and contingently determined than he predicted, and it is certainly less teleological and predestined than he wanted.

As Clementsian climax theory frayed around the edges, ecologists and land managers resorted to concepts like "range of natural variability" to bound their understanding of a system's innate potential. Let's say you are managing a forest that, because of fire suppression and other human influences, has become so grossly overpopulated with trees that it is primed for catastrophic fire, posing a threat to nearby communities, as well as to itself. Where might you turn for guidance in designing a thinning program, one that reduces danger to life and property while also returning the forest to a healthier ecological condition? An understanding of natural variability, expressed in basal area, stems per acre, or any number of other variables, might provide the parameters you need. As a concept, natural variability is flexible enough to accommodate the inherent dynamism of natural systems and inclusive enough (usually) to be relevant across the infinite range of situations encountered in the field. It also need not bog down its practitioners with questions of restoration (to what?) versus rehabilitation. The thrust is simply to return the forest to a condition in which it was known to function historically and to maintain it within the bounds by which that condition is described. For these reasons, "range of natural variability" has found much favor among managers of parks, refuges, and other conservation lands as a basis for formulating management goals.[16]

The trouble is, for such an idea to have utility, the *range* of variation must have reasonably fixed boundaries, which in turn are shaped by underlying soil conditions and by climate. When climate substantially changes, those boundaries weaken and dissolve, and "range of natural variability" wobbles into the ether like a satellite in unstable orbit.

AT BANDELIER NATIONAL Monument in the Jemez Mountains of northern New Mexico, Craig Allen, a lifelong student of vegetation change, monitors landscape conditions in a number of ways. One of these involves placing graduated metal bands around the trunks of living trees so that, as the tree grows, the enlarging circumference of the trunk draws the band through a register, recording the amount of growth. In 1991, Allen and coworkers banded three groups of ten ponderosas along the flanks of the mountains, at roughly the bottom, middle, and top of their habitat. A few years later he began to monitor piñons in a similar fashion.

Members of Allen's research team read the bands weekly and tabulate the results. The graphs they generate show the rhythms of tree life: stem expansion when conditions are good and, over the course of a typical year, a generally upward trend of diameter increase. But they also reveal episodes of contraction—periods of drought when the tree dehydrates and its wood actually shrinks, like lumber in a wood yard. The drought of the early 2000s is written in their dashed lines. For the ten low-elevation ponderosas, for instance, you see the averaged growth line slacken after the onset of dry years following 1996, as the trees react to moisture stress. The line nevertheless climbs modestly until August 2001, when the grip of drought tightens. Then the trees begin to shrink, and the line dives sharply. By this point, the trees are all but gasping. And then the line runs flat through the winter as they hold their breath for snow. But the snow only comes in token dustings. The dotted red line of 2002 sinks in February, steadies in March and April, and instead of showing growth, it plunges again in rainless May. In June and July it continues falling until, in a sense, it hits bottom. Not one of the trees survives the summer.

It wasn't only the banded trees that showed the effects of drought. Cottonwoods in the drier canyons died, as did patches of juniper and many grasses and shrubs. A wide range of plants at the vulnerable lower limits of their habitat succumbed, including even the Douglas-fir, a stalwart of the montane forests. Reading the landscape of the Jemez, Allen sees a wave of mortality that "worked its way up from lower [drier] elevations within any given species. Piñon was a little different because the bark beetle outbreak pretty much took out piñon at all elevations . . . but you could see [that] all up the elevation gradient the trees are growing more poorly as the drought goes on. The crowns are getting thinner, their leaf area is going down, and these trees are hurting."

The effects of the drought, which Allen compares to a pruning of the ecosystem, do not trouble him so much as does his sense that the drought of the early 2000s was only a step away from triggering extensive die-offs among all dominant tree species, that in fact this first "global-change-type drought" was a harbinger of colossal changes yet to come. Of the Jemez he says, "We were close to the flip point, the cusp of entire systems going down, the landscape reorganizing itself. I think the overstory of the whole upper forest zones of the mountains could have almost turned over in another year or two as drought stress pushed past that point of no return—that kind of non-linear *boom, gone* thing."

Allen is a peculiar mix of pure scientist and covert land manager. Throughout his career he has worked closely with park and forest rangers, applying his skills to practical questions like controlling erosion at Bandelier National Monument and assessing grazing impacts at the nearby Valles Caldera National Preserve. He has advised foresters on restoration treatments and the wildlife impacts of recreation plans. To all of these efforts, Allen brings a sense of urgency and intensity: *let's make this better, let's get it done.* Partly he is in a hurry. Chronically overbooked, he feels he is stealing time from the many unfulfilled research and programmatic commitments already on his calendar. But partly, too, he is driven by empathy for the land. He wants it to be healthier, more resilient, and more secure from the historically anomalous crown fires that compromise its productive capacity. The drought of the early 2000s and its attendant die-offs and fires ratcheted his sense of purpose to a new level. "People, scientists, we don't get it yet," he says. "Sure, stuff dies if it gets dry enough, but I don't think we see how close it was to becoming even more catastrophic."

In recent years Allen's professional activities have taken him farther afield—to research collaborations in Europe, Asia, and Australia. He is part of an expanding network of scientists monitoring the terrestrial effects of a changing climate. A paper he authored, cocredited with nineteen colleagues, on the steady worldwide increase in forest mortality has become the scientific equivalent of a best-seller; it is one of the most frequently cited new papers on the subject of forestry and climate change.[17] "There are examples of this stuff showing up all over the planet, particularly in semi-arid zones," he says. Wherever woody plants meet the limits of their range in terms of water stress, the nearly universal increase in temperature is stressing them further. "So, when you put a normal climatic drought on top of it…, you're getting die-offs."

These waves of mortality alone are not what worry him. They have emerged with a temperature increase of only 1–1.5°C. "What happens when the background increase is 2° or 3°? We're talking New Mexico probably 4°C higher by the end of this century. What happens then?"

Moreover, "those increases are projected mean increases; they are not the extremes. But we have good reason to believe that the extreme values are going to be more extreme, and that these extreme events will be the filters that determine the life or death of individual plants and possibly whole systems." For this reason, Allen believes most climate models are too conservative in projecting ecological effects.

Allen argues that the problem of broad-scale die-off is nearly invisible in most forecasts of vegetation response to climate change because of the difficulty of projecting tree death. "We don't know where the physiological thresholds of mortality are for most species, and even for those few whose thresholds have been determined experimentally in the lab, we don't know how to model what the actual stress is to plants on a landscape . . . much less the interactions with other disturbances, like bark beetles."

Allen has become a little messianic about die-offs and the need to address the scientific uncertainties that attend them. Since no one has mastered their modeling, climate change reports such as those issued by the IPCC (the Intergovernmental Panel on Climate Change—the most authoritative voice in the field) basically ignore them. Most of the research energy on the effect of climate change on vegetation goes instead to studies of what Allen calls "natality"—how vegetation types, even individual species, will adjust to cope with the new climate conditions. Will they "march" upslope following the temperature gradient, with upper and lower boundaries shifting in parallel, or will they merely "lean" toward higher elevation, growing denser and fuller at the high end of their range, with the range remaining in place?[18]

A prevailing idea postulates that plants will follow their climate niche to new places, assuming they can move fast enough—a worrisome assumption that a lot of people are studying in important ways. But as Allen points out, "What that view doesn't take into account is, What happens first? What happens to what's dominant there now, and how does that affect what can be there next?"

Allen confesses that, although he is working constantly to get people to think about climate change, he is also "a little schizophrenic about it." He hasn't yet put much effort into projecting changes for the landscape he knows best and cares about most. He hasn't tried, that is, because imagining the transformation of his beloved Jemez Mountains is too wrenching. The pain Allen anticipates, knowing what he knows, is something that more and more people will eventually share. It may be the pain of witnessing a treasured oasis dry up or of losing the inspiring grandfather ponderosas in an old-growth stand. It may involve a scenic vista turning gray with dead trees, or the loss of habitat for a familiar bird or mammal. Besides loss, there will uncertainty, akin to the irritating uncertainty that Darwin postulated, of plunging into an unmapped future. Although Allen and a group of colleagues have begun to model the changes coming to New Mexico's mountains, he admits, "I'd rather not have to think about it. What will it look like?…Maybe I live forty years yet. I am fifty now. What am I going to end up seeing? What will my children end up seeing?"

SOMETIME IN THE late 1970s I backpacked into an area of remote canyons in northern New Mexico. A forest of very old piñon covered the mesas, and giant ponderosas rose from the damp sands of the canyon bottoms. It was December, and very cold—so cold, in fact, that I awoke one morning much earlier than I wished and decided to get moving, even though the sky was only tinged with light. I had made my camp a half-mile from an ancient Native American shrine, and I headed toward it. The shrine consisted of a ring of large, upturned stones encircling other stones that were partly sculpted. There was an opening in the circle where the upturned stones flared out to make a short and narrow corridor, a kind of entry- and exit-way. I stood behind the shrine, admiring it, shivering a little, and rooting for the wan, deep-winter sun to top the horizon of blue mountains far to the southeast. The temperature was probably in the aughts or teens, and I knew the direct rays of the sun wouldn't bring warmth so much as the idea of warmth. Still, the idea alone would be consoling. I watched as the sun seemed to ignite a small orange fire atop the mountains that hid it, then burned its way free and rose into the sky, flooding the mesa with light. I was amazed to see that the rising sun exactly aligned with the entryway to the shrine. Its dawn rays poured down the corridor of upturned stones and bathed the sculpted boulders within. The scene was a unity of opposites: the hard rocks and the delicacy of light, the brittle, temporary morning and the timelessness of the ancient shrine, which nonetheless proved to be a kind of timekeeper. Days later, back at home, I realized I had visited the shrine on the morning of the winter solstice.

Or so I thought. As the decades passed I grew less sure of what I remembered, less sure of the rightness of my conclusion that the shrine functioned, in part, as a solar observatory. I decided to go back in December 2004 to visit it a second time at the dawn of winter solstice.

The point of bringing this up is not to talk about the silent, isolated shrine, the eyeball-aching cold, or the alignment of the upright slabs to the point on the distant southeast horizon where an orange disk bled into the whitening sky. (I can say, though, that my memory was not as decrepit as I feared—the alignment was precise and unmistakable.) The point is to say a word about the mesas and the canyons. This was the winter following several bad summers of bark beetles. The old piñons of the mesa tops were uniformly dead. So were most of the great ponderosas of the canyon bottoms. The odor of vanilla that used to hover between their plates of scaly orange bark was gone. The great green boughs that formerly soared overhead had vanished, and the thick limbs that had supported them lay shattered on the ground. The magnificent trees, some of which were already spreading their branches when Coronado marched up from Mexico, were now vertical corpses. They had presided over the canyons for hundreds of years, possibly as long as the shrine had been in place. But their era had ended, perhaps never to return, and a long, morbid transition was under way. It would take a few years for the piñons to fall over, creating jumbles impassable even for deer, and it would take years more for the big pines to break and topple, each crash like a geologic event, an avalanche of wood. I trudged up the last canyon, up the steep trail toward home, already planning to return—no, to be honest, more like hoping to be fit enough to be able to return—and visit the shrine in a decade or two, there to witness the new landscape that will have risen from the bones of the old, a landscape that may be like no other that human eyes have ever seen in this place.

IN THINKING ABOUT the ecological future, it is helpful to distinguish between linear and nonlinear change. In a linear event, things change evenly and in proportion to some controlling factor. Often that factor is time. Every year or decade or century, things change a certain amount, like silt

**DEAD PONDEROSA PINES, VICTIMS OF DROUGHT AND BEETLES, BANDELIER NATIONAL MONUMENT, NEW MEXICO, 2006.** *COURTESY OF CRAIG D. ALLEN, USGS.*

accumulating in a floodplain or spruce trees gradually replacing an old stand of aspen. Nonlinear change is completely different. Things happen abruptly— a threshold is passed; new forces take over—and the effect is out of proportion to the cause. The threshold of things bursting into flame is a good example. If fires were linear, a tiny match would make a tiny fire, but of course that's not the case. The Rodeo-Chediski Fire, born from sparks, brought sudden, nonlinear change to nearly a half million acres of Arizona forest. The exponential increase of bark beetles can mark another threshold, and trigger nonlinear die-offs for millions of acres of pinelands.

Nonlinear changes are part of what makes the future hard to predict. If the climate is warming, if multiyear drought returns to the Southwest, and if the boundaries of the "range of natural variability" melt away, then nonlinear change becomes highly plausible, even likely. It might take the form of epic fires, die-offs, or other transformations perhaps unidentified as yet, all of them interacting (as with fire coursing through an area of die-off).

Because the outcome is unpleasant, not to mention revolutionary in ecological terms, those initial three *ifs* warrant close examination:

Is the climate warming? A great deal of empirical evidence says that it is, and although anomalies and uncertainties will always exist, the case for a warming climate is about as solid as any scientific case will ever be.

Will multiyear drought return to the Southwest? Absolutely. It is as sure as death and taxes.

And as for the "range of natural variation," the breakdown of its comforting guideposts is assured if the first two propositions are correct.

So what happens if vast areas are profoundly transformed? Can one sequence of nonlinear change lead to subsequent, equally nonlinear cycles of erosion and repeat fire? Again, Craig Allen: "We have this sense kind of embedded in our collective consciousness that forests are relatively permanent features on landscapes, but we observe here in these landscapes that they don't need to be." For example, some of the big fires of recent years have blown holes in ponderosa habitat so big that the pines are ill adapted to recolonize them. "We may end up with huge areas of landscape dominated by really weedy kinds of stuff, stuff that migrates quickly, establishes well in

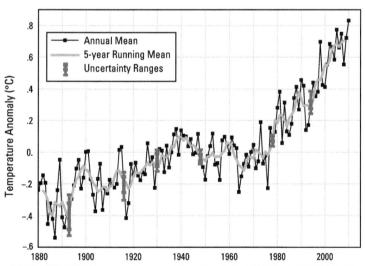

WORLD TEMPERATURE HAS RISEN MARKEDLY IN THE PAST CENTURY, ESPECIALLY IN RECENT DECADES. THIS GRAPH DEPICTS DEPARTURE FROM LONG-TERM NORMS FOR AN ASSEMBLAGE OF SITES FROM AROUND THE GLOBE. *NASA GODDARD INSTITUTE FOR SPACE STUDIES. DATA, GISS, NASA, GOV/ GISSTEMP.*

unstabilized areas, and reproduces in a short time period." Allen smiles, "On the other hand, I think a clonal shrub is a good life form to be thinking about if you were going to be something more resilient than trees in the coming century—something that can resprout, can reproduce sexually, and have seeds that are like an acorn. Like Gambel oak or wavyleaf oaks. You've got acorns, and birds can disperse those miles away, so you can get to new sites and establish. But once your clone is there, your top can burn in a fire or even die back in a drought, and then you resprout."

His smile is broader now, more boyish. He describes how the oaks he watched during the dry nadir of 2002 did just that: the tops died back during the heat wave of the summer and then sprouted for the first time that year after a spate of September rain: "I didn't even know they were phenologically that plastic, that they could do that. They were desperate to photosynthesize. That's another way to reduce your water risk, right? At the risk of starving yourself to death. For a deciduous plant, don't put your leaves on that year."[19]

The adaptiveness of the oak was a reassuring surprise. Allen expects there will be more. "My hope is that there are enough things out there that are part of the current mix, including some things with a fair amount of individual inertia—meaning some old things—that can make it a long time under the changes that we see coming. And maybe in part because we can't project it very well—I haven't wanted to go there—but in a general sense it's something that I worry about a lot, okay? ... So I think, yeah, again, a clonal woody sprouting species might be a good thing. I want to think positively about that, about the natural dynamism and resilience of individual species and ecosystems, but the reality is that I don't think we really give enough credence to how much things can change, how fast the natural order can change, beyond what we know and care about today."

We were sitting across from each other, at a table. When Allen finished what he wanted to say, we noticed that outside, beyond the window, a fresh wind was rattling the trees.

# 3

## SAND CANYON: VANISHING ACTS

IN THE SOUTHWEST the specter of climate change invites a long look into the deep past. For anyone who hunts for insights about the nature of the region and the trick of making peace with its aridity, the ubiquitous signs of vanished communities beckon irresistibly—in the ruins of Chaco Canyon, the empty cliff dwellings of Mesa Verde, and the mounded rubble of abandoned villages scattered near and far. The "lessons" they offer, however, are not always as clear as we would like them to be. Cautionary tales about the truths and errors of distant centuries can be easy to spin but surprisingly hard to reconcile to the complexity of the archaeological record, which is never static. As with any domain of science, the story told by the archaeology of the Southwest is always emerging, always gaining in heft and detail. When I went looking for someone who could help me read it, the trail I took led to the head of a rugged canyon, choked with piñon and juniper, in the far southwest of Colorado.

"There's a kiva, there's a kiva, there's a kiva," says archaeologist Mark Varien, who is vice president of programs at the Crow Canyon Archaeological Center, outside Cortez.

He points in succession to three circular depressions amid the rubble, signatures of the remains of subterranean rooms that once housed much of the life of the pueblo. Rough blocks of sandstone outline the space the kivas occupied, their roofs having long ago caved in. Wind has filled their cavities with the dust and litter of centuries. Now they bloom with cliff rose and sagebrush.

We stand just behind the kivas on a mound of half-buried building stones, which are canted at every angle—the remains of masonry rooms. To either side lie the mounds of more room blocks, their rear walls forming the perimeter of the pueblo, and the pueblo itself wrapping around the cleft of a rocky draw. The draw leads south and widens into Sand Canyon, a dry tributary of

McElmo Creek, which flows west out of Colorado and joins the San Juan River not far away in Utah. This place is called Sand Canyon Pueblo.

The trees here are big. Some of the junipers growing from the ruins are burly enough you'd have a hard time getting your arms around their trunks. But you would not want to try: they bristle with the jagged stubs of old branches. This is the kind of country where chaps, leather gloves, and a canvas jacket are mandatory for serious work on horseback. The woodland is everywhere armored with points and spines. Nothing bends; things either stab or snap.

The exception is a lone cottonwood, fat soft leaves aflutter, growing near the top of the draw in the center of the pueblo's embrace. The cottonwood marks the site of the now dry spring that was the lifeblood of the place seven and a half centuries ago.

"You don't really see a site until you map it," says Varien, moving quickly up the trail. "Mapping forces you to be more aware, to notice more." I scramble behind him realizing I am not noticing much at all. The foliage of the trees breaks the sunshine into spangles of intense light and deep shadow that obscure patterns on the ground. It is as though the ruins were dressed in camouflage. Near at hand, individual kivas and room mounds stand out, but a sense of the whole, amid the thick trees and mottled light, escapes me.

Varien has an unhurried manner but wastes little time. He moves and talks with economy, never pausing long from doing either: there is too much to explain. He ticks off the features that the mapping has revealed: 420 rooms, close to 100 kivas, 13 room blocks, 14 towers, plus other structures that qualify as "public architecture" and can't be captured in a phrase. Physically Sand Canyon Pueblo was several times larger than Cliff Palace, the largest of the famous cliff dwellings at Mesa Verde National Park, two dozen miles away. It had four times as many kivas, three times as many rooms. At the peak of occupation—within a few years of 1260, when the roof of the cathedral of Chartres was still new—as many as 500 people called it home. Sand Canyon Pueblo was the center of a community that spread across a broad landscape and included scores of outlying dwellings, shrines, satellite pueblitos—one of them tucked into a rocky alcove not far away—and extensive agricultural fields scattered over the surrounding uplands. The bones of the ancestors of the Sand Canyon people lay in those fields, which had been tilled for many generations. The pueblo did not exist in isolation, Varien explains. It was the hub of "a socially formatted landscape."

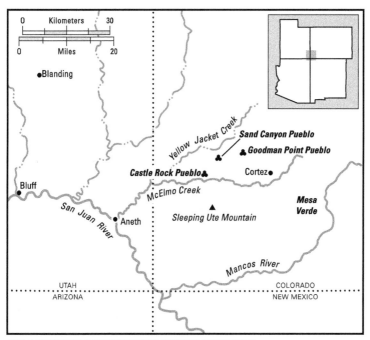

**MAP 4: THE CENTRAL MESA VERDE REGION, INCLUDING McELMO DOME.**

The startling thing about Sand Canyon Pueblo is the brevity of its occupation. Notwithstanding the enormous investment of human energy that went into building it and the fact that prior to its construction, the spring it enclosed had been the focus of a human community for at least several centuries, the pueblo itself was occupied only for about forty years. When the last of its occupants walked away, they departed not just from their pueblo but from their region, the world they had known. They abandoned its fertile soils and the trails trod by their ancestors. They also left behind many of their practices, perhaps even their beliefs, for their particular cluster of cultural equipment never reappears in its wholeness anywhere again. Of course it is possible—some would say likely—that the last inhabitants of Sand Canyon Pueblo did not walk away so much as steal into the night, refugees fearing for their lives. There are many signs of haste and disorder in the last days of the pueblo. Or perhaps they did not get away at all: the condition and posture of many of the skeletons unearthed at Sand Canyon Pueblo betray abundant evidence of violence.

But not everyone could have died. The environs of Sand Canyon Pueblo were inhabited at its peak by thousands of people—perhaps as many as 19,000 within the surrounding fifteen to twenty miles.[1] Why so many people, not to mention their brethren at better-known Mesa Verde, who departed at the same time, might have gathered up the few precious and holy belongings they could carry and set out for destinations beyond the horizon is the preeminent question of Mesa Verdean archaeology. For some archaeologists, it is a double question, asking not only what pushed them away but also what pulled them to the places where they went. In the case of Sand Canyon Pueblo, however, the *pushes* were shoves powerful enough to set people in motion. If pulling was important, it mainly mattered in where they elected to go.[2]

What happened at Sand Canyon Pueblo may have been more catastrophic than the final days of other communities in the ancient Southwest, but the issues of abandonment encountered there are not far different from those that shape the central questions of southwestern archaeology in general. Abandonment is a theme shared by all the Southwest's ancient civilizations—the Hohokam, the Mogollon, and the Casas Grandes peoples, as well as the vast group popularly known as the Anasazi, who include the Mesa Verdeans and whose petroglyphs, cliff houses, and ruined pueblos grace the land like a kind of weathered jewelry.

WESTERING ANGLO-AMERICANS IN the nineteenth century took note of the Southwest's windblown ruins and other vestiges of past grandeur; they liked to speak of "vanished races" and "lost civilizations." Many of them had read William H. Prescott's *The Conquest of Mexico*, which was a best-seller for its time. Prescott, who died in 1859, possessed either the sagacity or dumb luck to write a page-turning history of the Spanish conquest of Mexico just years before his countrymen embarked on a similar conquest—the Mexican War of 1846. The book made him so popular that in 1864 a committee of settlers elected to name their Arizona crossroads after him; for a time, the town of Prescott was capital of the territory. In subsequent years, when the beneficiaries of the second conquest of Mexico opened their copies of *Harper's Weekly* or *Scribner's Monthly* and saw engravings of cliff dwellings or Pueblo Bonito, it was not the sunbaked Utes or Navajos, living nearby, who came to mind, nor the somewhat distant, village-dwelling Pueblo Indians of Hopi, Zuni, and the Rio Grande. It was the Aztecs of Prescott's vivid pages, builders of an extravagant and powerful empire. Only Montezuma and his subjects measured up to the kind of antiquity Americans now wished to claim for their own. These notions were foremost in their minds as they set about renaming the landscape: Aztec, New Mexico, home to Aztec Ruins; Montezuma Creek and Montezuma Valley, both of which lie in Montezuma County, Colorado, a domain of 2,000 square miles currently occupied by about 25,000 people. The county seat is Cortez, named for the Spanish conquistador.

Eight hundred years ago, those same 2,000 square miles, which included Sand Canyon Pueblo, were fast growing toward a nearly equal population. It was a concentration of corn-farming people that even contemporary Iowans would have to admire.

Different stories resonate at different times. Evocations of a lost southwestern civilization—a sort of dehydrated Atlantis, its wind-haunted ruins dark with mystery—found nourishment in the romantic spirit of the Gilded Age, but that is not what excites people today. A friend of mine, a summer park ranger at Mesa Verde, says that the groups he guides through Cliff Palace or Balcony House seem to snap out of their vacation torpor when he tells them that the people who built those graceful stone villages *still exist.* The tourists, be they from Germany or Germantown, cap their water bottles and give full attention when he says that the descendants of Mesa Verde's cliff dwellers may be encountered on the streets of Albuquerque or Santa Fe, or in cities and towns across the nation, and certainly in the dusty plazas of pueblo communities from Hopi to the Rio Grande. They did not vanish; they moved away, scattering across the ancient world of the Southwest, in most cases joining existing communities, although the record is variously interpreted. Mark Varien's colleague Scott Ortman, the director of research at Crow Canyon, is a leading advocate of the theory that the Mesa Verdean people of the past became the Tewa-speaking pueblo people of the present, whose communities lie scattered along the Rio Grande and its tributaries north of Santa Fe. Ortman's highly innovative research combines a range of methodologies, including linguistic analysis, skeletal comparisons, and exhaustive exploration of oral traditions, something he calls "social memory."[3] Whether or not his views ultimately win acceptance, he has expanded the range of evidence and raised the standard of analysis that his critics will have to meet. In the meantime, the fact of a migration out of greater Mesa Verde and the absorption of those migrants into the larger Puebloan world is indisputable.

According to their descendants, the corn-farming people who built the villages of the Cliff Palace and Sand Canyon Pueblo are not *anasazi*, a Navajo word traditionally and inexactly translated as "enemy ancestors" or "ancient ones";[4] they are *ancestral Puebloans*. Most tribes accept this term, but not the Navajo, who feel *anasazi* needs no replacement. Fair enough. However one parses matters, today much of the fascination of southwestern antiquity derives not from worn-out nineteenth-century myths about disappearance, but from the saga of Puebloan continuity across oceans of time. It is a story as much about adaptation as loss, as much about tenacity and endurance as abandonment.

My friend the ranger says one of his favorite duties is accompanying tribal groups to the kivas at Spruce Tree House. The Park Service consults with twenty-four different tribes about the site, most of whom claim ties to it. The place was never abandoned, say the elders, for the spirits of those who lived there still remain. As long as that is the case, their descendants will never abandon Spruce Tree House or places like it. They might live somewhere else, but the old villages of Mesa Verde are still, in a way, *home*. The same might be said of every other ruin in the region, from Sand Canyon and Chaco in the north to the eroding walls of Casas Grandes deep in Chihuahua. From a traditional native point of view, the ruins of the region fairly brim with spirits. Logically, the presence of such beings at Spruce Tree House invites the occasional ceremony. On one occasion, however, a group of elders did a favor for the Park Service and its legions of visitors: they performed a ceremony to desanctify one of the underground kivas, so that the resident spirits would not be offended by the entry of the uninitiated. The idea was to make access to the kiva not only permissible but *safe*—southwestern lore abounds with accounts of curses incurred as a result of trespass into sacred spaces.

Every year, tens of thousands of visitors to Spruce Tree House climb down the ladder into the designated kiva. The few moments they spend in the cool, slightly musty confines of that windowless room become the climax, for many, of their visit to Mesa Verde. And as my irreverent friend says, thanks to the elders, "nobody has to worry about waking up in the middle of the night with a bone in his ear."

THE PEOPLE MAY not have vanished, but the cords of tradition and daily need that bound them to their villages frayed and ultimately gave way, like so much spider silk in the wind. In place after place, disruption overpowered continuity. We marvel at the rapidity and finality of their departure, and the reasons for our marveling are many. We want to know what went wrong when so much was demonstrably going right—the delicate but enduring architecture, the artful pottery, the unmistakable signs of subtlety and soulfulness in their worldview. We reverently acknowledge the impressive fact of their survival in an environment every bit as austere as it is beautiful. And mingled among the ideas and emotions that run through us as we contemplate their fate is a sense of kinship, a shared vulnerability that comes with living in this difficult land. It makes us want to know where things went wrong. Was it fate alone, bad cosmic luck, that drove the people to the desperate act of leaving, or did they use the land or each other in a way they should not have? If they erred, we want to know their error, lest we commit it ourselves.

TO UNDERSTAND WHAT happened at Sand Canyon, says Mark Varien, you must begin by asking the right questions. He points out that even if the people had not left, archaeologists would still ponder the pueblo's origins, for the fact of its existence embodies an enigma. People farmed corn in the uplands above the canyon, more or less continuously, for seven hundred years.

VIEW FROM SADDLEHORN RUIN SOUTH ACROSS McELMO CREEK TOWARD SLEEPING UTE MOUNTAIN. SADDLEHORN RUIN OCCUPIES A ROCK ALCOVE ABOUT A MILE FROM CASTLE ROCK PUEBLO ALONG THE TRAIL TO SAND CANYON PUEBLO. *AUTHOR PHOTO.*

Beginning in A.D. 780 the majority of the farmers gathered together in villages, but after about A.D. 900 they spread out again, making their homes instead next to the soils they cultivated, as Anglo homesteaders would do centuries later. They lived by their fields and walked to their water—drawing it from springs like the one at the head of Sand Canyon, and carrying it back to their homes in earthen jars. Then, for the last fifty to one hundred years that they occupied the uplands, they reversed their strategy: they lived at the water and walked to the fields.

The puzzle of Sand Canyon Pueblo's coming into being is no less important than the mystery of its abandonment. For centuries life on the uplands ebbed and flowed, reaching an early peak in the late 800s, then declining and building back to much higher levels in the thirteenth century. Life may not have always been fruitful, but by all indications it was sustainable. Even in the worst droughts some people stayed put and hung on, and they did this without bothering to congregate in crowded villages, without having to contend with the complications of lost privacy, community sanitation, and intensified interaction. In spite of that tradition, the people of the uplands radically reorganized themselves in the mid-1200s. For multiple reasons but with anxieties about common defense probably foremost on their minds, they elected to gather at the head of the canyon where the spring seeped forth and to build a village out of stone.

The making of Sand Canyon Pueblo was an orderly endeavor: residences were built in certain areas, communal and religious structures in others. A perimeter wall enclosed the whole. The layout, at least in its gross features, was planned. In many instances people dismantled their upland homes and carried the roof beams to the new pueblo, where they reused them. The mystery of village formation, however, has its own convolutions. Sand Canyon and nearby Goodman Point pueblos formed at roughly the same time, within years of 1250, and they lasted the same short span of roughly forty years. But fifteen miles to the north, the aggregation of Yellow Jacket Pueblo began a century and a half earlier, and in spite of a long and slow decline it lasted until the general depopulation of the area. On the matter of village formation, evidently, the medieval people of the Mesa Verde region held diverse, if not divergent, views.

What they could agree on, however, was that the land was good. As Varien and I drove out to Sand Canyon across open farmlands, Varien explained that geologically the uplands consist of loess soils blown in by the prevailing southwesterly winds. "If you imagine Monument Valley," he said, "and picture what had to erode to leave the Mittens and the other buttes standing as they do, well, a lot of what blew out of Monument Valley landed here." As soon as he says that, the rust-colored fields we pass begin to look even redder, like a John Ford panorama.

We stop and get out of the car to inspect a crop of beans, dry and ready for harvest. The soil is fine-grained, almost powdery, the kind that quickly inhabits the creases around your eyes and every crack in your skin. "It's been analyzed," Varien says. "The composition and size of the particles are considered optimal for retaining moisture. This is what made this area so good for dry-farming: even through a dry spring, the soil retained enough winter moisture to germinate the corn and keep it going until summer monsoons arrived."

Varien is blue-eyed and fair-skinned, a youthful fifty-something. His hair seems to have gone rapidly from blond to nearly white without stopping at gray. In all the pictures I have seen of him, he is squinting. Like me, he is a very untanned man in a sun-drenched world. He is squinting now. "This formation is called the McElmo Dome. It is a pretty interesting place. Underneath us is a huge natural reservoir of carbon dioxide. Over there, see that drilling rig? They are putting in another well. A company called Kinder-Morgan pumps out the $CO_2$, puts it in a pipeline, and sends it to West Texas for use in the oilfields. Something close to a billion cubic feet per day. It is one of the largest $CO_2$ deposits in the world."[5]

Varien gestures toward the plowed horizon. Apart from the drilling rig and a compressor station, almost every acre in view is devoted to agriculture. "After the Puebloans left, nobody farmed here until homesteaders came in the late nineteenth, early twentieth centuries. Then they, too, found out what this land could do." At one time the McElmo Dome's chief crop was pinto beans, which it produced even through the droughts of the 1930s and '50s—up until Mexico, a major purchaser, devalued its currency in the late 1970s. After that, farmers began to diversify, even finding success with dry-farming alfalfa, a notoriously thirsty crop. Somehow alfalfa manages to survive without irrigation in the remarkable soils of the McElmo Dome, and because of the way it concentrates nutrients under those conditions, dairies pay handsomely for it.

Clumps of brush blemish the regularity of the fields we pass. Plowed furrows, exact and parallel, abruptly stop at mounds entangled in shrubs, then resume on the other side. The mounds, explains Varien, are piles of masonry rubble, the remains of the homes of ancient farmers. For the homesteaders of more recent times those mounds were at times a resource, at other times a nuisance. Sometimes they recycled the sandstone blocks and used them in the foundations of their own houses, or to make a wall around a garden, or as steps to avoid the mud. For those who put the stones to use it was convenient that someone else, hundreds of years earlier, had troubled to muscle them up from the canyons. But just as often, if the new masters of McElmo Dome could muster enough energy or machinery, they bulldozed the rubble into a ravine to get it out of the way—which is to say, the many mounds to be seen today are a small sample of an earlier type of farmstead that was once ubiquitous across the Dome.

Most of the structures consisted of several rooms, built of stone and plastered with mud, with a kiva in front, usually facing slightly east of south, and a midden or refuse pile in front of that. This was the basic residential unit for a small, probably extended family. Today in the Pueblo world kivas tend to be reserved for ceremonial use, but centuries ago when people were not so far removed from the pithouses from which kivas evolved, they were an integral part of the domestic environment, housing countless utilitarian as well as ceremonial activities. Kiva roofs were no less important, contributing essential space as a kind of courtyard for households that worked and played more outside their walls than within them. Varien points out that if modern Zuni, the largest of New Mexico's pueblos, had the same ratio of kivas to population as Sand Canyon, it would need a thousand of them, when in fact the kivas in use at Zuni today can be counted on two hands.

When they moved from the uplands to the new village at the head of Sand Canyon, the people took their domestic architecture with them. Again they raised the stone walls of their rooms and made their kivas in front, except that instead of orienting the room blocks to the south, they angled them toward the spring that the village enclosed and toward the ravine down which the springwater ran. They left it to the crescent village as a whole, the sum of all the households, to face south, and without disturbing the overall reorganization, they also kept their kivas oriented to the south by in a sense swiveling them in relation to the room blocks. Through these adjustments they accomplished an orderly scaling up of their use of space, one that reinforced their essential alignment to the cosmos. Crow Canyon's Ortman likes to point out

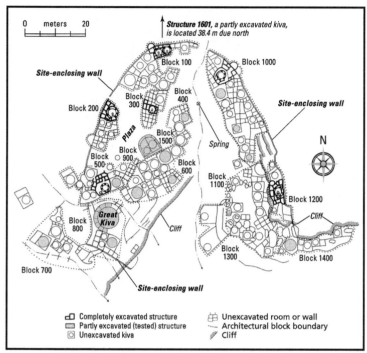

that the primary metaphors of the Puebloan world retain their meaning at
multiple scales. Etymologically, the Tewa word for "kiva roof" derives from
older terms that translate roughly as "a basket of timbers," and indeed the
cribbed roofs of the oldest kiva forms look a lot like a weaving of wooden
beams. The basket of the roof fits over the bowl of the kiva as the male basket
of the sky fits over the female bowl of the Earth. Each thing fit its opposite,
like yang to yin, and nested with others of its kind. A pottery bowl, for instance, bears nourishment, and it is made of earth. It nests within the earthen
kiva, a place of many kinds of nourishment, which in turn nests within the
village, which is also made of earth and which wraps itself around the spring,
forming the shape of a bowl, a bowl that is cupped into the earth at the head
of the canyon. Each unit of life is held by the next larger one, with no part fail-
ing to fit and no part left out, and with a male basket appropriate to each
female bowl also nesting in its place.[6]

BOWL, LADLE, AND MUG, SAND CANYON PUEBLO. THE CRAFTSMANSHIP AND ARTISTRY OF THESE OBJECTS EXPRESS THEIR MAKERS' HUMANITY ACROSS A TEMPORAL GULF OF NEARLY 800 YEARS. *PHOTOGRAPH BY WENDY MIMIAGA, COURTESY OF CROW CANYON ARCHAEOLOGICAL CENTER.*

Clues to the past are always fragmentary: a few scraps of oral tradition, the derivation of a word, correspondences between a certain style of pot decoration and the design of murals on a kiva wall. The world of the ancients becomes sensible to us only indirectly, but the puzzle, as it is pieced together, suggests a world resonant with symmetry and impressively complete. Varien says his sense of awe has deepened with time. He has come to view the philosophy of the people he studies as "a worldview based on acute observation of nature and people's place in it, and in that sense it is not far different from science." It is a philosophy, he says, "that meaningfully and effectively locates people in the cosmos." The things they made, from bowls to buildings, expressed a deep centeredness. Their quality of being rooted in place makes their decision to leave—and not to come back—all the harder to fathom.

"Maybe it helps to think of aggregation as a strategy to avoid leaving," Varien proposes. When people gathered to begin construction of Sand Canyon Pueblo about 1250, their world was shifting. Population was higher than it had ever been, reaching densities of nearly eleven people per square kilometer on McElmo Dome.[7] The land was crowded, and quite possibly not just because of natural increase. A flow of immigrants moving into the area, likely from the west, may have greatly complicated life on the Dome.

A plausible theory, based on tree-ring and pollen analysis, holds that sometime around A.D. 1180, the climate shifted, as it has done repeatedly over the centuries. The weather turned cold and stayed relatively cold for twenty-five years or more. As a result, lands less favored than McElmo Dome now failed to produce a crop. Such lands need not have been at high altitude, not even as high as McElmo Dome and Mesa Verde. They might have been canyon bottoms vulnerable to long shadows and cold air drainage.[8] In any event, if significant numbers of people left the places where they had been living and made their way to the red soils of McElmo Dome, the people who were already there would have faced a dilemma: should they embrace the newcomers, or some small group of them, as recruits to their community?

The arguments both for and against acceptance would be as numerous as they were unanswerable. Some people might have argued that the benefit of more willing hands to do the community's work would outweigh the increased drain on resources. In particular, they might have asserted that acceptance by the community of additional potential fighters would provide the margin of security to defend the pueblo against other waves of intruders who might soon come. But the intruders of the future were not the only threat. The pueblo no doubt already had enemies, who were perhaps absorbing other bands of newcomers, swelling their ranks and increasing the threat they posed. Moreover, arguments for recruitment need not have been solely defensive: some within the pueblo might have been hoping for new allies so that they might strike out against their existing enemies, lest they be struck. Yet others might have said such reasoning was nonsense; they were strong enough as they were, and they should drive off the newly arrived, impoverished, and homeless newcomers, even if by doing so they risked anger and retaliation, because their best gamble was to stay small; they could not afford more mouths to feed, more hearths consuming wood, more jars to fill with water, more needs to meet and arguments to settle.

The pressures of immigration surely generated powerful social stresses in the ancient world, just as they do today. Directly or indirectly, they may have provided the impetus for people to abandon their scattered upland farmsteads and build a rimrock village protecting its water source. Undoubtedly, other forces were also at work. Social turbulence takes many forms, violence being only one. If new people streamed into the area, others were likely departing, breaking old relationships and leaving "holes" in the socially formatted landscape. The builders of Sand Canyon Pueblo were among those who elected to stay. They would make a stand.

Hewn from the rocks at the head of the canyon, Sand Canyon Pueblo presented a formidable face to visitors. Its enclosing wall restricted access to the village core. The equivalent of loopholes allowed inhabitants to see without being seen. Two-story towers within the pueblo, as well as other towers and sites nearby, were visually linked and commanded a view of the surrounding area. Sand Canyon Pueblo may not have been a fortress, but it had a defensive character—and events proved that it needed such defense as it could muster, and more.

A group of Varien's colleagues at Crow Canyon Archaeological Center began to investigate Sand Canyon Pueblo in 1983. Fieldwork continued, off and on, for roughly a decade, but only a portion of the site—about 5 percent of the total, consisting of at least part of 111 structures—was actually excavated. Several rooms and kivas in Block 100, where Varien leads me now, were part of the sample. Varien explains that the usual expectation in such excavations is to find "no artifacts left behind because people move to a nearby place and they take everything with them." Even then, he says, the process is exciting. "When you're removing the fill, you know when you hit that floor, and you can actually take your trowel and twist it, and the fill above the floor just pops off. Uncovering those floors that haven't been walked on in seven hundred years is just a thrilling experience, an incredible connection to the past."

But at Sand Canyon, "because this site was occupied right up until the end and people were migrating a long distance when they left, the floors were just littered with artifacts. There were really, really rich assemblages, something that is not the rule in archaeology." Sometimes the surprises lay under the floors. When the field team removed the plaster floor of a kiva in Block 100, the archaeologists found a petroglyph carved into the canyon bedrock. It was worn and indistinct, but seemed to depict a flute player—the famous and enduring image of Koko Pelli, the humpbacked serenading seducer whose likeness has been pecked into rock walls across the Southwest.

In another room they found—buried in the rotted roof beams and the blown-in dust and litter—the remains of a man. "He was really distinct skeletally from the rest of the population. He was more robust," says Varien. "And he died from a blow to the head." This individual was officially recorded as Human Remains Occurrence 2, or HRO 2. Less officially, the archaeologists took to calling him Block 100 Man. He was relatively big for his time and place—nearly five and a half feet tall—and, to judge from his unusually thick collarbone, powerfully built. Kristen Kuckelman, another longtime Crow Canyon researcher, meticulously teased out Block 100 Man's story from the evidence of his bones. He was between forty and forty-five years old when he was killed, but by then he was no longer in his prime. He'd been close to death before, too. A healed depression fracture on the back of his skull showed that something, or someone, had once conked him very hard.[9] He suffered from dental abscesses and possibly other ailments of the mouth. He had arthritis in both elbows, and the pattern of wear and tear on his bones was that of someone who made a great many repetitive movements—scraping, grinding, smoothing, polishing. Together with the large number of stone tools found in the Block 100 midden, the condition of his skeleton suggests that Block 100 Man was a craftsman. Perhaps he made flutes like the one Kokopelli was shown playing in the petroglyph under his floor. Perhaps he processed leather, for his teeth were worn in a way consistent with the kind of chewing such an activity would have involved. Perhaps he was a jack-of-all-trades, handy at making all kinds of household equipment: flutes, arrows, effigies, sandals, whatever was needed.

A young woman died a few feet away from him in an adjacent kiva. She was "gracile," to use Varien's word, nearly as tall as the man, long-boned, and slender. Her bones were fragile and much curved, as polio or some other disease might have deformed them; she was quite likely frail. Anomalies in the formation of her teeth were similar to Block 100 Man's; perhaps she was his daughter or his niece. The cause of her death is unknown; it could have been murder by something other than a blow; it could have been illness or starvation.

The blunt trauma evident in the skull of Block 100 Man is a pretty good indication of violence. The weapons of choice in the ancient Pueblo world included stone axes and mauls, tools with which farmers cleared land and hewed building timbers. Such tools were handy in a fight. The farmers also had bows and arrows, possibly spears, but the wounds left by such weapons, even if fatal, were rarely recorded in the bones of their victims, at least not so graphically as a blow from a club. Another indicator of warfare is the lack of intentional burial. When archaeologists encounter the remains of a body that evidently lay slumped on the ground as it decomposed, showing no sign of having been carefully placed, and unaccompanied by funerary gifts, they fairly conclude that the last people in the company of the deceased were not his friends, or, if they were, that they lacked all opportunity to attend to him.

This appears to have been the fate of the robust craftsman of Block 100. When he was killed, he likely fell, or perhaps was thrown, into the room in which he was found. And there he remained, sprawled on his back and neglected, as his flesh joined the soil and his bones stayed behind. Many others at the pueblo suffered similar fates. Excavators found the remains of at least seven other individuals whose bones showed evidence of violent death, including scalping, as well as twenty-six others, causes of death unknown, who lacked burial. Moreover, the skulls of a number of these individuals showed evidence of healed cranial fractures, indicating that violence had been stalking the community for some time.

Because the remains of these thirty-four individuals were unearthed in the excavation of only 5 percent of the pueblo, the actual toll of violence was surely much higher. Interestingly, too, the fatalities at Sand Canyon included

THESE OLLAS, BOWLS, MUGS, AND OTHER VESSELS WERE RECONSTRUCTED FROM MATERIAL FOUND IN THE KIVA OF BLOCK 500, SAND CANYON PUEBLO. (OTHER MATERIAL MAY HAVE BEEN TOO BROKEN OR SCAT-TERED FOR REASSEMBLY.) PRESUMABLY THE ASSEMBLAGE REPRESENTS A SIGNIFICANT PORTION OF THE BLOCK 500 HOUSEHOLD'S MATERIAL POSSESSIONS. *PHOTOGRAPH BY ROBERT JENSEN, COURTESY OF CROW CANYON ARCHAEOLOGICAL CENTER.*

no men of fighting age. It may have been that the community was attacked when the members most responsible for its defense were away—perhaps on a hunt, perhaps on a raid of their own. Or, with brutal realism, they might already have migrated with their families to the south, leaving the old and weak unprotected.

Evidence of a cruel end to the pueblo is not restricted to the bones of its inhabitants. Valuable objects were left behind, sometimes scattered in unlikely ways, some of them clearly ceremonial and precious and also small, lightweight, and easy to carry. By all indications, the pueblo in its last days was the scene of a desperate confrontation, or series of confrontations, and the result was fatal not only to scores of individuals, but also to the community as a whole.

SAND CANYON PUEBLO had company in its suffering. A few miles down-stream, toward the confluence of Sand Canyon and McElmo Creek, another stone village, built to wrap around an imposing solitary butte, met a similarly bitter end. Crow Canyon archaeologists investigated Castle Rock Pueblo in concert with the excavations at Sand Canyon. Today, the remains of the stone houses that once climbed the sides of the butte are much less defined than those at Sand Canyon, and all evidence of the excavations and traffic of archae-ological work have been erased. The butte itself is nearly barren, a crumbling and wind-blasted geologic hulk that rises above the surrounding landscape like the keep of a ruined fortress. Natural stone terraces bound it on one side; an arroyo thick with piñon and juniper runs along the other. From the stand-point of a footsore human on a hot day, the mineral-heavy water of McElmo Creek, distant by at least a half a mile, seems a long way off.

Varien has written that "Castle Rock has become one of the best archae-ologically documented cases of warfare in the entire Southwest."[10] Crow Canyon's excavations revealed that at least 41 of the pueblo's roughly 100 inhabitants (estimates range from 75 to 150) were massacred in the commu-nity's last days. Human remains recovered there exhibit the unhealed fractures associated with wounds received at death. Interestingly, they also bear evidence of healed fractures, an indication that the people of Castle Rock, like their neighbors at Sand Canyon Pueblo, were no strangers to vio-lence. There is also evidence that the final massacre entailed a range of atroc-ities, including cannibalism, of which there is some slight suggestion at Sand Canyon Pueblo, too.[11]

Sand Canyon and Castle Rock were built and abandoned at roughly the same time, but Castle Rock sprang up in an area that, unlike the uplands above Sand Canyon Pueblo, had not been densely occupied for a long time. Castle Rock's environment was less advantageous; lacking a nearby spring, the villagers built a reservoir to capture runoff, and the soils they tilled were inferior to the loess of the uplands. It is possible that Castle Rock's inhabitants were relative immigrants to the region—perhaps an assemblage from different bands, who may have settled not so much where they wanted as simply where they could. It is also possible that they made their settlement at Castle Rock over the objections of nearby communities, thereby earning their enmity. The residents of Sand Canyon, Goodman Point, and other nearby pueblos might have resented the newcomers' placing additional pressure on wild game and other resources already depleted by centuries of human dependence. But the opposite possibility deserves consideration as well: that the Castle Rock people settled with the encouragement of those communities, whose members saw greater threats farther afield and sought advantage—and safety—in numbers. In either case, at Castle Rock the strategy of survival failed.

The identity of Castle Rock's attackers—and of the perpetrators of violence at Sand Canyon and throughout the region—is poorly understood and much debated. No archaeological evidence has yet been found to place the southward-moving Athabascan ancestors of the Navajo in the Four Corners region as early as the 1200s, and the same may be said of the ancestors of the Utes, who remain a major presence in Montezuma County today. But because the progenitors of both tribes lived lightly on the land, building no houses of stone and moving frequently and in small groups, for all intents and purposes they were archaeologically invisible. A firm date for their arrival in the region remains elusive. The oral traditions of both tribes, meanwhile, assert that they have been in the Four Corners area essentially forever.

Even supposing that early progenitors of either tribe might have reached the region by the thirteenth century, or that another outside rival—the Fremont people of central Utah, for instance—might have made incursions, it is tough to see how such people might have overthrown a fortified pueblo. The dynamics of military confrontation suggest that to storm and carry a strong defensive position, like Castle Rock, the attackers must substantially outnumber the defenders. But bands of roving hunters would have been unlikely to unite in numbers sufficient to mount such an assault; had they done so, they might well have left more sign of their presence. Nevertheless, the question remains open. Perhaps Castle Rock's main defenders were incapacitated, if not absent, as was apparently the case at Sand Canyon in its last days. Enough projectile points of distant origin have been found at both places to fuel speculation about invaders, but not enough to support a conclusion. In general, the preponderance of evidence directs the spotlight of culpability within the Puebloan world, and to the likelihood either that local pueblos were feuding with each other, or that interregional strife—conflict between, say, Mesa Verdeans and communities farther south along the San Juan River—had sickened the region with war.[12]

OTHER FINDINGS AT Sand Canyon Pueblo add detail to the picture of the community's last days. Analysis of the middens—the things thrown away—indicates that the proportion of corn among plant remains dropped precipitously as the pueblo approached its end and that wild plant foods, like goosefoot and lambsquarters, proportionately increased. Likewise, turkey bones diminished and the bones of rabbits, rodents, and other wild animals rose in frequency. Turkeys were the Puebloans' only domesticated animal besides dogs, and they added vital protein to the community's corn-based

CASTLE ROCK PUEBLO, 1874, BY WILLIAM HENRY JACKSON, VIEW SOUTH-SOUTHEAST WITH SLEEPING UTE MOUNTAIN IN THE BACKGROUND. RUINS OF THE PUEBLO EXTEND DOWN SLOPE FROM TOP OF BUTTE INTO FOREGROUND. *COURTESY OF HISTORY COLORADO (WILLIAM HENRY JACKSON COLLECTION, SCAN #20101440).*

subsistence. Complex computer simulations that take into account nutritional requirements, multiple hunting strategies, and numerous other factors have shown that the ancient people of McElmo Dome had significantly depleted the deer in their environment by about A.D. 900, even at relatively low levels of population. Only by feeding a portion of their corn to turkeys and thereby converting it into protein could they have sustained the population densities they achieved.[13] When the turkeys were gone or few—because they had been eaten or because they could not be fed—snaring packrats and rabbits could hardly fill the void.

Hunters from the pueblo evidently strove to fill the void another way. Some of the later accumulations at Sand Canyon include the bones of pronghorn antelope and bighorn sheep, neither of which was likely to be found within a day's or even two day's travel from the pueblo, suggesting that hunters from the pueblo were trekking increasing distances to bring home protein-rich meat. Such efforts may account for the pueblo's lack of defenders in its final agonies.

The decline of farmed foods, both corn and turkeys, and the evident increase in foraging indicate that Sand Canyon's crops had failed and that starvation stalked the last residents of the pueblo, a people who, in earlier years, by the evidence of their bones, had been well nourished. Crop failure, however, was a familiar problem in the ancient Pueblo world; by some estimates it may have occurred at least once every six or a dozen years, certainly frequently enough to fall within the experience of every adult within the community.[14] One way the ancient farmers of the region adapted to the vagaries of climate was by learning to store food effectively. In large pots and in dedicated rooms and granaries specially plastered to keep out rodents, they typically preserved enough grain to last them for up to three years, which seemed to be the practicable limit before the corn decayed. Well into modern times, many members of the Pueblo world continued the practice of keeping a three years' supply of grain always in reserve. It was a habit shaped by hunger, deeply engrained and long tested. If, in its last years, Sand Canyon's reserves were depleted, one can only conclude that the crops had failed for at least three consecutive years.

Bad weather can induce crop failure, and as will be seen, climate played a determinative role in the breakdown of agriculture in the Mesa Verde region. But the impact of climate can be expressed in multiple ways, not all of them direct. If because of raiding or other dangers farmers cannot sow or care for their crops, their farming will have failed as surely as if the crops had wilted from drought. The only success is in the harvest.

History has shown repeatedly that famines result as much from warfare as from lack of rain, and that wars are often the spawn of environmental stress. A recent and glaring example of this has inhabited the international news for over a generation, as the epicenter of starvation in sub-Saharan African has moved from Ethiopia in the mid-1980s to south Sudan in the 1990s to Darfur in the 2000s. It is as though a virus of violence were creeping across the region. The virus, in turn, seems to have drawn much of its strength from miseries induced by a drying climate—an expression, possibly, of an expanding or intensifying Hadley cell. In south Sudan and Darfur, for instance, competition for grazing land, water, and other subsistence resources underlay the efforts of Arabic-speaking Muslims from the desiccated north to seize the lands of black animist and Christian tribes to the south.

Something roughly analogous may have occurred in the subsistence agricultural economy of the medieval Southwest. Drought that led to crop failure may have also led to violence as hungry people attempted to occupy the lands or seize the goods of better-fed neighbors. As with the assassination of Archduke Franz Ferdinand in 1914 or the chimera of Iraqi weapons of mass destruction in 2003, a war's precipitating event may have little to do with underlying causes. Among ancestral Puebloans an exchange of insults might have grown into a feud, which led to a beating, which required revenge, and so forth, all in an atmosphere of want and fear and hair-trigger nerves. An increase in conflict—or even the threat of conflict—might then have resulted in smaller harvests from the maize fields than the climate might otherwise have permitted. The chain of cause and effect would have produced still more cohorts of hungry people whose reserves of stored grain were less than they should have been and who were understandably willing to fight in order to survive. And so as the noose of drought tightened, the cycle intensified.

Simulations and detailed analyses show that at Sand Canyon, human impacts on the environment were heavy, but not limiting. The constant need for cooking and heating fuel thinned the woodlands, but even at the last there remained wood enough, close enough. And the spring, too, seems to have provided sufficient water for as many thirsty mouths as depended on it. But the stresses in the land were not independent of each other. In a country filled with more people than it had ever before supported, each increasing difficulty frayed the fabric of subsistence. And then when the corn failed, even if it failed far away, the peace also failed, and the failures of peace and nourishment were two sides of the same coin.

IT SEEMS THAT, when the climate changed, it did not change in just one way. Perhaps it never does. For more than a century before Sand Canyon Pueblo was founded, from the end of the tenth century to the beginning of the twelfth, climatic variability was relatively muted in the Four Corners region. It was not "stability," perhaps; the graph of reconstructed precipitation is still jagged for that period, but the extremes are less pronounced than in most other intervals of comparable length. During this time a political and religious system based in Chaco Canyon rose to prominence. Some say the Chacoans ruled by might, even terrorism, but by whatever means, voluntary or autocratic, they harnessed a great amount of human energy over a long period of time for the pursuit of essentially public projects. Thanks to consistent and relatively bountiful growing conditions, the farmers of that era managed to produce surpluses, and surplus corn meant that labor could be devoted to projects other than agriculture. Public labor in turn made possible the canyon's monumental architecture, together with the life those buildings supported, dominated by a priestly elite.

A half-century of drought beginning about 1130 brought the Chacoan phenomenon to an end. In the reconstructed record of the region's climate, now 2,000 years long, no drought has surpassed it. As surpluses gave way to acute shortage, populations thinned. Some people stayed where they were; some dispersed. Some groups settled beside perennial rivers like the Animas at Aztec. Others migrated longer distances, trying first one place, then another. Some did not stop moving until they reached the Rio Grande, and when they did, they no doubt upset relations among the people already there.[15]

When the drought eased, conditions did not return to what they had been before. Analysis of pollens trapped in lake sediments (which record changes in vegetation) suggests that sometime around 1180 the weather grew wetter but also colder, shortening the growing season and possibly triggering an exodus, mentioned earlier in this chapter, from the cooler limits of the farming frontier.[16] Some of the people uprooted by the change probably migrated to the lands of the McElmo Dome, adding measurably to population pressure and social tensions. Then, around 1250, roughly coinciding with the construction of Sand Canyon Pueblo, the climate appears to have shifted in a way it had not done in the previous six centuries—or in more recent centuries. The weather, which in the Southwest is always capricious, became markedly and, from a human point of view, destructively more so. Its instability, coming at a time when human societies were pushing the limits of the land's carrying capacity, trumped the long-tested agricultural adaptations of the ancestral Puebloans and took those societies to their breaking point.

It is not a simple story. Jeff Dean and Gary Funkhouser of the Laboratory of Tree-Ring Research at the University of Arizona teased out its main features from an immense quantity of tree-ring data drawn from twenty-seven locations and going back 1,500 years.[17] The basic procedure of tree-ring study is well known: most trees accumulate annual growth rings, which are thicker in wet years, narrower in dry ones. By assembling chronologies of tree rings for a given location and calibrating those chronologies in various ways, the location's precipitation history can be reconstructed.

Dean and Funkhouser looked for shared patterns among their twenty-seven chronologies, drawn from sites scattered across the Southwest, and they found that the chronologies divided into two geographical groups with roughly the same ups and downs. One group included the Mogollon Rim of Arizona, the southern Colorado Plateau, and the mountains of northern New Mexico. The second group consisted of everything south and east of the first. An S-shaped line separated the two zones, and it wobbled around a good deal from century to century, but never too far. The two zones remained separate and consistent for the entirety of the 1,500-year chronology—except, that is, for the 200 years from 1250 to 1450.

The two zones reflect the two primary patterns of southwestern precipitation known from historical times. The southeastern zone experiences what Dean and Funkhouser call a "unimodal pattern," in which summer precipitation, mainly consisting of thunderstorms, dominates. By contrast, the northwestern zone is "bimodal"—precipitation from both winter storms and summer monsoons tends to be equally important.

Dean and Funkhouser's discovery was that the stability of the bimodal and unimodal weather zones broke down around 1250. Instead of manifesting two zones, the data erupted into four, and for some periods six, groupings. From one decade to the next, the zones shifted shapes and moved around, like cells in a Petri dish. The worst of the fragmentation was in the northwestern zone, centering on the Four Corners region, which included Sand Canyon Pueblo and its neighbors. The breakdown of the weather zones meant that the climate grew chaotic. Places that were relatively close together experienced markedly different weather conditions. Few places experienced conditions consistent with those that had dominated their weather for the previous six centuries.

Something big had changed, but it is hard to say exactly what. Some scholars hypothesize that the summer monsoons failed—they arrived late or not at all, rendering attempts at dry-farming futile.[18] Perhaps the region had cooled and could not generate convective air currents strong enough to draw in summer moisture. But perhaps not. Dean, who knows the data best, declines to speculate on the exact nature of the climatic breakdown beyond saying that "it would have had profound adaptive implications for the populations of the affected area (the northwest) and important secondary impacts on the people of the unaffected southeast who had to put up with immigrants from the affected area." He says that the monsoons might have failed, but that the data could also "reflect a failure of the winter (frontal) component, which would be just as disastrous to farming in the area."[19] In either case, the breakdown of regularity in the climate was a crushing blow to the people of the Four Corners region.

Interestingly, the same period saw an explosion of population in the Galisteo Basin southeast of Santa Fe, an area that until then had lain beyond the limits of the Puebloan corn-farming world. But conditions now seemed to favor it, and people thronged there, forming small villages at first, then gradually aggregating into larger and larger communities, no doubt in response to a rise in intercommunity conflict. By 1400, the Galisteo Basin was home to some of the largest pueblos ever formed, running to thousands of rooms and housing thousands of people. Had the pueblos of the basin been built less with adobe and more with stone, like communities at Chaco and Mesa Verde, they would surely be the jewels of a national park today. Their chief building material, however, turns to mud in the rain, and long ago, the walls of San Lazaro, San Cristobal, Pueblo Blanco, and their cousins collapsed into unphotogenic mounds that sparked little appeal in the popular imagination.

The decline of Mesa Verde and the rise of Galisteo appear to be opposite manifestations of the same change in climate, but in social terms they do not appear to be closely related. Some Mesa Verdeans may have found their way to Galisteo, but the basin seems to have drawn its population mainly from nearer sources along the Rio Grande. Still, the arcs of growth and decline in the two regions ended similarly. The climate phase that favored the Galisteo Basin did not last. Shortly before the Spanish arrived in the mid-sixteenth century, the productivity of maize cultivation in the basin began to decline, and the pueblos declined with it. They were already in a weakened condition when the authorities of New Spain began to tax them for food and labor. Increasing predation by Apaches and later by Comanches hastened their end.

Eventually the people of the Galisteo pueblos, notwithstanding Spanish efforts to hold them in place, moved away. As was the case in the Mesa Verde region, they did not move all at once, nor all to the same hoped-for refuge. But, in time, the area emptied out. Today, anyone who might attempt to dry-farm corn in the Galisteo Basin, as thousands of Puebloans in the fifteenth century successfully did, would be branded a lunatic.

FOR AS FAR back as one can look, the vagaries of climate have shaped the ebb and flow of human occupation of the Southwest. Rarely has climate remained consistent for more than two or three centuries at a time, although *consistent* hardly means *unchanging*. Rather, it means that variability operates within a consistent range, with bad times including good years and good times including bad. Short-term variability is the rule, not the exception. People plan for it, and adapt to it. But the long-term variations are something else. No one sees them coming. When they do, human cultures undergo wrenching change.

The florescence of the so-called Pueblo I period ended with a period of regional cooling and drought in the A.D. 800s, forcing a cycle of dispersal and shrinking population as people generally moved to lower altitudes. Then came Chaco, which flourished through the A.D. 900s and 1000s, a period shown by Dean and Funkhouser's analysis to have been "especially favorable." The great drought that began in 1130 triggered the decline of Chaco and forced another dispersal and regrouping, but when that drought eventually eased, settlement on the McElmo Dome and throughout the Mesa Verde region gained momentum, reaching its apogee in the middle of the thirteenth century—about the time Sand Canyon Pueblo was founded. This period of growth was in turn brought to an end by more climatic instability, including a crushing drought from about 1276 to 1299, termed "the Great Drought" by

A. E. Douglass, the founder of dendrochrology. The Great Drought was the final blow to a civilization already at the point of collapse. A sequence of very bad years in the early 1280s destroyed the hopes and prospects of the last holdouts. Probably by 1285, certainly by 1300, the exodus of farming people from the Four Corners region was complete.

The changes went on from there—through the growth and decline of the Galisteo Basin pueblos, through the Pueblo Revolt of 1680, for which decades of drought were an underlying cause, and through other cycles of relative plenty and want—and on to the restless present. If anything, the modern era has been a long departure in both climatic and cultural terms from the kinds of conditions that caused Fray Atanasio Dominguez to describe the Pecos Indians in 1776 as wretches "tossed about like a ball in the hands of Fortune."[20] Notwithstanding the trials of the Dust Bowl era, which launched a mass emigration of 2 million people from the Plains states, the ability of modern society to shield its citizens from the harshest impacts of environmental change has yet to be fully tested.[21]

Eventually it will be. Jonathan Overpeck says he's not sure who first coined the term *megadrought*. It may have seen its first published use in a paper he wrote with his colleague Connie Woodhouse on the climate history of the Great Plains. They used it simply to refer to a decades-long drought. The ninth-century drought that, combined with cold weather, brought the Pueblo I period to an end was a megadrought. So was the twelfth-century drought that ended the florescence of Chaco. The quarter-century-long drought that descended on the region in 1276 and delivered the final blow to Sand Canyon Pueblo was yet another.

It turns out that in the long view of world climate, such droughts are common. They tend not to show up in the relatively recent record documented by rain gauges, thermometers, and other modern equipment, but according to Overpeck, wherever scientists have reconstructed the prehistorical climate of semi-arid regions around the world, they have found evidence of "these very long droughts, much longer than anything we see in the instrumental record." More than the prospect of 130°F days in Phoenix and other potential manifestations of climate change, it is the specter of megadrought that worries him. "I'd say if anything is scary, that the scariest is the possibility that we could trip across a transition into a megadrought."

The triggers for the megadroughts of the past are not well understood. Probably, at a minimum, they had to do with the currents and temperatures of the oceans, the kinds of phenomena climatologists reference by their acronyms: ENSO, the El Niño Southern Oscillation; PDO, the Pacific Decadal Oscillation; and AMO, the Atlantic Multi-decadal Oscillation. It was likely some unhappy conjunction of these cycles that tripped the climate of the Southwest into the megadroughts of the past. Whatever the trigger may have been, it certainly did not include a 42 percent increase in atmospheric $CO_2$ over the preindustrial baseline, which is what we have now. We must hope that the trigger for the next megadrought will not be so delicate that the cluster of effects brought on by greenhouse gases might set it off.

Unfortunately, there's a possibility that the next megadrought has already begun—we just don't know it yet. The character of a drought becomes clear only retrospectively, and every long drought includes wet years that break the pattern of sustained dryness. A megadrought, by definition, manifests only over a span of decades. In a way, our decisions for the future should be the same, no matter whether we are a few years inside a megadrought or lucky enough to have decades of relative abundance ahead of us. Deep, crushing cycles of drought are part of the natural history of the Southwest and, for all practical purposes, they always have been. Building resilience against drought into the region's water systems and cultural practices would be a wise course, irrespective of the cause or timing of the next emergency. Perhaps the dangers now arising from anthropogenic climate change will goad us into doing the things we should have been doing all along. This is especially urgent because the next megadrought will pose unique challenges. "You can probably bet your house," says Overpeck, "that unless we do something substantial about these greenhouse gas emissions, the megadroughts of the future are going to be a lot hotter than the ones of the past. So their impacts are going to be a lot more dramatic."

AT SAND CANYON Pueblo one day, Mark Varien had a kind of epiphany. He was testing a small site nearby, and nearly every day at lunch he would take the Sand Canyon map and go over to the ruins and walk among the mounds and crumbled walls, visiting the kivas. He was trying to absorb the place, letting it sink in. Eventually he stood in every kiva, every circle on the map. They were all small, just a few paces across. One day he walked between some trees and rubble into an open space between Block 500 and Block 800, and then just stood there, looking around. He noticed four masonry piles evenly spaced in the clearing. He noticed also that a small arroyo, not much more than a ditch, came into the space as though to run through it, but was deflected by something and took a strange turn. "It had to hit something that stopped it." Something hard and permanent. He went over to investigate. The arroyo seemed to follow a buried structure around the edge of the space. Whatever it was, the structure was circular, and it was big. Then it hit him: he was standing in a great kiva.

Anyone who has visited Chaco Canyon has seen a great kiva. Some of them are more than sixty feet across, perfectly round, and as much as fourteen feet deep. They are among the most spectacular features of that spectacular place, and they are a kind of signature of the Chacoan worldview, an architectural embodiment of the religious and political system that was based there. Days before finding the great kiva at Sand Canyon Pueblo, Varien had gone with a friend to the ruins at nearby Goodman Point. It was a kind of busman's holiday: an archaeologist on his day off, pondering an ancient site. One of the things he pondered at Goodman Point was a clearly defined great kiva. He wasn't looking for one when he walked into the clearing at Sand Canyon, but perhaps the image was in the back of his mind, waiting to be awakened. In any event, he rushed back to tell his colleagues at Crow Canyon of his find, and there was excitement, followed by discussion, inspection, and more discussion. In due course, a grant was solicited, money became available, and the trowels began probing anew.

What they found was both more and less than they expected. More in that the kiva offered surprising insight into the story of the pueblo, but less in that it was missing something. The kiva was large—its interior space roughly forty-six feet in diameter. It had a firepit, a floor vault, a stone bench around the periphery of the room, and other necessary accouterments. The four masonry piles Varien had initially noticed were the remains of pillars, oriented to the cardinal directions, which ordinarily would have supported the primary beams of the roof. But Varien's colleagues at Crow Canyon, much to their surprise, found no roof material collapsed into the space. They also determined that the masonry piles, had they been reassembled into pillars, would have been too short to support a functional roof. They soon concluded that the Sand Canyon great kiva, unlike its older counterparts in Chaco Canyon, was open; it had never had a roof. The undersized pillars may have been symbolic. The roof they supported was metaphorical: the basket of the sky.

Because the great kivas at Chaco were roofed, they were also private. Access was restricted. Presumably, only the chosen were allowed in. Had the great kiva at Sand Canyon been roofed in the fashion of its Chaco predecessors, the structure might have accommodated about 100 witnesses to its ceremonies. One row of people might have sat on the stone bench, with another row sitting on the floor in front of them. But if there were no roof, if the kiva were an open amphitheater instead of a closed chamber, then a third circle of viewers—or participants—might have perched atop the exterior wall, and still another ring might have stood behind them on the roofs of the adjacent buildings—much in the manner that rooftops provide a kind of grandstand for dances at many pueblo plazas today. With the roof omitted and the sacred theater opened to the sky, the great kiva at Sand Canyon would have been democratized; it might have held an audience of 250, fully half the population of the pueblo at its peak.[22]

Such a structure invites multiple layers of interpretation, all of them speculative. The truth in such situations is almost always different from what we imagine it to be, and our imaginings keep changing as new information comes to bear. But putting caveats aside, consider for a moment, the following scenario.

The Mesa Verdean world in the first half of the thirteenth century is turbulent. There have been enough good years on the McElmo Dome that families are large, and there are many of them. Yet bad years elsewhere have brought outsiders streaming in. People fight over farmland and access to springs. Ordinary tasks grow harder. A woman walks twice as far as she used to for an armload of firewood. A man is constantly on guard to protect the family's stored grain from theft. Feuds grow hotter. There is raiding, violence, murder. Families who have farmed the red loess soils of the Dome for generations begin leaving, perhaps joining relatives in the south. Their exodus demoralizes those who stay behind. And still the strangers keep arriving. Those who elect to stay decide to defend themselves against the unceasing turbulence by banding together and building a village around the spring that has always nourished them. Being pressed together so closely in the new place adds more stresses to those that brought them there. They have to become more organized, more of a unit, both to face the challenges of the outside world and to manage the internal affairs of their lately compressed community. They reach back into their past for ideas and examples. Urged by certain among them who claim special knowledge, they come to agreement: one of the first things they will build is a great kiva, and they will follow a body of practices that go with it. But they will depart from the ceremonies, the hierarchies, and the architecture of the distant past in at least one important respect: they will take the roof off the kiva. They will let everyone in.

It has been argued that Sand Canyon Pueblo is the site of a Chacoan revival, a revitalization movement aimed at surviving the dangers of the present by reverting to the lifeways of the past.[23] Varien has another view. "I don't think Chaco ever went away," he says. "I don't think you have to have a period when it goes away and then gets revived. I think you can see its influence throughout the post-Chaco period."

The great kiva is not the only element of Chacoan architecture at Sand Canyon Pueblo. The overall D-shape of the village conforms to the topography at the head of the canyon, but it also evokes the shape of Pueblo Bonito, Chetro Ketl, and other centers within Chaco Canyon. At some level, perhaps, when the builders of Sand Canyon Pueblo looked at their site, it looked like home. More concretely, a prominent and evidently public building within the pueblo is D-shaped and features doubled walls, a Chacoan style: two lines of careful masonry with earth and rubble between them.

There are other links, but the most tantalizing concerns Block 100 Man, the robust but aging craftsman who was bludgeoned and died in the last days of the pueblo. It turns out that Block 100 Man had six toes on his right foot—a rare condition under any circumstances, but one, it turns out, with a particular link to a powerful place. On the face of a sandstone cliff behind Pueblo Bonito in Chaco Canyon there are three petroglyphs depicting six-toed feet: two right feet, one left. In the early excavations at Pueblo Bonito a collection of bones was removed from an area presumed to be the location of elite burials. The bones included a right foot with a sixth toe. Anatomically, there are at least six ways for a sixth toe to attach to a foot; in the case of Block 100 Man and the bones from Pueblo Bonito, the manner of attachment was the same: the metatarsal of the little toe was Y-shaped, and the sixth toe branched from there. One might argue that the similarity is merely coincidental, the result of random genetic variation. Or one might speculate that Block 100 Man was biologically related to a cadre of six-toed leaders at Pueblo Bonito, perhaps directly descended from them. Going a step further, one might imagine that, more than his contemporaries, he carried the traditions of Chaco Canyon within him.

The past supplies vocabulary, but people structure their sentences in the present. The inhabitants of Sand Canyon Pueblo, be they the great-grandchildren of Chacoan priests or the descendants of ten generations of McElmo Dome farmers, were trying in every way they could to solve the puzzle of survival. The great kiva, the D-shaped building, the long-distance hunting, and innumerable other adaptations, including the aggregation of the pueblo itself, were all tools toward that end. The fact that they "took the roof off" their kiva shows not a slavishness toward the past but a willingness to experiment, to use their cultural toolbox flexibly in search of answers to the challenges that beset them. It might have worked. The pueblo was organized around 1250. As the 1250s gave way to the '60s, construction and expansion of the community continued. The people of Sand Canyon Pueblo might have prided themselves that they had crafted a survival strategy that fully suited their place and time. They were at least holding their ground, waiting for better times.

They waited in vain. The drought of 1276 was deep, and the next year's was deeper. Cold temperatures made things worse. As the dry years continued, climate effectively trumped the people's efforts to wrest life from the red loess soils of the uplands. The drought was nightmarish, a megadrought, but even so it was neither as deep nor as long as the drought of the previous century, which had humbled Chaco and forced a reorganization of human life throughout the region. Back then, farming people had managed to stay in the region—perhaps not all, but many of them. This time was different: everyone left, or died in place.

Perhaps the past droughts had been more predictable; perhaps their austerity fit a pattern that was more susceptible of adaptation. Perhaps the present drought was somehow more chaotic, more impossible to outwit. In any event, the hostile weather, combined with a crescendo of violence and the general depletion of the land, made staying impossible.

When the Sand Canyon people and all their neighbors in the Mesa Verde region departed, they left a lot behind, including their penchant for building towers, their style of making mugs and kiva jars, and their approach to domestic architecture, which produced a very high ratio of kivas to ordinary rooms. All these were signatures of a way of life that never reappeared elsewhere. Perhaps the people of the region had suffered too much trauma to haul their past along with them, either physically or conceptually. Perhaps there were too few survivors, and the bands leaving the Mesa Verde region lacked the mass or the unity to re-create their world in new places. Questions like these are the subject of active research and hot debate among archaeologists.

It is likely that the Mesa Verdean diaspora went not to one place but to many, scattering group by group from Hopi to the Rio Grande. Perhaps a majority, as Scott Ortman, Mark Varien, and others at Crow Canyon argue, were Tewa speakers who eventually made their way to the valleys of the Rio Chama and upper Rio Grande, where the Tewa Pueblos persist today.[24] In time the people who departed Mesa Verde and the McElmo Dome were absorbed into communities already formed or still forming, communities they eventually changed by their own presence.

The people of Sand Canyon Pueblo did not leave their stone village because of the drought alone, for no big thing happens for just one reason. If their population had been fewer, the land less worn, and the violence less constant and severe, the outcome might have been different. They saw their situation and adapted to it to the best of their ability. The brief flourishing of the pueblo suggests that they almost succeeded. But in the end, climate in the guise of both drought and the breakdown of established patterns played the card that ended the game, and they walked away, probably in successive groups over a period of time. After the massacre in which Block 100 Man and many others perished, perhaps no one was left to bury the dead—or perhaps there was no time. The danger and the sorrow made leaving urgent, and possession of the stone village and its life-giving spring was ceded to the wind and the sand.

# 4

## JANOS: A MIRROR IN TIME

THE RAINS HAVE forsaken El Cuervo for nearly a year, and the mountain-ringed plain that used to be a prairie is as naked as a parking lot. Not a blade of grass is in sight, scarcely a bush. A few low mesquites, defoliated and dormant, hug the parched ground, the wind having packed into their thorny embrace the dried-out stems of last year's tumbleweeds. Except in the burrows of the kangaroo rats, nothing can be hidden here. A lost coin or key would shout its presence, much as the potsherds do on the mounds of the ancient pueblo by the arroyo. Every edible thing has been consumed, every plant nipped off at the level of the ground. Even the soil is leaving, blown away, tons to the acre, by winds that sweep down from the Sierra Madre, a dozen miles to the west.

If you were to make your way to the top of one of the chipped-tooth peaks of the sierra (no small task), you would be able to look down into great canyons. One of those canyons belongs to the Río Gavilán, where in 1936 Aldo Leopold glimpsed a kind of ecological heaven that no longer exists. From atop the peak you would also see for great distances, certainly as far as Janos, the crossroads and market town through which nearly every visitor to this northwest corner of Chihuahua passes, and on a dustless day you might see the gritty penumbra of Ciudad Juárez and El Paso, far on the northeastern horizon. The air is dry, and here it is empty of pollution, which makes El Cuervo and its environs a good place for looking long distances, even into the past.

One way to understand changes in the land is to visit a place that shows how things used to be. That's what Leopold realized when he visited the Río Gavilán. He saw it as a fragment of the Southwest that had escaped the pressures of white settlement, and he recognized it as a mirror of how Arizona and New Mexico used to be, back in the days when the Apaches still roamed their homeland in freedom.

Down on the naked, cowburnt plains of El Cuervo, it strikes me that these grazing lands provide a similar, if less happy, source of comparison. They pretty well represent what a lot of the U.S. Southwest looked like a century ago. Vernon Bailey, chief field naturalist of the U.S. Biological Survey, was looking at land like this when he described the public rangelands of New Mexico in 1908: "The best grasses have been killed out and fully half the range rendered almost worthless. Mile after mile of bare ground may be found in many of the valleys or of ground bearing only weeds and worthless vegetation that stock cannot eat."[1]

Bailey was then roughly halfway through a career of open-air camp life that lasted more than forty years. His job was to appraise and catalogue the land and wildlife of most of the western states. When he wasn't doing that, he was not infrequently exterminating wolves. He was small, wiry, and owlish, and he had a good eye for landscape, although he didn't like the condition of much of what he saw. Neither did his colleague E. A. Goldman, who visited the southwest corner of New Mexico in the same year Bailey produced his report on public rangelands. Goldman reported, "The stock

"PASTURE" AT EL CUERVO, JULY 2008. KANGAROO RAT BURROW IN FOREGROUND. *AUTHOR PHOTO.*

ranges depend on the winter snow to make early grass for feed, and when little or no snow falls, as happened last winter, many cattle are apt to die of starvation."[2] The land Goldman visited lies only about sixty miles north of El Cuervo, and the odor of death he inhaled in 1908 was no different from the stink of decomposition that hangs in the air here. A short while ago, we saw the corpses of five cows at the dry water tank we passed a half-mile back. Another five or six stood listlessly by, poised to join their defunct sisters, unless rain comes quickly.

But Ed Fredrickson is not dismayed. Although he agrees that the present state of affairs in El Cuervo mirrors the U.S. Southwest's dismal past, he believes its future can be brighter. "There are more of the pieces still here," he says. By *pieces* he means ecological elements, in particular creatures like kangaroo rats and, pre-eminently, prairie dogs. Aldo Leopold was thinking about pieces like prairie dogs and people like Ed Fredrickson when he said, "To keep every cog and wheel is the first precaution of intelligent tinkering." If Fredrickson's business card were more honest, it would have the word *tinkerer* on it, but that's not how he's identified. Officially he is a research rangeland scientist at the Jornada Experimental Range of the USDA's Agricultural Research Service. He has an office in a sleek and modern building on the campus of New Mexico State University in Las Cruces. His preferred habitat, however, is the Jornada Experimental Range itself, where he retreats to a permanently moored travel trailer, or the grasslands south and west of Janos, which include Ejido San Pedro, a remote farming hamlet where UNAM, the Universidad Nacional Autónoma de México, the country's top school, maintains a biological field station, and where Fredrickson rents a small house.

One of the goals of Fredrickson's tinkering is to prevent what happened at the Jornada Experimental Range (JER) from taking place in the Janos grasslands. The JER consists of 302 square miles of mostly flat desert two dozen miles north of Las Cruces, New Mexico. It takes its name from the Jornada del Muerto—the journey of death—a famously waterless and Apache-plagued stretch of the old Camino Real, which grazes the northwest corner of the JER. In Spanish colonial days the Camino Real was the lifeline of the fragile colony of New Mexico, linking Santa Fe to Chihuahua City, Mexico City, and other commercial and administrative centers to the south. In the mid-nineteenth century, when the Mescalero Apaches were still a potent force in the area, the Jornada was "a great expanse of grass with only isolated spots of mesquite."[3] Today, virtually none of Jornada's 193,394 acres is free of brush, and no one would mistake any major part of it for grassland. Mesquite and its shrubby cohorts, creosote and tarbush, have taken over the place, and they show no sign of giving it back.

Tens of millions of acres of the Southwest, including huge portions of west Texas, have shared the Jornada's fate and have succeeded from grass to woody shrubs.[4] (There is no agreement on the exact acreage affected—the infinite gradations between a pure stand of grass and a pure stand of shrubs make precise accounting impossible and probably meaningless.)

Brush encroachment was already well advanced when the JER was established in 1912, the year New Mexico became a state. It was hoped, in fact, that the Jornada would be a place where the dynamics of brush encroachment and other afflictions of the ranching world would submit finally and fully to human understanding, and then, with enough luck and science, the means might be found to reverse them. Unfortunately, while understanding has advanced enormously, reversal of brush encroachment—at a scale of intervention short of bulldozing and poisoning the land to a fare-thee-well—remains doggedly out of reach. For all practical purposes, the shift from desert grass to desert scrub is a one-way flip from one ecosystem to another. No one has figured out how to run the process backward. That's one reason Fredrickson is down in Chihuahua tinkering with the cogs and wheels that he hopes will keep mesquite and its thorny cousins from marching into the Janos prairies.

Climate change compounds the problem. It is hard to improve the vegetation of a landscape when rain doesn't fall. In times of drought plants and animals draw inward. They grow little and reproduce less, sometimes not at all. But drought can be beneficial, too, because it also slows down the negative changes one hopes to avoid. Fredrickson points out, "Drought is just as critical to the function of this ecosystem as the rain." He says the trick in dealing with drought, or with the hotter temperatures of a changed climate, is to strive for resilience—the capacity of an ecosystem to experience disturbance without losing its essential character and becoming something else. It stands to reason that a grassland with a diversity of grasses—some that flourish with fourteen inches of rain, some that prosper with just eight—will fare better through fluctuating conditions than will a monoculture of a single species. The lesson is the same across ecosystems and at all scales: biodiversity supports resilience. If biodiversity declines in a grassland, a forest, an ocean, or across the entirety of a planet, the odds rise sharply that the affected environmental system, under the stress of climate change or some other "disturbance," may crash.

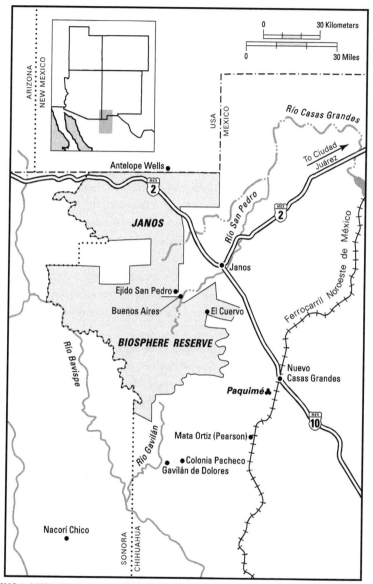

**MAP 5:** NORTHWEST CHIHUAHUA, SHOWING THE JANOS BIOLOGICAL RESERVE.

ED FREDRICKSON IS big. He stands a head taller than I do, and I nudge six feet. In a clothing store, he would shop the sizes defensive linemen wear. He wears a big hat. He drives a big truck. We went down to Chihuahua in a Ford pickup so high off the road that I almost had to chin myself to get in. (And if it hadn't been so big, we might not have gotten home, at least not without a lot more wear and tear—but that's a story for later.) Although he's handy at growing cattle (operations at the Jornada include a working ranch, which Fredrickson oversees), he doesn't fit a cowboy stereotype. Years ago, swathed in Lycra, he rode his bike from Las Cruces to Albuquerque, a distance of over 200 miles. His recreational interests depart from most people's idea of "normal." He and his wife Susie, who is Indonesian, think a pretty good weekend involves camping in an isolated corner of the desert and collecting scorpions, tarantulas, and other arachnids, which they carefully identify by genus and species and take home to add to their menagerie. The collection is big and taxonomically complex, and it commands a large portion of their domestic square-footage. Ed says, "If one of our shelves falls over, we would have a lot of species running around the house." Not to mention that they would alarm the faint of heart.

Fredrickson also thinks big. Ecological tinkering is only part of his agenda in the Janos grasslands. Economic development and education figure prominently, too. "When you look at an ecosystem," he says, "the people are part of that ecosystem. So we're not going to solve the problems that we see in how we interact with those landscapes until we start dealing with people." I asked him: if he were appointed czar of greater Janos and could initiate land rehabilitation of any kind, where would he begin? He did not pause. He shot back, "Education."

In Mexico, he explained, primary education is free but after that school attendance entails costs—for books, uniforms, and the like—and in a place as cash-starved as Ejido San Pedro such costs can be prohibitive. So kids don't learn science and they don't get much chance to learn about the broader world. That's a tragedy in its own right (repeated often enough in the United States, although not so much because of cost), but it is a double tragedy in San Pedro and neighboring ejidos, because, says Ed, "people know they need help. They're really looking for answers." Their openness to change is one of the things that keeps him coming back. Sometimes with his own money, sometimes with fragments of grants, Fredrickson covers educational costs for a few of the kids, but not as charity. "They pay us back by assisting us in conservation efforts—remediation projects, also with research. So we put them one-to-one with researchers in the field. We have them entering data. They're a really integral part of that research program." The impacts on the students have been much to Fredrickson's liking. He says he's seen changes in the kids' attitudes, maybe in their worldview. They hang around the house he rents tinkering with the microscopes and browsing the books he leaves out for them. One mother took her son to get shoes and offered to buy him cowboy boots. She thought the child would be overjoyed. But, no, he didn't want cowboy boots. He told her, "I want biologist boots."

If the means were available, Fredrickson would expand his work-for-study outreach into a kind of Youth Conservation Corps. He dreams of building a cadre of land rehabilitators in San Pedro the way a doctor in a lonely clinic might dream of having a team of physical therapists. In the meantime he and the graduate students he advises at the UNAM field station do what they can.

ED FREDRICKSON, CENTER, WITH GRADUATE STUDENTS LALO PONCE, LEFT, VERONICA SOLIS, AND DANIEL ACEVES AT RANCHO EL UNO. THE LOW FENCES BEHIND THEM ARE PART OF AN EXCLOSURE EXPERIMENT INTENDED TO TEASE APART THE RELATIVE IMPACTS OF CATTLE AND PRAIRIE DOGS ON VEGETATION. *AUTHOR PHOTO.*

When the school bus broke down, for instance, they became morning carpoolers, carrying the San Pedro kids in their battered field trucks to the school in Tres Alamos, several miles away. After that, they went back to their computers or out to the field.

IT IS CLOSE to sundown and we are on our way to an improbable task, tracking porcupines in the desert. Months ago, two of the graduate students, Lalo Ponce and Rodrigo Sierra Corona, radio-collared some of the big rodents, and it is time to monitor how they are doing. The young men complement each other nicely. Lalo is cool and reflective. Rodrigo brims with energy and jokes. Both are smart and dauntless. One afternoon they were stranded in the field station's cramped Nissan pickup by a sudden flood, which filled an arroyo they needed to cross. Evening gave way to night and night to morning before the water receded enough for them to get back to the station. All they had was a single can of Coke and a Monster energy drink. The experience led Rodrigo to suggest that in the future they always travel with an emergency kit—"some jerky and a bottle of tequila," he says grinning, "with the jerky optional." I ask if the truck we are in is so equipped, and Rodrigo assures me it is. "But the emergency will come later."

We head west toward the blue sierra. Lines of tall, drought-stressed cottonwoods mark the dry streambeds that angle across the plain. The road twists through the headquarters of a ranch where José Bustillos and a couple of helpers struggle with an ancient drilling rig. They've deepened their water well by forty feet and reached a reliable flow, they say, but the dropping of the water table worries everyone. Bustillos blames it on heavy logging in the mountains and the ceaseless groundwater pumping that drives the sprinkler irrigation of nearby Mennonite farmers.

The road snakes past a corral enclosing a pile of manure twice as tall as the horse that stands beside it, and then meanders over rutted clay and cobbles toward a stand of cottonwoods where two Harris hawks shriek. Ed Fredrickson's concern for the education of the children of San Pedro meshes with his sense of what the lands around Janos need, but it might also have something to do with his own educational experience, which was as rocky as the kidney-jarring road we are bouncing down. He grew on Abiqua Creek, near the town of Scott's Mill, Oregon, which is smaller than what most people would consider to be a town. Hunting, fishing, and roaming around the woods occupied the place in his childhood that television and mall visits fill for most kids today. He was self-reliant enough to leave home at fourteen, or at least thought he was, and found high school, which he departed with a GPA of 1.8, decidedly uninspiring. The woods, along with the increasing importance of the opposite sex, may have provided some distraction. He went to work in the woods as a faller, cutting trees, handling other jobs, and meanwhile wondering what logging was doing to the land and how he ought to feel about his role in it. Then he got hurt. A log rolled and pinned him. All that day, awaiting rescue by his coworkers (who wouldn't miss him until it was time to go home), and then for weeks afterward recovering from his injuries, he had even more time to think. He concluded education might be a good thing. He went to Oregon State University wanting to study plant ecology but was told men couldn't get jobs in that field; he should study range science instead. So he did, and somewhat to his surprise, he became an academic hotshot. A master's degree at Montana State and a doctorate at New Mexico State eventually followed.

Range management is a conflicted discipline. It is a "science" burdened with an undercurrent of romanticism, at least for many of its practitioners. The allure of a fabled cowboy past is hard to resist. And the accompanying scenery is usually pretty good. Fredrickson, however, seems to avoid the romanticism pretty well, along with the value-laden vocabulary that depicts the transformation of grass to shrub and range to parking lot as a struggle between good and evil. Instead of casting mesquite and its woody cousins as "invaders," he expresses admiration for their toughness. He also questions how grasslands got established in the first place and how "natural" they are.

"If you look at the plant that's most adapted for this country, it's mesquite," he says. Mesquite has a taproot that penetrates eighteen feet or more into the ground to harvest deep water, and aggressive laterals, spreading far beyond its crown, to harvest water in the upper levels of the soil. It can withstand extended droughts. The foliage and interior of the plant, which comprise the engine of metabolism keeping it alive, is protected by sharp stiff thorns that hold browsing animals at bay. But the plant is not stingy. It produces a high-protein fruit, packaged efficiently in pods that hang on the outside of the plant—freely offered to any creature that would spread the seeds across the land. Giant ground sloths and camels helped handle the job 20,000 years ago, and rabbits, horses, and Herefords do likewise today. The seed itself is a bean hardy enough to pass intact through the digestive tracks of most mammals, at least much of the time, so that it lands well fertilized and ready to sprout, often far from its mother tree. But it doesn't have to sprout right away. It can lie dormant a long time, easily years, sometimes decades, until the environmental cues are right. And then mesquite suddenly appears where before there was none.

WE ARE WALKING across an overused pasture toward a clump of celtus trees in which we hope to find the porcupine whose transmitter is beeping into Lalo's headphones. Grazing has been heavy here, and grass is sparse, but so is mesquite. Most of the plants are cactus or forbs that are unpalatable to a cow. Ed bends down and picks up a lump of dry horse manure, and cracks it open. Inside it is packed with mesquite beans. "What's this pasture gonna be in twenty years?" he asks, chucking it away.

It used to be thought that shrub encroachment was directly the result of overgrazing. After the Jornada Experimental Range was established in 1912, the managers excluded the cattle from some of the pastures and continued to graze the rest. In the years that followed they were alarmed to see the ungrazed pastures succeed to mesquite at more or less the same pace as the grazed. The first technical paper identifying transport of mesquite beans by cattle as a factor in brush encroachment appeared in 1929, but for a long time the idea failed to get traction.[5] In some quarters it still doesn't. Ed says he's gotten tired of suggesting to ranchers, "Why don't you hold your cattle off mesquite when the beans are ripe?" It doesn't seem to work. "They look at you like you are out of your mind. 'It's the best feed I've got this time of year,' they say. So there go more pastures to mesquite."

Considering all that humans have done to introduce mesquite to new habitat, arguing that it is an "invader" is like putting signs around your neighborhood that you are having a garage sale and then calling the people who show up "intruders." Ed explains that most of what people did to expand ranching across the ranges of the Southwest was a boon for mesquite. With the enthusiastic support of the government, and sometimes with outright subsidies, ranchers "developed" water—drilled wells, installed windmill-driven pumps, and built earth dams for stock tanks wherever surface water ran. The main thrust was to make possible the occupation of more land by cattle, which is to say that the industry inadvertently encouraged the dispersal of mesquite beans over the whole landscape. Up to that time, other creatures like jackrabbits, kangaroo rats, and coyotes had been dispersing mesquite a half-mile, three-quarters of a mile at a turn. "But when we brought in cattle, we brought in a more effective dispersal agent. We were moving seed five, six miles away from water. So we ratcheted up the speed of this transformation."

In the United States the well-meaning, Progressive-spirited application of technology to resource management also removed from the ecosystem possibly the most significant force keeping mesquite in check. At considerable expense and with a great deal of effort, the U.S. government and its various public and private allies poisoned out the prairie dogs from over 95 percent of their habitat. No serious thought was given to the ecological role of prairie dogs or to the scores of other species, like ferrets and burrowing owls, that depend in some way on the habitats that prairie dogs create. The prairie dogs were considered varmints, pure and simple, because it was believed they consumed grass that would be better used to fatten cattle. Therefore they were wasteful. And therefore they were sentenced to death by strychnine.

The Jornada Range, true to its intended character as a range management showcase, got rid of its prairie dogs early. In 1916 and 1917, as part of a hasty attempt to increase red meat production during World War I, government agents successfully exterminated prairie dogs throughout the Jornada Basin.[6] Other public and private lands followed suit as money and manpower became available. Range science, amply funded through the West's land-grant universities, guided the effort. "We focused on the typical things that you can manage in range management: What type of animal do you use? What about maximizing distribution and achieving proper forage utilization? And the timing, intensity, and frequency of grazing. We looked at those things and we forgot that we were operating in a complete ecosystem."

In Mexico conditions were different. With less capital and fewer incentives for intensive management, even the *ricos* did not get around to developing water or poisoning rodents until comparatively later, and the poorer *ejidatarios* never had the money to get started. That's why, from Fredrickson's point of view, there are more cogs and wheels to work with in Chihuahua.

Prairie dogs have a lot of predators, from eagles and mountain lions to coyotes and rattlesnakes—anything that eats meat will enthusiastically eat them. Their best defense is to share sentry duty and maintain an open field of view. They don't tolerate mesquite close to their colonies. They girdle the stems, and defoliate the leaves. Fredrickson thinks they probably chew away the roots underground. After a long, good rain, when mesquite beans sprout, they gobble up the seedlings with relish. Prairie dog colonies do not necessarily stay fixed in place. They expand here and contract there, responding to changes in their environment. They suffer abandonments and produce migrations. Over time, they contribute mightily to the diversity of the landscape. They are like the white pieces on a chessboard, and mesquite is like the black. Where one is, the other isn't, but in the long span of the game, both sides surge and retreat repeatedly across the squares. The result is a restless mosaic, a landscape in motion through time.

Fredrickson wants to use that restlessness constructively. It turns out that prairie dogs prosper amid unobstructed views to such a degree that they multiply excessively on overgrazed ranges. Indeed, in the day of the great prairie dog exterminations, their prevalence on overgrazed ranges won them undeserved blame for the overgrazing. Fredrickson says the best way to keep prairie dog numbers in check is to graze judiciously, which is to say that prairie dog colonies represent a continuous and tangible referendum on the quality of land management. Just as important, the dogs rarely prosper where there is no grazing. Without the help of livestock or wild grazers to keep down the grass, prairie dogs will eventually move on. Fredrickson therefore theorizes that it might be possible to steer a prairie dog colony across the landscape, moving it slowly and incrementally, by excluding grazing on one side of the colony so that the taller vegetation provides a "push," and by using fire and grazing animals to generate a "pull" on the other. Fire tends to fertilize and invigorate grasses, and the new shoots that rise where a fire has passed will be both tasty and nutritious. If the corner of a large pasture has been burned, eventually the cows will linger in that corner, keeping its grasses trimmed down, always favoring the new growth, creating an environment attractive to prairie dogs.

Ranchers have traditionally focused on running as many cows and growing as much meat as possible. Over the years the more progressive among them have shifted to viewing themselves as grass producers, leaving animal production to follow. But even that, says Fredrickson, is not enough. "You've got to have the larger context always in view," which means not focusing on one creature or subset of variables, but understanding the behavior of the whole system and working with all of it, or at least as much as you can figure out, and doing so deliberately through space and time. And with patience. Out of such an understanding comes a vision of land use that has little in common with clichés about stirrup-high grasses. Within the bounds of that vision, some land is grazed hard, some lightly or not at all, some of it has prairie dogs, some hasn't; mesquite is there, as it has to be, in tension with the rest, and the complex of the whole is in continuous flux, restless, diverse, and, because of those qualities, resilient.

IN THE "BOSQUE" pasture of Rancho El Uno, Albino Parra Hererra gestures toward a sea of mesquite. "When my father was young, sixty or seventy years ago, all this was *navajita,*" he says. Navajita is blue grama, the durable and nutritious mainstay of shortgrass prairies from Mexico to Manitoba. It gets its Chihuahuan name from the appearance of its seed spike, which projects horizontally from the stem like a half-open *navaja,* or clasp knife. "Había unos pocos mesquites allá," he says, pointing. There used to be a few mesquites over there. Now mesquite is all you can see. During the same span that mesquite was advancing, the wealthy owners of Rancho El Uno were poisoning their prairie dogs, even until recently. Fredrickson, however, thinks the population can be brought back. He'd like to see colonies reestablished in the Bosque pasture to hold the advancing edge of the mesquite in check. "Now the ranchers are going to have to show the prairie dogs some love," says Rodrigo.

THE GREATER JANOS grasslands support close to 300 species of birds and scores of mammals, reptiles, and amphibians. They also provide a home for the only free-roaming bison herd in Mexico—yes, bison, in what most people think of as desert. (Unfortunately, the herd's future remains in doubt; its ability to roam back and forth across the border from the bootheel of New Mexico, as it has done for most of the twentieth century, has been compromised by border barriers erected by the United States.) Notably, the Janos grasslands harbor the second-largest black-tailed prairie dog colony still extant, which is a key to their biodiversity. The prairie dog tunnels provide burrows for other creatures; their digging loosens the soil for a host of invertebrates; their cropping maintains habitat for short-grass obligates; and with their bodies the prairie dogs provide sustenance for predators of all kinds: some that slither, some that fly, and some that travel on their feet. Recognizing this diversity, members of the Mexican scientific community, particularly at UNAM, began in 2002 to press for protection of the area as a biosphere reserve. In 2005 the Nature Conservancy boosted the effort with its purchase of the 46,000-acre Rancho El Uno as a kind of beachhead of land preservation. For a time, it enlisted Pronatura Noroeste, a Mexican conservation group, to manage the ranch. Year after year, the lobbying continued, the scientific reports and surveys accumulated (with the graduate students at the Ejido San Pedro field station much involved), and the coalition of partners supporting the designation steadily expanded. Finally, on December 8, 2009, a presidential decree formally established the Janos Biosphere Reserve, encompassing more than 1 million acres of northwest Chihuahua.[7] The designation gives environmental advocates new leverage. Many hope it will lead to enforcement of pumping regulations and protection of groundwater resources, which the sprinkler-irrigated farms of Mennonite colonists at Buenos Aires are depleting at a great rate.

Here on Rancho El Uno, Lalo Ponce is embarking on his doctoral research. With Fredrickson's assistance and supervision, he will attempt to move prairie dog colonies across the landscape using the push and pull of fire, grazing, and nonuse. If all goes well, his experiments will have more than passing interest for outside observers, because the grazers he is using are bigger, shaggier, and more mythic than cattle. Just weeks before the biosphere reserve declaration, twenty-three bison were set loose on El Uno (they came from Wind Cave National Park in South Dakota, not from the feral herd that plies the border).

No one expects bison in the desert, but then again, deserts are lands of surprise. Their diversity runs from moonscapes where little grows to prairies like the grasslands of Janos, which abound in birds, mammals, and grasses, and which, if unabused, are highly productive. Some authorities argue that the Janos prairies (and the adjacent high desert grasslands of the New Mexico bootheel) are biologically a dislocated fragment of the southern Great Plains and should not be classed as desert. The matter is open to debate, as is the question of whether bison were a continuous or ephemeral presence in northwestern Chihuahua prior to European settlement.[8] The small herd that has lived along the border since at least the 1920s, which likely descend from captive animals that escaped or were released, have proved beyond question that they and their habitat get along rather well. Hunting on both sides of the border appears to have kept their numbers in check, and today the greatest threat they face may be the "Normandy" barriers—trestles of steel beams like those with which the Nazis defended D-Day

BISON ROAM THE GRASSLANDS OF NORTHWEST CHIHUAHUA AND FAR SOUTHWEST NEW MEXICO, A
FACT ATTESTED BY THE NM DEPT. OF TRANSPORTATION, WHICH IS RESPONSIBLE FOR HIGHWAY SIGNS.
RECENT CONSTRUCTION OF "NORMANDY" BARRIERS ALONG THE BORDER HAS FRAGMENTED BISON
HABITAT AND LIMITS MOVEMENT OF THE HERD. THIS VIEW LOOKS NORTH FROM THE INTERNATIONAL
CROSSING AT ANTELOPE WELLS. *AUTHOR PHOTO.*

beaches—that the United States has erected along the border. The barriers, which prevent passage by vehicles, also halt the movement of bison, antelope, and deer, effectively fragmenting their habitat, constraining migration, and interrupting access to water.[9]

Yet another surprise has attended the new interest in prairie dogs in the Janos grasslands. A conservation group that goes by the simple name Rare teamed up with ProNatura and the Nature Conservancy to organize what it calls a Pride Campaign—an effort to rally support for conservation by arousing in people a sense of pride in the uniqueness of the place where they live. The prairie dog was chosen to serve as mascot of the Janos campaign, and activists costumed as prairie dogs appeared at elementary schools and community halls and carried on in various ways intended to excite interest in grassland appreciation and protection. They also plastered posters of prairie dogs on power poles and buildings throughout Janos and the surrounding ejidos.

Then something strange happened. Within days of the posters' going up, they disappeared. The images seemed to vanish as fast as they were displayed. Cynics said it was an indication of the local people's resistance to change and their hostility to the idea of protecting a critter they'd long regarded as a pest. The disappearance of the posters was taken as proof that real conservation, where poverty was so great and education so limited, was impossible.

But further investigation told a different story. It turned out that people had not torn down the posters. What they actually did was to remove them carefully from their exposure to the elements and take them inside their houses to decorate their walls. It seemed that nearly everybody wanted one, and there weren't enough to go around. Far from carrying a grudge against prairie dogs, the people of Janos seemed to like them and think they were cute, and they considered the posters nice to look at. All in all, they were proud that one of the largest remaining prairie dog colonies belonged to their world.

That kind of openness in the local spirit makes Ed Fredrickson optimistic. He'd be the first to say that although prairie dogs are important in controlling mesquite, they are not half as important as humans are: "If we want to maintain the grasslands, we need to ask how they got here in the first place." Through the analysis of fossil pollens and other traces of the ancient past, researchers have learned that mesquite and other shrubs dominated vast areas of the Southwest, including the Jornada, 2,500 years ago. Since then, those ecosystems gave way to the grasslands that early explorers encountered. What turned the tide? Fredrickson thinks that fire, climate, prairie dogs, and other factors may have nudged the shift along, but that humans accounted for most of it. "As I go back and look at the early ethnobotanists, some of them claim that mesquite beans made up 50 percent of the diet of the native people here. They would have been a trade item. One thing about mesquite, it consistently produces seed, except maybe at the end of long-term droughts. So it's a reliable food."

"One of the things that I've been playing with is the idea that for a longer period than we've been here, from about 900 to 1450, the people around the Jornada and other places used mesquite heavily and they essentially depleted it." They consumed the beans for food and the wood for fuel, which meant that they both curbed the mesquite's reproduction and also physically removed it, along with other shrubs, from the landscape. They also had turkeys; by some reports, they had lots of them. "It was *the* domesticated animal of the time. So, as you have families moving turkeys around the landscape, you can imagine what that would do to those seed banks, if in fact turkeys are eating those seeds. The impact can be enormous. So the grasslands may not have been here for any other reason but the fact that the mesquite was over-used—exactly the way we have overused the grasslands, but in reverse."

If this is true, then far from being an invader, mesquite and other shrubs are the default vegetation of large portions of the Chihuahuan desert, at least under the climatic regime of the last thousand years or so. Grasslands replaced them largely as a result of sustained, long-term aboriginal exploitation.[10] You can think of the seed- and fuel-gathering activities of native people and their animals as the lid on a pot of mesquite beans. When the lid was removed, the beans spilled out and began to spread, receiving a lot of help from the live-stock industry. They are still spreading.

Fredrickson concludes that maintaining the grasslands that still exist may require an approximation of those ancient patterns of use. He wants to enlist some of the youngsters of Ejido San Pedro in growing turkeys to see how the birds fare, how much mesquite seed they eat, and whether they might give a worthwhile boost to the local diet and economy. He'd like to interest the kids and their elders in cutting mesquite to smoke meats and make *machaca* (a spicy shredded beef) from the cattle they graze. He'd like to see humans and prairie dogs in a kind of interspecies collaboration to stem the tide of mes-quite encroachment and to maintain the diversity and productivity of the Janos grasslands.

The stakes are high, not just for the ecology of the region, including its enormous diversity of birds and other creatures, but for the people of San Pedro and every other village on the Janos plains. Their predicament is the predicament of rural Mexico. If ranching and small-scale agriculture cannot sustain them, they must either live on remittances from relatives in the States, or leave. (The dark alternative of smuggling, both of people and of drugs, draws closer every day, too.) A final reckoning is coming, and it is coming irre-spective of climate change, although climate change will likely hasten it. The Mennonites in nearby Buenos Aires are draining the aquifer to grow potatoes and other crops with center-pivot sprinkler systems a quarter-mile long. When they've exhausted the resource, they will move on, as they've done before, to Belize or Nicaragua or points farther afield, and the *ejidatarios* will be left behind with depleted lands and a hotter and likely drier climate. In San

Pedro the societal reset button is relentlessly descending toward contact. People are going to have to live a different way, if they are going to live there at all. That's the subtle emergency that holds Fredrickson's attention. It is an emergency without boundaries. Not only do El Cuervo and other battered grasslands in the vicinity of San Pedro give a glimpse of the past; they also offer a taste of what may be the future for large areas of the Southwest.

THE INTERRELATIONSHIPS OF mesquite, humans, prairie dogs, grasslands, and many other agents and actors constitute a kind of ecological riddle about aridlands. Getting to the bottom of the riddle, as Ed Fredrickson is trying to do, matters at more levels than simply the economics of communities like San Pedro and the biological future of some impressive grasslands. It is part of the long-running story of how two societies, in Mexico and the U.S., have struggled to learn the trick of living with aridity. Where mesquite has been concerned, both were wrong in their early science and slow to adapt as the science got better. The need to do better would be great if the climate were stable. It is all the greater as the aridlands grow more arid.

If you made a list of the thinkers who have deeply pondered the problem of living in arid land, you would have to put Aldo Leopold near the top. The Southwest shaped his understanding of the natural world in fundamental ways, and Chihuahua provided some of the finishing touches. Leopold did not come to Chihuahua and the Río Gavilán expecting to travel into the Southwest's ecological past, but once there, he realized he'd traveled as much in time as he had in space.

Leopold's 1936 hunting trip turned out to be a journey of discovery, one of the greatest in the history of North American conservation. He was a university professor at the time, on holiday, but by no means taking time off from the great vocation of his life, which was reading landscape and deciphering its story. Wherever he went, close to home or far away, he was a bit like an old-fashioned doctor on a house call, always taking down a history, diagnosing ailments, probing for insights into character. But the Río Gavilán, far from being a "patient," came to represent the opposite. It became for Leopold the living embodiment of land health, a concept that crystallized from his Gavilán experience.

The more Leopold climbed the Gavilán's mesas and hunted its canyons, the more he believed it to possess the kind of ecological integrity he had been looking for all his life. It was ironic, he wrote, "that Chihuahua, with a history and a terrain so strikingly similar to southern New Mexico and Arizona, should present so lovely a picture of ecological health, whereas our own states, plastered as they are with National Forests, National Parks and all the other trappings of conservation, are so badly damaged that only tourists and others ecologically color-blind, can look upon them without a feeling of sadness and regret."[11]

None of this might matter to us today except that Leopold was then also embarked on a second vocation as a nature writer and ethicist. In 1936 he was still finding his voice—and his core ideas—and the Gavilán helped him do both. In originality, depth of insight, and grace of presentation, Aldo Leopold soon joined the highest ranks of those who have written on the natural world.

At the University of Wisconsin he taught in a field of study new to the university and new to the country—*game management*, soon to be broadened to *wildlife management*, and Leopold initially represented the whole of the field's professoriate. In a way, he had moved to the other side of the lectern from his own days as a master's candidate at the Yale School of Forestry, which was then a similarly pioneer institution legitimizing another embryonic field. In 1909, when Leopold received his degree, the Yale forestry school, the nation's first, was less than a decade old, and forestry as an academic discipline and a profession was still establishing its credibility.

In 1905 Gifford Pinchot, an intimate of President Theodore Roosevelt and one of the founders of the Yale school, had taken charge of the newly created United States Forest Service. His intention was to apply a utilitarian brand of conservation, through the vehicle of forestry, to as much of the United States as he could win the authority to manage. With a large institution to build and millions of acres of unmapped land to inventory and bring under administration, the new agency's appetite for trained manpower was voracious. Leopold was an early recruit. He'd gone to Yale hoping to work for the Forest Service, and he was soon hired. Within weeks of graduation, he reported for duty in Albuquerque and was assigned to the Apache National Forest in southeastern Arizona as a forest assistant.

Early on, like many young men in their early twenties, Leopold gave evidence of more enthusiasm and ambition than judgment. His first major assignment, in which he supervised a small crew in a summer-long survey of 65,000 acres of timber, proved a failure on multiple levels. A touch of arrogance and a reluctance to admit or address his own errors led to dissension among the crew, accusations of mismanagement, and an investigation. In the words of one of Leopold's biographers, "It was clear to all that he had taken a serious mismeasure of both his men and the land."[12] But Leopold made good on his next assignment, and his supervisors, recognizing his promise, stuck by him. In time, he "made a hand," mastering the skills required by his profession in both the field and the office. His ascent through the organization was rapid. By the age of twenty-five, in 1912, he was supervisor of Carson National Forest in northern New Mexico, responsible for the management of a million acres. Seven years later, after a horseback journey through a winter storm broke his health, requiring a long recovery, he returned to the Forest Service as assistant (southwestern) regional forester for operations. His responsibilities took him to every forest in Arizona and New Mexico and gave him abundant opportunity to observe the hard-used terrain of the Southwest at the close of pioneer days.

At first, understandably, he viewed the land through the lens of his education and training. He had been taught, for instance, that all types and varieties of fire were harmful to the forest; there were no exceptions. Even the intentional use of light, low-intensity fire, for which Leopold in one essay adopted the derisive term "Piute forestry," was considered an enemy of the forest, because it killed small trees that the forest would later "need," as well as for a number of other reasons, all of which, years later, Leopold would come to see as specious. But if fire was bad, predators like wolves and mountain lions were worse. A devoted hunter, Leopold campaigned energetically for the extermination of all predators that consumed deer and other preferred game species. As late as 1920, he told delegates to the Sixth American Game Conference in New York City, "It is going to take patience and money to catch the last wolf or lion in New Mexico, but the last one must be caught before the job can be called fully successful."[13]

Leopold's alertness to the evidence of his senses was too lively and his intelligence too critical for him to carry such ideas without examination. If his admirers—and they are legion—sometimes present him as an environmental saint, he is at least, like Paul or Augustine, a saint with an imperfect and therefore humanizing past. Perhaps the great lesson to be drawn from his example concerns less the Delphic wisdom he seems to have possessed at the end of his life than the habit of restless observation and questioning that he nurtured along the way. This restlessness appears to have intensified, rather than diminished, with age, and was coupled with a compulsion to wrestle into coherence the sometimes conflicting lessons of what he observed.

The Río Gavilán was one of three great sources of revelation along Leopold's philosophical journey.[14] Another derived from the decades of labor he devoted to the restoration of a depleted farm property on the banks of the Wisconsin River that he and his family called "The Shack." The third was a 1935 trip to Germany during which he toured forests so excessively "managed" that every understory plant and fallen branch was removed from the all but sanitized forest floor. Deer in those forests had to be fed artificially because there was nothing for them to eat among the rows of plantation trees.

If the German forests were a vision of ecological hell, the Río Gavilán was a glimpse of heaven. In a lifetime of avid hunting, it was the only place where Leopold observed a population of deer that was both abundant and in healthy balance with its environment. Nearly every other deer range he had known had been either depopulated or overstocked and subject to irruptions—sudden explosions of population, like that which afflicted the Kaibab Plateau in the mid-1920s and led ultimately to extreme overbrowsing and mass starvation. Leopold wrote that in nine days of "hard hunting" on the Gavilán, he and his companion, Ray Roark, a fellow professor, saw "187 deer, fifty of them bucks of two or more prongs. Deer irruptions are unknown. I doubt whether the lion–deer ratio is much different from that of Coronado's time." But much more than the deer herd impressed Leopold: "To my mind these live oak–dotted hills, fat with side oats grama, these pine-clad mesas spangled with flowers, these lazy trout streams burbling along under great sycamores and cottonwoods, come near to being the cream of creation. But on our side of the line the grama is mostly gone, the mesas are spangled with snakeweed, the trout streams are now cobble-bars."[15]

Leopold and Roark had journeyed by train to El Paso, where they met their guide, Clarence Lunt, a Mormon settler from Colonia Pacheco, the village from which they would set out for the Gavilán. From El Paso's cross-border twin, Ciudad Juárez, they rode the Mexico Northwestern Railroad to Pearson, a former logging town at the foot of the Sierra Madre. In his journal Leopold noted the condition of the country through which the train passed: the eroded slopes behind Ciudad Juárez, the mesquite flats as the desert opened out, and the sacaton-covered playas at Laguna Guzman, where the train stopped for lunch, "full of doves and mosquitos."[16]

"South of Laguna Guzman we passed over great stretches of range, all overgrazed, part of it fenced, and stocked with Herefords. The draws are eroding. The flats have thin sacaton grass." It was a landscape like innumerable others he had seen north of the border, hard-used and far less vital than it once had been. Ever watchful for wildlife, Leopold saw "a few jackrabbits between Casa Grande [*sic*] and Pearson."

Casas Grandes is the Mexican town adjacent to the ancient ruins of Paquimé, which was formerly a vital exchange and ceremonial center of the medieval Southwest. The word *Paquimé* is Nahuatl for "big houses"—the *casas grandes* now in ruins. The people of Paquimé traded in large and brilliantly colored macaws, the show-offs of the parrot family, and may even have bred them within their adobe village. When their pueblo flourished, in the thirteenth and fourteenth centuries, it numbered over a thousand rooms, and the countryside around it abounded in satellite villages. Leopold would savor a taste of that lost world. In the foothills of the Sierra Madre and across the Continental Divide in the canyons of the Río Gavilán, he marveled at what he called "check dams," ancient barriers of dry-laid stone behind which soil accumulated. There were hundreds of them, virtually everywhere: "On the dry tops of the highest mesas, in the bottoms of the roughest and wildest canyons, anywhere in fact where a short watershed is intercepted by a ledge, dyke, or other favorable spot for impounding soil."[17]

The dams, called *trincheras* in Spanish, became emblematic for Leopold of the human potential for improving land. The soil they held back nurtured grasses, oaks, and ponderosas hundreds of years old. Deer seemed to like the luxuriant flats that formed behind them. Leopold deduced that "prehistoric Indians" had built them to create "little fields or food patches." Quite likely, the builders were farmers from the days of Paquimé.

Leopold, Roark, and Lunt disembarked the train in Pearson and clambered into Lunt's wagon for the slow trip west to Colonia Pacheco and thence to the Río Gavilán. The "road" they followed was an abandoned railroad grade, and the hills through which it rose were cut over and much abused. The town's incongruously Anglo name of Pearson commemorated Fred Stark Pearson, a Canadian investor who spearheaded the consolidation of several railroads in 1909 as the Ferrocarril Noroeste de México, the Mexico Northwestern Railway. At the same time, he also acquired the Sierra Madre Land and Timber Company, along with concessions of public timber from the Mexican government.[18] His companies soon built a sprawling saw- and planing mill in Pearson, exporting their lumber to the United States. The voracious appetite of the mill for timber would no doubt have devoured the forests of the Río

Gavilán and destroyed the vision of "land health" that later inspired Leopold, except that the Mexican Revolution broke out in 1910. Rebels seized Pearson early in 1912, after which the mill operated only intermittently. By 1920 it had closed, never to reopen. A few years later the town jettisoned the name Pearson and called itself Mata Ortiz, after Major Juan Mata Ortiz, a hero of the Apache wars.[19] If you go to Mata Ortiz today—it is a busy pottery center, famous for designs that evoke the artistic traditions of Paquimé—you may note not only the denudation of the surrounding hills but the general absence of wood in the buildings of the town. As is the case in lumber towns throughout Latin America, Mata Ortiz is built with masonry; the wealth of its forests, reduced to boards and beams, rattled away on flatcars to distant markets.

Westward from Pearson, Leopold bounced along in Lunt's wagon under a bright moon until 1:00 a.m., meanwhile noticing "many 'agrarian' homesteads recently granted by the government to settlers"—a continuing legacy of the revolution and a harbinger of the hard fate that lay ahead for the Gavilán as the goats, sheep, and cattle of a growing and hungry population followed logging roads into the backcountry.

ALDO LEOPOLD BOWHUNTING ON HIS SECOND TRIP TO THE RÍO GAVILÁN COUNTRY, IN 1938. *COURTESY OF THE ALDO LEOPOLD FOUNDATION (WWW.ALDOLEOPOLD.ORG).*

Leopold's arrival in Colonia Pacheco gets little mention in his trip notes, but this Mormon village, settled in the late 1880s, was already in steep decline. Like its sister colonies scattered through the region, Colonia Pacheco had been founded by members of the Church of Jesus Christ of Latter-Day Saints, who were fleeing strictures against polygamy in the United States. In its early decades, the people of Colonia Pacheco supported themselves with lumbering, ranching, farming, and cheese-making. By 1909 the community had a population between 300 and 400 and had discovered a new business: guiding hunters from the United States into the rich gamelands to the west. But the revolution cut short its prosperity. In 1912 all Mormon colonies in Chihuahua were evacuated, Colonia Pacheco among them, and their inhabitants left with what little they could carry. Those who returned to resettle Colonia Pacheco in 1918 found the village burned to the ground. They rebuilt, but times were hard. By 1940, sixty settlers remained; by 1951, there were six.[20]

From Colonia Pacheco, a few hours on horseback carried Leopold and his companions up and over the ridge of the Continental Divide and into the watershed of the Río Gavilán.[21] They left behind the tree stumps, the gullies, the guttering roads, the grassless pastures, and the dull gazes of the cows that hung at the edge of the settlement. Once across the divide, they rode through country rich with deer and turkey, both of which species kept fat on the mast of the oak thickets. They saw sign of wolf and lion. They feasted on native trout that teemed in the creeks. Leopold noticed the effects of recent low-intensity fires among the big pines. He asked Lunt and Lunt's neighbor "Harl" Johnson, who came along to assist,[22] how frequently the fires burned. They told him, "Every few years"—which is to say that the natural fire regime of the Río Gavilán's pine forests was undisturbed in 1936, whereas throughout Arizona and New Mexico the natural frequency of fires had ceased by about 1890.

Of fire on the Río Gavilán, Leopold wrote, "There are no ill effects, except that the pines are a bit farther apart than ours, reproduction is scarcer, there is less juniper, and there is much less brush, including mountain mahogany— the cream of the browse feed. But the watersheds are intact, whereas our own watersheds, sedulously protected from fire, but mercilessly grazed before the [National] Forests were created, and much too hard since, are a wreck."[23]

The key to the intactness of the Río Gavilán was not so much the absence of humans as it was the absence of humans' goats, sheep, cattle, and horses, along with humans' saws and machinery. Leopold pondered the ancient *trincheras*, their soil-rich terraces, and the positive role that an earlier, less heavily equipped culture had played in the ecology of the mountains. Lunt and Johnson now told him that the key to keeping settlers from the mountains in recent times had been another group adept at living lightly on the land—the Apaches. Geronimo had surrendered in 1886, whereupon he and the rest of his band were shipped off in boxcars to Florida, but hostilities dragged on, especially in the Sierra Madre, as the Apache Kid and successive bands of fugitives from the San Carlos reservation continued to roam the borderlands, raiding intermittently, sometimes brazenly, for years more.

Although Leopold's journals are silent on the subject, Lunt and Johnson might have shared with him stories of Apache depredations in the mountains west of Colonia Pacheco. At least one old-timer would later remember that "the danger in the mountains lasted until 1900, when several Apaches were killed by a group of Mormons, and this was the last heard or seen of the Apache in northwest Chihuahua."[24] But that wasn't the whole story. Immediately across the border in Sonora, if not sporadically in Chihuahua, cattle rustling and kidnappings (of Mexicans by Apaches and of Apaches by Mexicans) continued into the 1920s and '30s. There were murders, too, the most famous of which was the 1927 killing of Maria Dolores Fimbres and the kidnapping of her son Gerrardo near Nacorí Chico, fifty miles southwest of Colonia Pacheco. The aggrieved father and husband, Francisco Fimbres, kept up a desperate search for his son for two and a half years, combing a large swath of the Sierra Madre and finally encountering his son's captors in April 1930 in the wild country between Nacorí Chico and Colonia Pacheco. He opened fire immediately. The ensuing gunbattle cost the lives of two Apache women, a mother and her daughter, as well as the leader of the band, known as Apache Juan, who attempted to defend them. Uncounted others, including Gerrado, disappeared into the brush. Fimbres evidently scalped or beheaded his victims—a photograph exists of him with grisly trophies— and he left the corpses where they lay. Unfortunately, the tragedy had not ended. Days later, two Mexican men came upon the scene and found the three Apaches neatly buried. Nearby lay the body of Gerrardo Fimbres, dead at seven years of age. Accounts are imprecise, but it seems the surviving Apache women tied him to a tree and stoned him in retribution.[25]

EVEN AS APACHE hostilities faded, Chihuahua's remorseless land tenure policies limited settlement in its far northwest. People in the United States, to the extent that they think historically of frontier life, conceive the frontier as a westward-moving boundary that yielded steadily to the expansion of settlement and that was never static for more than a generation. Northwest Chihuahua, by contrast, was a frontier for two and a half centuries, held tenuously at outposts like Janos, originally a Franciscan mission, which later became one of the largest and most isolated presidios on the northern frontier. It was part of a chain of forts defending "a line more than twice as long as the Rhine-Danube frontier held by the Romans, from whom Spain learned her lesson in frontier defense."[26]

In 1776 Janos and several other towns were appointed "military colonies," and the settlers of each received a communal land grant of 112,359 hectares— almost 300,000 acres. In exchange the colonies furnished men for campaigns against the Apaches. Their people, although poor, necessarily developed a fierce spirit of independence and self-reliance. For as long as the Apaches remained a menace, the *ricos* of Chihuahua, owners of giant haciendas, allied themselves with the people of Janos and other frontier settlements. As soon as the threat—and the need for mutual self-defense—was removed, however, the *hacendados* who controlled Chihuahua's political system betrayed the soldiers they formerly had commanded. They passed laws in 1884 and 1891 depriving towns like Janos and other political subdivisions of the right to elect their own officials. With the mayors and district administrators subsequently appointed by the governor of the state, the path was clear for large portions of the old land grants to be sold to neighboring *hacendados* or other investors. A lucrative market for Mexican cattle in the United States helped to drive the process of expropriation. It also triggered curtailment of the long-standing custom of allowing villagers to graze their small herds of livestock on neighboring haciendas. These attacks on the viability of peasant life did more than limit the expansion of settlement. In Chihuahua they led directly to the outbreak of revolution in 1910, and many aggrieved descendants of the militiamen of Janos soon placed their rifles in the service of Pancho Villa.[27]

The anarchy and violence of the revolution restricted settlement for another decade, and it also slowed the westward march of logging operations. A few timid attempts at land reform gained traction in the 1920s, and even more took hold under President Lázaro Cárdenas in the '30s, producing the "agrarian homesteads" Leopold noticed between Casas Grandes and Colonia Pacheco. Most of these occupied land better watered than the Janos grasslands and less isolated than the canyons of the Sierra Madre. Even so, Leopold saw what was coming. At the close of a story about a mule deer buck, bedded above a *trinchera*, that he shot at and missed, he wrote, "Some day my buck will get a .30-.30 in his glossy ribs. A clumsy steer will appropriate his bed under the oak, and will munch the golden grama until it is replaced by weeds. Then a freshet will tear out the old dam, and pile its rocks against a tourist road along the river below. Trucks will churn the dust of the old trail on which I saw wolf tracks yesterday."[28]

LEOPOLD'S PREDICTION WAS correct, except for the prophecy of a tourist road. The integrity of the Gavilán did not last long after his 1936 visit. He came back for a second hunt at the close of 1937, bringing his brother Carl and eldest son Starker, and this experience was as enchanting as the first. He also urged in various quarters that the area be preserved as "an international experiment station" to serve "as a norm for sick land on both sides of the border."[29] But nothing came of his proposal. Game began to decline in the mountains in subsequent years, and the outfitting business in Colonia Pacheco was all but dead by 1941.[30] Shortly after Leopold's death in 1948, his son Starker returned to the Río Gavilán hoping to initiate some of the ecological studies his father had dreamed of, but the paradise he remembered from a decade earlier was gone. Logging roads threaded through the headwaters and livestock had followed. Many stretches of the formerly moss-hung river had become a scour of cobbles.

The downward ecological spiral of the Río Gavilán continues to the present.[31] In the all-too-familiar cycle that Leopold in his early days documented throughout Arizona and New Mexico, logging and overgrazing bared the land to the violence of storms. Deprived of its defense of grass, litter, and living roots, the soils of the slopes yielded to the force of downpours, and floods tore every soft thing from the rivers and creeks, leaving rocky gutters in their place. The damage was a kind of sin against the land, which Leopold captured in one of his best metaphors: "Somehow the watercourse is to dry country what the face is to human beauty. Mutilate it and the whole is gone."[32]

MANY ARE THE accounts of southwestern watersheds that unraveled. In one of his essays on Mexico, Leopold cites Will Barnes's telling of the destruction of valley lands along the San Simon River in Arizona. In "Pioneers and Gullies" (1924) Leopold himself describes the loss of 90 percent of the cultivated land along the Blue River in Arizona's White Mountains. Elmer Otis Wooton, in a seminal 1908 monograph, *The Range Problem in New Mexico*, recounts how deepening arroyos cut the life out of the Mangas Valley in the Burro Mountains. These and other reports point unmistakably to human agency as a trigger for widespread erosion, if not through logging and grazing then as a result of roads, trails, and other disturbances that guttered runoff and initiated the formation of arroyos. Just before his first trip to the Gavilán, Leopold drafted an extensive discussion on this topic, which he titled "The Erosion Cycle in the Southwest." Dutifully, he sent the essay out for review, including a copy to the eminent Harvard geologist Kirk Bryan, a native of New Mexico who had closely studied arroyos in Chaco Canyon and other southwestern locations. Bryan and many of his colleagues maintained that climate, not human activity, governed arroyo formation and that, independent of human influence, the Southwest had repeatedly undergone alternating periods of degradation—arroyo cutting and stream entrenchment—on the one hand, and aggradation—the building up of alluvial lands—on the other.

Bryan's critique evidently prevented Leopold from settling his mind on the subject of arroyo formation, and "The Erosion Cycle in the Southwest" was never published. But he did not let the matter rest. He urged his son Luna to go to Harvard to study with Bryan and get to the bottom of the issue.[33] Not many families pass intellectual conundrums from one generation to the next, but the Leopold family was in a class by itself. All five of Leopold's children became accomplished naturalists, and three, Starker, Luna, and Estella, achieved election to the National Academy of Sciences.[34] Luna's dissertation, "The Erosion Problem of Southwestern United States," was not completed until after his father's death, but the conclusion he reached neatly bridged the points of view of his father and his mentor. Near the end of his own life, he restated it in a book, *A View of the River*, that sums up his life's work as a hydrologist: "The deep gullies cut by erosion in pre-Columbian time took less than 200 years to evacuate a large part of the early Holocene fill. In the period 1880–1920, overgrazing and climate change repeated the events of A.D. 1200–1400 in a period of less than 50 years."[35] Luna Leopold had concluded that climate and geology control cycles of large-scale erosion, but also that human activity, like an enzyme in a biological reaction, accelerated the process markedly.

The relative weights accorded to climatic and human causes of arroyo formation are important. If climate is the sole cause, then clear-cutting a watershed or grazing it to the point that it can't hide a golf ball might alter its biota and ruin its productivity, but the gullies that subsequently take form *would have formed anyway.* Such a view is nonsense, as both Leopolds well knew, and the Río Gavilán was their proof. The watershed of the Gavilán did not unravel during the arroyo-cutting heyday of 1880–1920. It came apart later, after its timber was cut, its grass removed, and its slopes braided with roads and trails that concentrated the energy of storm runoff on newly exposed targets. The most reasonable conclusion, in fact, might be stated less conservatively than the restrained scientist Luna Leopold was willing to do: namely, that the way people treat the land influences arroyo-cutting considerably, and that the influence is especially great when climatic factors favor downcutting and degradation.

Two key factors are important to understand: effective aridity and storm energy. In the period 1880–1920, when most of the valley arroyos in the western states were cut, total precipitation did not differ markedly from other periods of similar length, but more precipitation appears to have come from heavy storms, and less as light rain.

A light rain—what some Pueblos call a female rain—falls slowly and gently. It soaks the ground, and a large proportion of its water becomes available to plants. Under a regime of abundant light rain, which would normally mean many mild storms, the vegetative cover of the land tends to increase. By contrast, heavy rains—male rains—quickly saturate the topmost layer of the soil. During the downpours, the water falls so fast that it soon exceeds the absorptive capacity of the soil, and most of it runs off, sometimes in violent flash floods. Only a small portion remains behind to nurture plant growth. As a result, a period with many violent thunderstorms and few light rains is effectively more arid than a period in which light rains produce the same total amount of moisture. As aridity increases, the protective clothing of the land—in the form of vegetation, leaf litter, and other organic matter—decreases, making the soil more vulnerable to the increased energy of those big, wild, violent male rains.

If all this sounds familiar, it should. The prospect of bigger, more violent storms coupled with increased effective aridity is part of the bundle of effects predicted for the Southwest as a result of climate change. In the early 1990s, Luna Leopold put the matter bluntly: "The climatic changes of the past suggest that if the trend toward a warmer and more arid climate actually continues in the coming decades, the erosion of alluvial valleys seen in the thirteenth century, and again in the nineteenth, will be repeated in many of the semiarid areas of the planet where the rainfall is primarily of the thunderstorm type."[36]

AT LAST THE rains have come to El Cuervo and San Pedro. They arrived as a steady sprinkle after we returned one of Lalo's porcupines to its celtus tree. The walk back was long enough that we were soaked to the skin when we reached the truck, an outcome that Rodrigo deemed an emergency, justifying immediate deployment of the emergency kit. To no one's surprise, it lacked a supply of jerky.

The next day dawned cloudy and drizzled intermittently as we paid a call to the Hacienda San Diego, a ruined palace of tile and adobe, once owned—together with twenty-some similar estates scattered across the region—by Chihuahua's wealthiest and most powerful man, Luís Terrazas. At various stages of his extraordinary life, Terrazas was a commander of militia, governor of Chihuahua, intimate (and later enemy) of successive Mexican presidents, and cattle baron at a spectacular scale. The lowest estimates of the number of cattle Terrazas once grazed across northern Chihuahua cluster around half a million. The high end of the range is stratospheric. (Two oft-repeated anecdotes suggest the imprint he has left in popular memory. Asked by an ingénue if he was from Chihuahua, Terrazas answered, "I am not *from* Chihuahua; Chihuahua belongs to me." And when an American meatpacker inquired if he could deliver a hundred thousand head of cattle to the border, Terrazas supposedly replied, "In what color?")

Terrazas abandoned the Hacienda San Diego soon after the revolution broke out—it is said he hired a train to haul his belongings from Pearson to the safety of El Paso. Today the hacienda is like a beached and ransacked treasure ship, its grand rooms empty, its murals fading, the atrium where executions are said to have taken place overgrown with weeds. A family descended from *vaqueros* Terrazas once employed has colonized a suite of rooms, which are bitter cold in the winter and drafty against the wind. They say they are protecting the place in hopes its glory might one day be restored.

Some miles farther along, the rain was pouring down by the time we got to Paquimé, site of the ruins of the big houses, the Casas Grandes, that some scholars think were a kind of southern cousin of Chaco Canyon. We took shelter in the modern and well-appointed interpretive museum, and listened to rain leak through the roof, puddling in the exhibits. We gazed through the outside windows as though into an aquarium, the rain falling in sheets, the air almost solid with water, quickly generating little creeks that ran chocolate-colored down the alleyways of the ruins.

The rains continued for the better part of the next two days, more often harsh and male than gently female, at least in the way they registered on the mind. Soon all of northwest Chihuahua was soaked.

The drought had broken, and, biologically speaking, the coming of the rains should have ushered in a period of rest and recovery. Drought in arid and semi-arid land is like the bear market that assures an eventual bullish run. During drought the food chain thins. If the metaphorical chain was as

THE RUINS AT PAQUIMÉ (CASAS GRANDES), CHIHUAHUA, AS THE RAINS FINALLY ARRIVE, JULY 2008. *AUTHOR PHOTO.*

thick as a ship's cable in times of plenty, it becomes a dog leash now. Plants grow less or die back, and the herbivores that consume them, from mice to antelope, decline as well. Inevitably the predators that prey upon the grazers grow fewer, too. When rains come, the plants rebound first, much faster than the herbivore populations can build back up. They sprout, flower, develop root reserves, increase leaf surface, and send out suckers in a world where creatures that chomp and nibble are scarce. "Usually after a drought we rush in with cattle and bring them back too soon," says Fredrickson. "The plants, grasses especially, need a long period of recovery, and we don't let them have it."

Premature restocking is the likely fate of the bald prairies of El Cuervo, depending on how soon the landowners buy new cattle to replace those that starved in the drought. Long before the new cattle arrive, the few slat-ribbed survivors that we saw teetering by the water tank will shake off their lethargy and devour the green shoots of Russian thistle and other weeds that will be among the first plants to sprout. Some of them, unable to digest the sudden bounty, will bloat and die. Those that don't will stagger on, metabolically crippled, unable to grow much meat, the heifers among them likely never producing a calf. They will remain, in fact, economic liabilities, consuming resources but capable only of maintaining themselves.

Not far away on Rancho El Uno the rains will renew the tug-of-war between grasses and mesquite. The resurgent colonies of prairie dogs that Fredrickson and his colleagues have nurtured will partly arbitrate the result.

But still the rain pours down. After dark, a dog huddles for shelter on the mats and rugs we've piled to keep water from flowing under the front door and into the house. The dog flees when we open the door to peer into the night. Rain streaks the yellow light of a solitary streetlight. The dirt road has become a gleaming chain of puddles all the way to the corner, and the intersection is flooded in all directions. A din of frogs has erupted. They croak and warble, wail and bleat, and the drowned streets are alive with them. The streetlight shows at least a dozen frogs leaping through the puddles, some disappearing into darkness, others coming into the light, hustling down the road this way and that, all of them on errands of evident urgency. It is a frog jubilee, a night of orgiastic copulation making up for a year of lying inert and dehydrated under layers of dust. If drought is essential to desert plants, gully-washing rains are no less essential to desert frogs.

Next day we prepare to leave. We pay a call at the field station, which is now moated by water four inches deep. We pick our way gingerly. We say goodbye to Lalo, Rodrigo, and the others and fend off the affections of the splashing, hyperactive stray dog they have adopted, which they alternately call Obama or Osama to get the greatest rise from their audience. Farewells completed, Fredrickson takes the wheel of the big Ford and we roll wetly down the dirt lanes of San Pedro and then along the crowned gravel roads of Buenos Aires, the Mennonite settlement, where for once the sprinkler rigs in the potato fields are silent.

This portion of the Janos grasslands is a plain with ill-defined drainage. Water has pooled in the plowed fields and open pastures. But except within the carefully tended confines of Buenos Aires, the pastures and croplands stand higher than the bulldozed road. For long distances, water continues to drain from the fields into the depression of the narrow lane we are driving. Fences hem us in on either side, so there is no way to outflank the long, narrow pools. Besides, the saturated soils beyond the roadbed would mire us in an instant.

Fredrickson guns the truck through the first linear pond. It is only fifty or sixty yards long, and the center ridge between the ruts crests the surface of the water like the back of a crocodile. The truck jolts through an unseen hole and grinds on to safety. No problem. The next pond is deeper and a little longer. The truck roars and bucks its way to hard ground again. And another pond, and another. Traction has been good, but we have miles to go to reach the paved highway. No homes are in sight, only pastureland and Mennonite fields. If we mire or the engine quits, we will have a long way to wade and a longer way to posthole through deep mud, finally to beg some farmer to pull us out with his tractor. After the first mile, the ponds graduate into lakes so deep that the forward charge of the big Ford sends plumes of dark water high up the sides of the cab. The first plume drenches us both, for it is July in Chihuahua and we have the windows down. They will be down no longer.

Fredrickson was nonchalant about the water at first, but now his jaw is set and his eyes are in a squint. His knuckles on the steering wheel look a shade whiter than they did before.

The lakes grow still deeper, and now the truck is pushing a bow wave. We assume that every wire and fitting under the hood must be drenched. It seems improbable, but the Ford keeps roaring on. Soon we are looking at a stretch of flooded lane at least a quarter mile long and straight as an arrow between two lines of four-strand barbed wire fence. There's no turning back—and for that

END OF DAY IN THE JANOS GRASSLANDS, WITH RAIN APPROACHING AFTER A YEAR OF EXTREME DROUGHT, JULY 2008. *AUTHOR PHOTO.*

matter, there is no room to turn the giant pickup around. "Here goes," says Fredrickson, and we plunge in, going deeper, deeper. We feel the resistance of the bow wave. The truck's rear end begins to fishtail. Either the road beneath the water is a quagmire, or we have lost traction to flotation; maybe both. Fredrickson pushes the RPMs to the top of their range, and all four knobby tires are clawing and arguing with the water and mud. Froth flies from the truck in all directions. Engine roar fills the cab. Suddenly we go deeper still, and the bow wave breaks over the hood and washes up the windshield. The truck staggers, lurches, keeps grinding forward. If the engine quits now, we'll have to swim. Still the truck churns, creeping onward, engine howling. We rise an inch, then several inches. The bow wave recedes. Now we are like an LST crawling onto a beach. Gradually we become a truck again, no longer a boat, growling through the sodden aridlands of northwest Chihuahua, a landscape riddled with riddles, as we prepare for another immersion.

# 5

## LAVA FALLS: THE BLOOD OF OASIS CIVILIZATION

RIGHTLY OR WRONGLY, everything challenging on a whitewater river in North America gets compared to the booming rapid that culminates, in space, time, and difficulty, a river trip through the Grand Canyon. Say "Lava" to anyone who has tasted whitewater, and the association to Lava Falls Rapid is automatic. A friend who guides trips on some of Alaska's wildest rivers bristles when she hears the name *Lava North* applied to the most sphincter-tightening, life-or-death rapid on the mighty Alsek River. "It shows how Grand Canyon–centric the rafting world is," she says. Advocates for other rivers say much the same. Still, Lava is king, and the Colorado River, for which Lava is a mere riffle in its eons of canyon carving, is the most mythic of river kingdoms. If you have a weakness for wild rivers, eventually you float the Colorado, and eventually you make your way to the Grand Canyon and to Lava.

The drop at Lava Falls is thirteen feet almost immediately, followed by fourteen more in a few hundred yards. At most water levels, Lava earns a difficulty rating of ten, on a scale of ten. It confronts you at mile 179 on the 226-mile voyage from Lees Ferry to Diamond Creek, an incomparable outdoor adventure. The trip has the shape of a well-crafted novel. It establishes its themes in the red-rock stillness of Marble Canyon. It tests its characters in the churning whitewater of the Inner Gorge. Then Lava comes exactly where a novelist would place the climax, about four-fifths of the way through the saga. All the way down the river, you have had Lava in the back of your mind. Everything that precedes it feels like lead-up. Everything that follows is coda, resolution, release, perhaps recovery. The crux of the tale, the defining moment, resides in the hurricane waters of Lava Falls.

By the time our small flotilla got there, we'd courted disaster at Crystal, dodged the rock horns of Horn Creek Rapid, ridden the roller coaster of Granite, and thrashed and crashed our way through scores of other rapids. Our baptism was long and wet, and it taught us that the river has many voices. We might have thought we knew them all, but we didn't. Half a mile upstream from Lava, floating in the eerie calm of the slack water backed up by the falls, we heard one that was new and unexpected. It existed outside the range of the other voices of the river, a bass rumble at the lower limit of hearing. You felt it as much as heard it. It was both majestic and menacing, like a naval bombardment on a faraway coast. Its vibration trembled in the tubes of the boat. The new folks turned a shade paler under their sunburns. The old hands, if they spoke at all, spoke in low tones. They had the set-jaw look of a rider about to mount a horse known for its nasty habits.

We brought our paddleboat to shore on river right and trekked up a promontory of heat-soaked basalt to scout the rapid. At first the cauldron of froth and roar looked so chaotic I couldn't make sense of it. Then the noise in my head, if not in my ears, quieted a bit, and I began to pick out its parts. On river left lay a maze of jagged rocks with no workable line through them. A left run would be possible only at higher water when most of those rocks were drowned. A center run, on the other hand, was a worse idea. It would begin with a plunge into the maw of a giant hole, formed where the river pours over a series of basalt boulders—the remnants of a debris dam that once held back the entire river. If the hole did not flip us, the run would end instantly in collision with a wall of water the size of a mobile home. The standing wave would churn our boat and all of us until every person and part came loose from everything else. Nothing would come out the way it went in. Not a good alternative.

The only feasible route began on river right, around the end of a long pour-over ledge. The trick for us would be to float down to the ledge, nip in under its right corner with a burst of velocity, and dig with our paddles for the center of the river. Then we'd slam into a big wave like a whale's back just below the pour-over, and the collision would veer our boat downstream, aligning it (we hoped) to take on a massive V-wave more or less at its apex. The V-wave channels you into a convergence of water eruptions that blast like fire hoses from all directions. It's an orgy of aquatic fury, which one wit has called "the world's biggest car wash." At that point you put your head down, hold on to your breath and the boat, and try to keep your paddle from flying loose and remodeling your own nose or that of your neighbor. A lot of the rapid still lies ahead, but from there on, pretty much your only option is to place your trust in dumb luck and chaos.

Our calculation, as we stood on the scouting rock, was a series of *ifs*: if we tucked in quickly under the ledge, if the whaleback wave turned us just so, and if we hit the V-wave close to the apex, then we just might punch through the ten-foot wave at the end of the car wash and be in position to paddle away from a boulder like a lost remnant of Stonehenge at the end of the run. If we missed our entry, well, a lot of things could happen. It all came down to having the right position, angle, and velocity at the outset.

YOU MIGHT SAY that the West, together with most of the rest of the world, is somewhere in the flat water above the rapids of global change. Not that we haven't already felt some of the early effects of an altered climate, but the big excitement lies ahead. Our trip through the hazards will be a first run— which is to say we will have the benefit of no one else's experience. Still,

LAVA FALLS RAPID, GRAND CANYON, ARIZONA. THE SCALE OF THE RAPID IS SUFFICIENT TO CAUSE AN EIGHTEEN-FOOT RAFT TO BRIEFLY DISAPPEAR IN ITS WHITEWATER. *AUTHOR PHOTO.*

enough good science exists to constitute a decent scouting. Looking down-stream, for instance, we know that the Southwest will become hotter and drier, with greater extremes of both storm and drought. Increased aridity is assured by higher temperatures, even if precipitation does not decline, which it is likely to do. A greater proportion of that precipitation will come as rain, less as snow, and runoff from winter snowpack will peak roughly a month or so earlier than it used to. These are some of the big rocks and ledges we know about. Other hazards—waves of hyper-powerful forest fires, ecological die-offs, and dust storms—lie downstream, but we are less sure where; we only know to look out for them. Additional and as yet unidentified dangers may also crowd our path, but their present invisibility could be just as well: if we seriously attend to what we already know about, our hands and our agenda will be full.

Lester Snow, director of the California Department of Water Resources from 2004 to 2010 and later secretary of natural resources, thinks that part of his job as a public servant is to scout these rapids, and he has devoted consid-erable thought to his point of entry. He knows that given the length of time it takes to plan and build a reservoir, a pipeline, or a flood protection levee, the decisions of today will ramify onward indefinitely. "I hate to use a clichéd phrase," he says, "but there is a tipping point somewhere [past which] you can't catch up, and if you don't make decisions now, you may have made a decision you can't recover from."[1]

For purposes of future planning, Snow and his staff assumed a 40 percent reduction in Sierra Nevada snowpack by 2050 and engineered new levees and other water projects to accommodate a fifty-five-inch rise in sea level. The current rate of sea-level rise is about 3.3 millimeters per year, roughly double what it was for the majority of the twentieth century. It will likely continue to accelerate.

An increase in sea level will have immense impacts on California's inland waters as well as the coast. The San Francisco Bay and the deltas of the Sacramento and San Joaquin Rivers constitute a vital nexus for the state's economy, environment, and hydrology. All of it lies within a few feet of sea level. Flood hazards, endangered species, and other issues critical to agricul-tural and domestic water supplies converge there, affecting the fortunes and security of millions of people. Ignoring a probable rise in sea level is not an option. "There are numbers all over the place," Snow explains. "We just picked

---

fifty-five inches. At the same time we are commissioning scientists to come up with a range that could either verify or undermine that target. At least when I pick a number, then I have given the engineers something that they can engineer."[2] And adjusting to a revised number, if necessary, is a lot easier than starting from scratch.

Compared to other states, California is at least attempting to grapple with the implications of climate change. It ought to. Its vulnerability to floods, especially along the Sacramento River, and the dependence of tens of millions of its people on fantastically elaborate water-supply systems make it vulnerable even to small shifts in existing conditions. It may be counterintuitive, but in times of crisis, sloppiness is a kind of grace. Paring away at waste is always the first strategy for dealing with shortage. It's easy and produces fast results. Fix a leak and things get better. California, however, has been paring away at waste for a long time, and most of the easy work is already done. Although a certain amount of sloppiness remains, there's no longer enough of it to provide the flexibility and potential for rescue that 20 or 30 million people in a tight spot are likely to need. The Los Angeles metropolitan area, for starters, is the largest desert conurbation in the world. It receives its water from a complex of aqueducts long enough to reach from southern California to Rapid City, South Dakota. Every drop of that water, you might say, is filtered through a web of contracts and agreements that requires an army of lawyers to defend. One bad day in court, and the consequences can be devastating. The fact that faucets in Malibu flow with snowmelt mixed from the Wind River Mountains of Wyoming, the Colorado Rockies, and both sides of the Sierra Nevada is a daily miracle. And the tighter and tauter the system stretches, the harder it is to keep delivering the miracles.

Miracles also abound in Phoenix, Las Vegas, Tucson, and Denver, even to the point that molecules of $H_2O$ from the same Colorado snowdrift that slakes LA's thirst may similarly appear in those precincts. The replumbing of the rivers of the American West, especially the brawny Colorado, makes daily marvels of water delivery not just possible but imperative. The abundance engineered by a previous generation has had its intended effect: a powerful society has risen unconstrained by the limits of the land, wealthy and populous. But now nearly everything that can be used is being used; every new tap and faucet adds strain to the system, and abundance has metamorphosed into

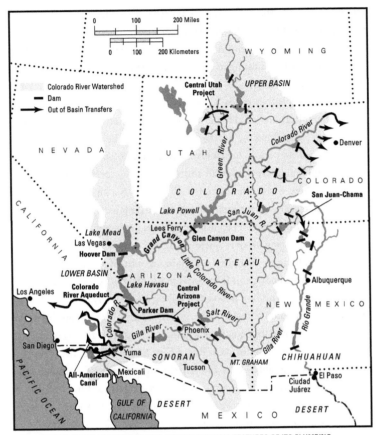

**MAP 6:** THE COLORADO RIVER WATERSHED AND THE PRINCIPAL FEATURES OF ITS PLUMBING.

scarcity. The good news for the rest of the Southwest is that it may have a little more time than California to order its affairs in preparation for climate change. It is not stretched quite as thin. At least, that's what Brad Udall thinks.

Udall directs a program at the University of Colorado called the Western Water Assessment. The National Oceanic and Atmospheric Administration funds it, and its purpose is to deliver science-based climate information to western water managers in a form and manner they can use. Udall spends most of his time on issues relating to the Colorado River. That suits him fine, for a couple of reasons. The Colorado, in the words of Marc Reisner, "is the most legislated, most debated, and most litigated river in the entire world."[3] Nearly 30 million people in the United States and Mexico (roughly equivalent to the combined populations of Illinois, Indiana, and Ohio) depend on it for all or part of their water. If you are involved in western water—and if, like Udall, you want to be where the action is—you want to work on the Colorado.

Udall also has the river in his blood. His great-great-grandfather was John D. Lee, for whom Lees Ferry is named (the apostrophe washed away long ago). Lees Ferry is the launch point for boat trips into the Grand Canyon. It is also the site of the single most important gauge along the 1,300-mile length of the river.[4] It lies in northern Arizona a short distance downstream of what is today the Utah state line. Ostensibly, Brigham Young dispatched John D. Lee to settle at that lonely crossing in 1872 in order to build a ferry and ensure transport across the river in a location vital to the growth of the Mormon colony. But Lee's assignment was more complicated than that. It was also a banishment. Brigham Young wanted to get Lee out of sight because Lee had played a large role, years earlier, in the Mountain Meadows Massacre, in which Mormon militia attacked a California-bound wagon train, killing close to 120 men, women, and children, sparing only the infants. Unfortunately for Lee, even the desolation of a lonely red-rock crack in the Colorado Plateau failed to shield him from the attention of the law. Ultimately, he was brought to trial and, after lengthy legal proceedings, sentenced to death. His executioners took him back to the Mountain Meadows, where a firing squad carried out the order of the court. A scapegoat for at least a score of guilty, John D. Lee was the only perpetrator of the massacre to be punished.[5]

Brad Udall's genealogical connection to the river doesn't end with his great-great-grandfather, nor even with his great-great-grandmother, the redoubtable Emma Lee, who was John D. Lee's seventeenth wife and who continued operating the ferry for several years after her husband's death. His father, Morris Udall, represented Arizona in Congress for thirty years and helped craft the devil's bargain that produced the $4 billion Central Arizona Project (CAP), which brings Colorado River water across the breadth of Arizona to Phoenix and Tucson. Without the CAP, Arizona's furious Sunbelt growth would have been impossible. Morris's brother Stewart—Brad's uncle—served Presidents Kennedy and Johnson as secretary of the interior and played an even bigger role in brokering the birth of the CAP. Stewart Udall also presided over innumerable other projects touching the energy and water resources of the West, including the final construction and commissioning of Glen Canyon Dam, just upstream from Lees Ferry, which impounds Lake Powell. Both Stewart and Mo Udall embodied the values of their time when they entered public office, and they advocated for water development on a massive scale. But both left office much changed by their responsibilities. By the time their public careers ended, each had compiled a record placing him among the leading American conservationists of the twentieth century. Nevertheless, their legacies are by no means unblemished, even in their own appraisal. Looking back, Stewart came to see his support for Glen Canyon Dam as one of his greatest mistakes.[6]

Unlike his brother Mark and cousin Tom, both now serving in the U.S. Senate, Brad Udall chose a career outside politics. He became an engineer, though not too quickly. As a young man, he indulged his affection for wild water and briefly pursued a career as a Grand Canyon river guide, learning the moods of the Colorado River in an intimate, tactile way. Udall is lean and rangy, and it is easy to imagine his long arms levering the oars of a raft. It is easy, too, to picture his considerable native energy released in the unconstrained environment of the canyon, less easy to visualize it contained behind a desk. When he talks about the river and climate change, which lie at the core of his current work, he speaks in fast long paragraphs, the words streaming almost on top of each other, so that he sometimes runs short of breath and finishes his sentences on empty lungs.

Udall has the sense that the 30 million people who depend on the Colorado River are poised at the top of a rapid, and he has the sound of the whitewater in his ears. "What people don't get," he says, "is just how out of the ordinary these times are right now in human history." His diction is a touch unusual, a little like that of the old-time movie actor Jimmy Stewart. He plays with the tempo and pitch of words as though he were trying to wring something out of them that other people don't hear. "Sustainability," he intones, stretching out the syllables, "it's the IQ test for humans in the twenty-first century. And I am not certain that we're going to pass the test."[7]

Passing the test will be as hard in the watershed of the Colorado River as anyplace in the world, which is one reason the Western Water Assessment exists. The river's basin is rugged terrain, not just physically, but also legally and politically. Most of Udall's colleagues are scientists, but most of his "customers"—the policymakers, administrators, and elected leaders whom the Western Water Assessment is purposed to inform—are not. It typically falls to Udall to fill the role of translator, and he plies his trade in both directions: explaining to policymakers what the scientists have learned, and helping the scientists appreciate what the policymakers need to know. Udall has won a reputation for representing each group to the other in a way that both find accurate and trustworthy.

The world of the Colorado is as complex as it is big—the river's watershed sprawls over nearly a quarter-million square miles, representing one-twelfth of the continental United States. "What you see on the climate models, which makes the Colorado River Basin a little challenging to figure out, is [that] the basin's so big north to south that parts of it sit on this dividing line where it's likely to be dry south and wet north, and yeah, you can scratch your head and go, what does this really mean?" For Udall, the likely meaning includes gradual expansion of the Hadley cell across the Lower Basin and the risk of reduced precipitation in the areas thus affected. It means winter storm tracks moving northward, shorter winters, and generally less snow in lower elevations, "although we've got to be kind of careful here because this more powerful atmosphere can dump more snow in shorter periods of time." The clincher is that higher temperatures, producing "longer summers and drier soils," will result in "5, 10, 15, 20 percent reductions in flow of the Colorado River by 2050, and potentially a lot more by 2100, depending on what goes on with humans and greenhouse gases."

Lester Snow likes to say, "All models are wrong; some are useful," and Udall agrees. He points out that science is still in the "Model T" stage of modeling: "Are we going to get a whole lot more clarity than we have now? Probably not in the next five years. But I think we've got more clarity than some people might think." Part of the problem is that people fixate on the models' lack of precision—"Is it 15 percent or 18 percent, and which one of these numbers do we plan for?"—rather than on the central message, which describes the direction and approximate magnitude of change. "It's kind of like the old famous story of the drunk looking for his keys under the porch light [when in fact they were dropped in shadow]. We've gravitated to these models because they can answer these kinds of questions, but we don't step back and say, all right, how much bigger is the uncertainty than what the model has already told us?"

A CASE IN point is the debate over the future of Lake Mead, the giant reservoir formed by Hoover Dam. In 2008 Tim Barnett and David Pierce, researchers at the Scripps Institution of Oceanography in San Diego, published a paper in which they postulated that Lake Mead, under the one-two punch of climate change and high rates of water consumption, had a 50–50 chance of going dry by 2021.[8] The implication was that in terms of water supply, the cities of Los Angeles, San Diego, Phoenix, Tucson, and Las Vegas, all of which depend on releases from Mead, were in peril. Their lifeline was running dry, and no rescue was in sight.

The news was jarring but, for anyone who had been paying attention, not wholly unexpected. After years of plenty during the 1980s and most of the '90s, which left the reservoir brimming, drought gripped the region through the early 2000s. In 2002 the Colorado River shrank to its puniest flow on record—only 25 percent of the historical average.[9] By the time the Barnett and Pierce paper appeared, both Lake Mead and Lake Powell stood at less than half their capacity. In less than a decade they had become forlorn pools shrunken from the inland seas that they used to be. Their formerly busy boat ramps (excepting those few that were lengthened at great expense) ended high and dry at white, chalky bathtub rings marking where the water used to be.

Officials of the Bureau of Reclamation, which operates Hoover Dam and Lake Mead, as well as other water managers downstream, knew the situation was bad, but Barnett and Pierce had done something that the river's water establishment, surprisingly, had not: they crafted an analysis of water availability that took climate change into account.[10] The Scripps team modeled the inflows and outflows of both lakes, Mead and Powell; they factored in a decline in river flow under various climate change scenarios; and they included the standard estimates for increased Upper Basin water withdrawals, as water rights in Colorado, Wyoming, and Utah are progressively put to use in the years ahead. It was a budget analysis: so much in, so much out, how much left?

Barnett and Pierce were like auditors who had shown up uninvited to examine the finances of a bank, and they plied their craft more comprehensively than the bank's management had bothered to do. It is hard to say to what degree the managers' neglect was owed to the press of daily business or to the George W. Bush administration's suppression and avoidance of climate change science. Whatever the case, the auditors' verdict was simple: the bank was unsound. It was going to fail.

By the standards of water management and infrastructure construction, the year 2021, when the bank would have only a 50–50 chance of solvency, was not much farther off than the day after tomorrow. And worse, the crisis Barnett and Pierce were predicting was not something the region could quickly engineer its way out of. When there is no water to put in a new pipeline, it doesn't matter whether the pipeline gets built or where it goes.

The prediction made big news. Media from near and far picked up the story, and local boomer editorialists quickly branded the study alarmist. Trotting out the usual Sin City clichés, the Las Vegas *Review-Journal* boasted, "We'd love to buy some action on the odds provided by Mr. Barnett and Mr. Pierce. They can name the amount at stake. Are they willing to put their money where their mouths are?"[11] Water managers from the Bureau of Reclamation and the Lower Basin states,[12] meanwhile, protested that they would never allow the lake to go dry, pointing to a shortage-sharing agreement, known as the "Interim Guidelines," that they had collectively negotiated the previous year. Barnett replied that it was precisely the shortage-sharing agreement that had prompted him to undertake the study. The reductions in consumption that it mandated, he said, were too small. At the most they would require cutbacks of 6.6% in Lower Basin consumption, or 500,000 acre-feet. It would be too little too late for a system that already (independent of climate change) was living beyond its means to the tune of more than a million acre-feet of water a year.

THE DEBATE OVER Lake Mead grows from nearly a century of hard bargaining, colossal engineering, and wishful thinking. The catechism of the river begins with the Colorado River Compact of 1922, which divided the river at Lees Ferry into upper and lower basins, and apportioned use of the river evenly between them: each basin would be entitled to consume, on average, 7.5 million acre-feet of water per year, an acre-foot being a quantity of water sufficient to cover an acre to a depth of one foot. (In the Southwest at normal levels of conservation, an acre-foot, roughly 326,000 gallons, will support four single-family homes for a year.)[13] The states of each basin would allocate their respective portions of the river among themselves as they saw fit, which is to say, as the politics of power, cajolery, litigation, and self-interest dictated.

Fulfilling the promise of the compact required some of the world's greatest feats of construction. Lake Mead, which began filling in 1935, following construction of Hoover Dam, would serve as the Lower Basin's primary source of storage. Glen Canyon Dam, completed in 1963, would impound Lake Powell, whose purpose was to enable the Upper Basin to store above-average flows, balance them against the deficits of drought years, and meet its compact requirement of delivering a rolling average of 75 million acre-feet of water to the Lower Basin every decade.

In addition, under the 1944 Water Treaty with Mexico, the United States agreed to deliver 1.5 million acre-feet (maf) per year at the International Border. This obligation, plus the 15 maf divided between the two basins, brought the principal legal entitlements on the system to 16.5 maf per year.

These bald facts sound simple, but they hide complicated histories. That of the 1944 Mexican Water Treaty, which had as much to do with geopolitics as with the merits of how to share a river, provides a good example. After decades of entreaties, the United States finally consented to assure Mexico a share of the Colorado River, even as Mexico made parallel assurances about the contributions of the Rio Conchos to the lower Rio Grande. The treaty was unpopular among the water establishment of California because California, already using its full allotment (and more) of the river, stood to gain nothing by it. The other compact states, however, welcomed the certainty about downstream obligations which the treaty would provide and which they (and most of Congress) viewed as essential to justifying the big water projects they desired. The federal government held the trump card: with World War II drawing to a close, it wanted to secure a treaty in order to "tidy up loose ends" in the uncertain postwar world and as a way of demonstrating the United States' lofty approach to international leadership.[14]

The hydraulic history of the Colorado River, southern California, and the rest of the Southwest constitutes one of the defining stories of twentieth-century America. Like many great stories, it has a dormant but deadly problem planted at the beginning, a kind of pistol on the mantelpiece. In this case, the shadow of future tragedy takes the form of a fatal mortgage embedded in the

Colorado River Compact. It results from the fact that the men who drafted the compact (they were all men) believed their knowledge of the river to be true and sufficient to their task. It wasn't. The initial decades of the twentieth century, when the river's flow was first reliably gauged, up to the 1922 negotiation of the compact, happened to be one of the wettest pluvial periods in the last fifteen centuries; the high flows of those years led to assumptions about the long-term dimensions of the river that were misleadingly optimistic.

It is not clear what the members of the compact commission assumed the flow of the river to be. Some evidence points to a figure of 17.2 maf per year; others say it was 16.4 maf; more traditionally 17 maf is used.[15] Quite possibly, different members may have worked from different assumptions. In they end, they managed to persuade themselves, erroneously, that the river was reliably generous enough to provide at least 15 maf for the two basins and that any future allocation to Mexico might come out of what was left.

Even without worrying about evaporation, system waste, or environmental uses of water, the system was oversubscribed. By the mid-1970s, with the century's initial wet period long past, the average of the gauged record had declined to 15 maf. But worse news was still to come. In 1976 Charles Stockton and Gordon Jacoby published a reconstruction of river flows based on tree-ring analysis going back to 1520. They concluded that the average flow at Lees Ferry over the past four and a half centuries was about 13.4 maf.[16] Subsequent studies have produced estimates ranging from a low of 13 maf to a high of 14.7 maf and have pushed back the starting point for the reconstruction chronology to the year A.D. 762.[17] In addition, all of the studies confirmed Stockton's and Jacoby's second troubling discovery, which was that droughts longer and more severe than any known from the historical period have occurred repeatedly in the Colorado River watershed. In other words, a state of sufficiency within the Colorado River system, as defined by the sum of today's legally ordained water entitlements, would be utterly abnormal. What is normal is a much smaller and quite variable river whose low-flow periods are lower and more protracted than anything record-keeping humans have yet known.

DISCUSSIONS OF THE Colorado River, including this one, tend to brim with an excess of numbers. Here are the two most important to hold on to:

Total legal entitlements to mainstem Colorado River water (Upper Basin, Lower Basin, and Mexico): 16.5 maf/yr.

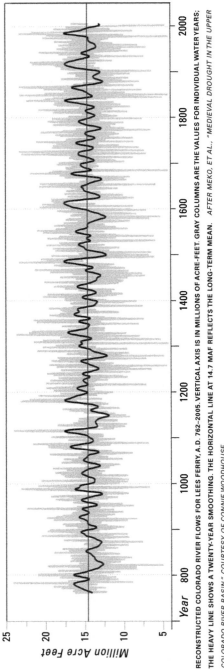

RECONSTRUCTED COLORADO RIVER FLOWS FOR LEES FERRY, A.D. 762–2005. VERTICAL AXIS IS IN MILLIONS OF ACRE-FEET. GRAY COLUMNS ARE THE VALUES FOR INDIVIDUAL WATER YEARS; THE HEAVY LINE SHOWS A TWENTY-YEAR SMOOTHING. THE HORIZONTAL LINE AT 14.7 MAF REFLECTS THE LONG-TERM MEAN. *AFTER MEKO, ET AL., "MEDIEVAL DROUGHT IN THE UPPER COLORADO RIVER BASIN," COURTESY OF CONNIE WOODHOUSE.*

Best-case (and probably most thorough) estimate of natural river flow at Lees Ferry: 14.7 maf/yr.

These two figures recapitulate the basic math of the Colorado River Compact with updated numbers. Clearly, a problem exists: putative outlays exceed revenues. But of course the actual facts on both sides of the ledger are more nuanced. Note that the figure for river flow is obtained at Lees Ferry. Below that point, even excluding the Gila River (which the "Law of the River" holds separate for accounting purposes), inflows from the springs and rivers of Grand Canyon, the Bill Williams River, and other downstream sources approach 1.56 maf/yr, a boon to the Lower Basin.[18] These contributions lift "revenues" to 16.26 maf/yr.

For a moment, the books of the river might almost seem balanced. Unfortunately, as with any enterprise, there are operating costs to be considered. From 1971 to 2004, Lakes Mead and Powell together evaporated an average of 1.2 maf/yr, and Upper Basin reservoirs gave up to the air an additional 630 kaf/yr (kaf = thousand acre-feet). In addition, consumption by Lower Basin vegetation and unavoidable overdeliveries to Mexico account for a little more than 650 kaf/yr. Add these figures together and total operating cost is roughly 2.5 maf/year. Combining these costs with entitlements of 16.5 maf/yr lifts total potential expenditures to *19 maf/yr.*

Now comes basic arithmetic: annual income of 16.26 maf less expenses of 19 maf leaves red ink at the bottom of that balance sheet. The scarlet letters declare a potential annual deficit of *2.74 maf.*

And that's before we account for flow declines attributable to climate change, which fairly soon might amount to 20 percent of gross income, and continue to grow after that.

If this were your personal profit and loss statement, you would have a hard time obtaining a mortgage, buying a car, or even getting your hands on a credit card. You'd have to talk fast, make the most of every opportunity that fell your way, and practice creative accounting just to get by from year to year. In short, you'd have to learn to do business like the Colorado River water establishment.

THE GAP BETWEEN entitled consumption and actual wet water has been finessed two ways. Historically, not all the entitlements were put to use. Arizona, for instance, was much slower to develop than California, and the CAP, which delivers Colorado River water to Phoenix and Tucson, did not come on line until 1993. Similarly, the Upper Basin has been slower to make use of its water than the Lower Basin, which has allowed California and more lately Arizona to develop profligate habits. But times and consumption habits have changed, and they will change some more. Arizona now diverts its full 2.8 maf allocation, while the Upper Basin diverts (and evaporates) somewhere between 4.3 and 5.1 maf annually. It is on track to raise that number to 6.04 maf by 2060, if it can find that much water in the river.[19]

The Upper Basin faces the future from a stronger position than its neighbors downstream for at least two reasons. First and most obviously, its slower growth prevented it from becoming as addicted to unrealistic flow expectations. Second, it long ago abandoned the fiction that 7.5 maf was available for its use. The Upper Basin developed its internal agreement, the Upper Basin Compact, in 1948, after witnessing the drought of the 1930s. Sobered by evidence of the river's variability, the Upper Basin divided its waters by percentages, not absolute amounts. Thus the state of Colorado gets 51.75 percent of what is available to the Upper Basin but is explicitly on notice that the amount of water represented by that share may vary.[20]

The Lower Basin, by contrast, regards its 7.5 maf allocation as Holy Writ. It fails to budget for evaporation, vegetative consumption, or a proportionate share of the treaty obligation to Mexico. Its old strategy of using the Upper Basin's leftovers to cover up its deficits, however, will no longer work. Together, Upper Basin development and climate change will guarantee the same result: there aren't going to be any leftovers.

The second strategy for dealing with the gap between consumption and actual water has been to store surplus flows to offset the deficits of drought years. This works only as long as cumulative surpluses equal or exceed cumulative deficits, as was the case through the wet period of the 1980s and 1990s—a happy circumstance for promoters of unlimited growth. Counting on such a strategy over the long term, however, is the same as saying you don't believe in the law of averages. In the long run, if evaporation and outflow exceed inflow, a reservoir will go dry, no matter how big it is. Lakes Mead and Powell together have the capacity to store 50 maf; other reservoirs on the Colorado bring total potential storage to 60 maf— the rough equivalent of four years of native flow. This immense capacity successfully sustained the Lower Basin through the drought of the first years of the twenty-first century.

By 2010, however, most of that capacity was depleted. Barring a return to the anomalously wet conditions of the 1980s—an eventuality that both paleo-climate reconstructions and climate change models argue against—it is unlikely that either reservoir will refill to capacity again. At the present rate of consumption, even without climate change, the region faces a future of long-term water deficits. With the arrival of climate change, however, the long term has grown markedly shorter, and the potential for a water-supply catastrophe looms ominously. What was once possible is now probable. Barnett and Pierce stopped short of saying "certain," but not far short.

"I THINK BARNETT ultimately did everybody a favor," says Udall. "He put out this paper and raised some issues out there that needed to be raised." Considered responses from scientific peers, not just editorial broadsides, were not long in coming. Barnett and Pierce had missed a few things. They'd failed to factor into their analysis the inflows to Lake Mead downstream of Lake Powell. Rivers like the Pariah, the Little Colorado, and the Virgin, which enters the Colorado from Nevada, as well as the spring-fed creeks of the Grand Canyon itself—Havasu, Tapeats, Shinumu, and other jewel-like waters—contribute well over 800,000 acre-feet per year, a substantial resource and an important palliative for an insolvent water budget.

Barnett and Pierce had also treated evaporative losses from the lake as a single, steady value, when in fact evaporation declines as lake level drops and surface area diminishes. These and other issues of fine-tuning became the substance of a dialogue between the Scripps researchers and a group of Udall's colleagues at the University of Colorado.[21] In the end, the conclusions of the two groups pretty well converged. As Barnett and Pierce put it in a more exhaustive 2009 follow-up paper, "Overall [the model improvements] delay the onset of problems compared to our earlier work by roughly 4 to 10 years . . . depending on the particular scenario."[22]

This was a stay of execution, not a commutation of sentence. The central message did not change. The models show that the system is unsustainable if either of two things happens: if mean annual flows approach the paleoaverage or if climate change reduces streamflow by as little as 10 percent. If both even-tualities should arise, as the odds suggest they may, the degree of adaptation required among water users would be simply stupendous. Doomsday might not arrive as early as 2021, as Barnett and Pierce first postulated, but, says Udall, "Risk just skyrockets after 2026 because of declining river flows and increasing demand.... There isn't really that much risk till then. It's not zero but it's not huge. But beyond it, by gosh, we've got a real problem."

> Risk assessment for a paddleboat at the top of Lava Falls Rapid,
>     assuming an experienced captain and reasonably competent crew:
> Risk of drenching: certain.
> Risk of exhilaration: high to certain, depending on personality type.
> Risk of one or more members of crew having out-of-boat experience:
>     moderate.
> Risk of boat flipping (and everybody swimming): low to moderate.
> Risk of bruises, abrasions, etc.: same.
> Risk of serious injury due to collision with rocks, gear, paddles, flailing
>     bodies: low.
> Risk of death by drowning: very low, but (especially if the putative victim
>     is you) never insignificant.

Our group's five oar-boats ran the rapid first. They were eighteen feet long, weighted with up to a ton of gear and passengers, and had side tubes two feet thick. The whitewater seemed to swallow them whole. They would burst from the froth like breaching whales and then disappear into the next set of waves. Each had a good run, although one ran close enough to the big rock at the bottom of the rapid to pin an oar and threaten a crash. A cushion of water piled by the current against the rock held off the boat, and it floated safely by. One by one, the oar-boats pulled into eddies below the main rapid, spaced well apart on both sides of the river, and turned to watch us. They formed a human net designed to catch anyone who took an involuntary swim, or all of us, if we flipped.

Compared to the oar-boats, our paddle-craft provided a much thinner membrane of protection: fourteen feet long, with tubes (and freeboard) half that of the oar-boats, its total weight was not much more than the sum of the seven of us wedged within it. Everyone checked the carabiners securing their day-bags and water bottles, and gave an extra tug to the cinches of their life vests. No more risk assessment now; it was time to go.

PROBABILISTIC RISK ASSESSMENTS cause Patricia Mulroy's eyes to roll, and such is the force of her personality that, when her eyes roll, everyone in the room feels dizzy. She is petite, attractive, and laughs like a man. Today she is

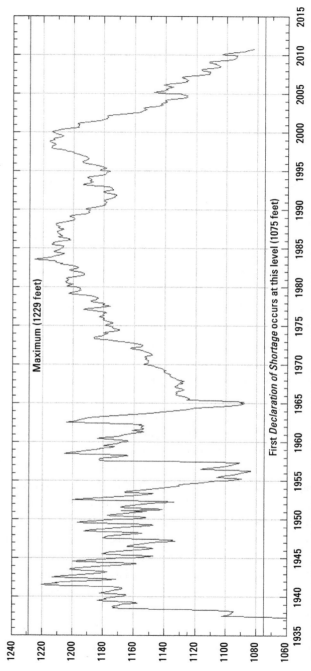

GRAPH OF LAKE MEAD WATER LEVEL, 1937–2010. IN OCTOBER 2010 LAKE ELEVATION DROPPED BELOW THE LOWEST POINT OF 1950s DROUGHT. DECLINE OF 1963–1965 REFLECTS FILLING OF LAKE POWELL.    *IMAGE AND DATA COURTESY OF P. LUTUS, HTTP://ARACHNOID.COM, AND THE U.S. DEPARTMENT OF THE INTERIOR, HTTP: WWW.USBR.GOV.*

wearing an off-white, tailored pantsuit and gold lamé ballet flats. She has a voice with a touch of gravel in it. It is the kind of voice that might sometimes intentionally abrade its hearers, and you expect nothing less from someone who has survived the rough-and-tumble of Las Vegas water politics for twenty years.

Mulroy is general manager of the Southern Nevada Water Authority, and she is not a fan of risk assessment. "Probabilities are like a bookmaking operation," she says. "What the water professionals have done for so long is take false comfort in high probabilities of no risk. It makes life easy. If there's only a 5 percent chance, why should I worry about it? I've got a 95 percent chance it won't happen. The problem is when the 5 percent does happen. So unless you can tell me it's impossible, I have to assume it can happen."[23]

In fact, there is a zero percent chance that you will tell her it is impossible, because after she finishes speaking she fixes a stare on you that feels a lot like a half nelson. Her grip does not loosen as she further explains that one of her greatest professional regrets is to have believed the rosy forecasts of the 1990s, when groupthink on the river prophesied full reservoirs to the limit of the foreseeable future. In 2000, "we had a fifty-year, reliable water supply. By 2002, we had no water supply. We were out. We were done. I swore to myself we'd never do that again. We can't be in that hole. It was such a stark awakening experience for me to have relied so heavily on those probabilities of the '90s. And it made me mad not to have seen it."

Nowadays Mulroy's agency provides water for 2 million residents and 40 million annual visitors to greater Las Vegas. Things were different in 1989, when she was appointed head of the Las Vegas Valley Water District, having had no previous experience in the water business. At that time her utility served only 118,936 active customers. It had little clout and was beholden to upstate centers of political power. Not for long. Within two years, Mulroy had welded the Las Vegas Valley Water District and a half dozen utilities serving the rest of the city and neighboring suburbs into the Southern Nevada Water Authority, providing water to all of Clark County's then-835,000 residents. To the surprise of no one who knew her, she emerged as general manager of the new entity. Greater Las Vegas now spoke with one voice on the subject of water, and the voice was hers.

Within two more years Mulroy had gained control of a majority of seats on Nevada's Colorado River Commission, which represented the state on matters relating to the Colorado River Compact. Now Mulroy sat eye-to-eye with the water czars of the other six Colorado River states. She represented Nevada on issues pertaining to the compact, the federal government's management of the river, and the ceaseless pursuit of more water for Las Vegas and its environs. From that point forward, through year after year of go-go Las Vegas growth, her authority and influence, within Nevada and beyond it, steadily increased.

Mulroy was born in Germany of an American father and a German mother. She grew up outside the traditions of the rural West, which have shaped not just western water law, but also the personalities of the (nearly always male) water buffaloes who interpret the law, administer the projects, and cut the deals. Mulroy's perspective is different. She came to public service as a twenty-five-year-old who needed a job and hoped to go to graduate school in German literature. She found work at the Clark County Manager's office. Her charm, toughness, vision, and general competence carried her on from there.

Mulroy's constituency is nontraditional, too, as different from the big-hat and belt-buckle power brokers of the past as the lobby of the Sands is from a cattle range or the factory fields of Imperial Valley. She points out that the Las Vegas Strip consumes only 3 percent of Nevada's water and is the state's largest employer and the greatest single contributor to its gross economic product. Perhaps because cattle, cotton, alfalfa, and other leading consumers of southwestern water don't swing votes within the SNWA, or perhaps simply because she's smart and hires smart people, she's conjured water (or water storage, which is sometimes the same thing) out of places nobody thought to look before. She's paying Arizona to pump water owned by Nevada into the ground so that at some future date Las Vegas can pump some of Arizona's water out of Lake Mead and trade the underground water back to Arizona. She's building a reservoir in California to capture surpluses the Imperial Valley used to release to Mexico so that she can later swap for Imperial Valley water in Lake Mead. She swung a deal to claim the flow of the Virgin River for Las Vegas, separate from the compact, by threatening to build, but never building, a reservoir that no one needed. She swung another deal to solidify return-flow credits for the river of waste her city produces. She paid her customers $110 million to tear up their lawns and replace them with desert vegetation, reducing water demand by 18 percent.[24]

Some of her moves have been more traditional and plainly ominous. During her first year as general manager of the Las Vegas Valley Water District, even before the SNWA was formed, she filed for rights to unclaimed water in thirty hydraulic basins across northern Nevada. Mulroy's "rural water grab" has been a high-stakes drama ever since, evoking inevitable comparisons to William Mulholland's draining of Owens Valley for Los Angeles in the 1920s.[25]

Notwithstanding that in 1994 Mulroy called the notion of tapping northern Nevada groundwater "the singularly most stupid idea anyone's ever had,"[26] the SNWA now has "at least 37 billion gallons lined up for the project,"[27] a liquid bonanza that it will tap once it builds a pipeline and other necessary infrastructure to deliver the water to Las Vegas at the whopping cost of $3.5 billion. Mulroy asserts that groundwater pumping in the valleys of the Great Basin can proceed without drying desert springs or producing other environmental damages. She also says that the project will enhance the water security of nearby rural communities. These are, of course, obligatory statements, which almost no one believes. In her heart of hearts, perhaps not even Mulroy believes them.

She is meanwhile churning forward on another nearly billion-dollar project, the goal of which is to tap the deepest waters of Lake Mead by boring a three-mile tunnel under the lake and installing something like a bathtub drain at its bottom. This will be Intake 3. Its ostensible purpose is to sustain the Las Vegas water system after Intake No. 1, at an elevation of 1,050 feet, is high and dry. Intake No. 2, at 1,000 ft., is not capacious enough to meet the city's needs alone. It is worth noting that Intake No. 3 is not being installed at or close to the elevation of Intake No. 2. At great cost and enormous technical risk, it will tap the lake as close to its bottom as an intake can be placed. Las Vegas's "third straw" will keep sucking from the lake even when Intake No. 2 no longer touches water, and it will keep on sucking from Mead's "dead pool," which is the water that remains when the spillways and outlets of the dam are stranded above the surface of the lake.

Construction of the intake will entail extraordinary difficulty and danger. A tunnel-boring machine manufactured in Germany will make the tunnel, and a vertical intake column, built on shore, will be connected to it by means of underwater excavation. The hydrostatic pressure of both lake water and saturated sediments bearing down on the tunnel will exceed that of either the Chunnel or the Big Dig in Boston. "OSHA is going crazy," says Mulroy. Intake 3 will access lake water at 860 feet above mean sea level, effectively the bottom of the lake. To understand the significance of that number, it helps to know a little about the elevational way stations that stand above it. (See box, below.)

---

## The Levels of Lake Mead

1,219.6 ft. above mean sea level: Official "full capacity" of Hoover Dam, although the lake can rise somewhat higher.[28] Capacity of Lake Mead is roughly 28.5 maf and exceeds three years of obligation to Lower Basin (7.5 maf/yr) and Mexico (1.5 maf/yr).

1,214.26 ft.: Level in January 2000, prior to drought of early 2000s.

1,083.81 ft.: Level at end of water year 2010. (Water years begin October 1, after summer demand slackens and before winter snow accumulates.) Only one year in the previous eleven (2005) interrupted a continuum of precipitous declines.

1,075 ft.: Level that triggers the first declaration of shortage under the 2007 "Interim Guidelines." Water delivery to Lower Basin will be reduced from 7.5 maf to 7.167 maf (4.44%). Mulroy says that construction of the pipeline to northern Nevada will "need to start the minute we hit 1,075."[29]

1,050 ft.: Second declaration of shortage. Water delivery to Lower Basin is further cut to 7.083 maf. Below this level Hoover Dam's generators can no longer produce electricity—historically, what has been about 4 billion kilowatt-hours per year, providing power for more than 1 million consumers and for lifting Colorado River water over the mountainous spine of California so that it can flow to Los Angeles and San Diego.[30] Declining water level has limited power production in recent years.

SNWA's Intake No. 1 is also sited at elevation 1,050. As to its vulnerability, Mulroy says, "We expect to lose the upper intake. We fully expect to lose it."

*Continued*

*Continued*

1,025 ft.: Third (and last) declaration of shortage, under the Interim Guidelines. Water delivery to Lower Basin drops to 7.0 maf, a total reduction of 0.5 maf, or 6.67 percent.

1,000 ft.: Level of SNWA's Intake No. 2. Says Mulroy: "The cuts that would have to be made in the Lower Basin to keep that reservoir at an elevation above 1,000 are just unbelievable."

895 ft.: Dead Pool. Lake elevation falls below lowest outlet of Hoover Dam. River flow and downstream water deliveries cease. About 2.0 maf remain in the lake.

860 ft.: Level of SNWA's Intake No. 3, completion expected in 2013.

LAKE MEAD FROM ATOP HOOVER DAM, JANUARY 31, 2010. THE TALL STRUCTURES ARE WATER INTAKES FOR POWER GENERATION. NOTE THE CONSIDERABLE "BATHTUB RING" OF WHITE MINERAL DEPOSITS. THE ELEVATION OF THE LAKE SHOWN HERE IS 1,100 FEET. *PHOTO COURTESY OF JOHN WEISHEIT.*

Our paddleboat drifted toward the right-hand limit of the ledge. We angled the boat toward the center of the river and held ourselves poised, ready to sprint. The flatwater ended, and we felt the river drop. The current accelerated. One stroke with our paddles, and we ducked neatly left into a curl of water below the hissing ledge. The river boiled. We dug hard for midriver. Two, maybe three fierce strokes and we slammed into the whaleback wave, crashing up its shoulder and sliding, half spinning, into the froth below. For a split second we caught a glimpse through the splash of the waves downstream. We were headed for the apex of the V-wave. We grabbed flip lines and D-rings, anything secure; we ducked our heads, pressed our paddles to the tubes. Then we were in it. The river roared and gulped us down. A thousand liquid fists punched from every direction, and water and spray closed overhead; the raft shuddered, bucked left and right, up and down; the furious wetness was louder than a gale. And then, as quickly as it had swallowed us, the river spat us out. We shot into the light and the air and saw that we were still aligned with the current, right side up, and on course to speed by the big rock at the bottom of the rapids. A good thing, too, because we were full of water and wallowing uncontrollably. We paddled a few strokes to speed the work of the self-bailing drains. As we slid past the Stonehenge rock, we raised our paddles and let out a soggy cheer that didn't seem to go away, even when we stopped cheering. We looked around: our echo was the applause of the oar-boats. We'd had a perfect run. We couldn't have hit our entry better.

INTAKE 3, THE bathtub drain in Lake Mead, is a point of entry, also of exit. It is an endgame strategy for when the pawns, knights and bishops are gone and the king is on the run. The water quality in Lake Mead is already a subject of concern. Counterintuitively, it is worst near the surface. As the volume of the lake shrinks, the quality of its tepid dregs will become increasingly controversial, heavy with metals and salts, endocrine-disruptors, and the residue of prescription drugs that people have thrown away or excreted. The joke about Las Vegas is that eventually no one in that high-strung city will need anti-depressants. People will get all they need from their drinking water.

But, as Mulroy says, Las Vegas can't not build the intake. The city is too vulnerable, especially with its ultimate lifeline, the pipeline to the ground-water of northern Nevada, still unbuilt, still challenged in the courts, and a long way from completion. Yet the real reason it has to build the intake—

and the real reason the water supplies of other metropolitan areas are strained—is not a shortage of water, not the prospect of drought. It is overall-location and ever-rising human demand. Part of that demand is for energy. The region produces vast amounts of water-cooled, water-consuming electricity in order to air-condition its sweltering cities and pump artificial rivers over mountain ranges and across deserts. A typical 1,000-megawatt coal-fired plant evaporates about 10,000 gallons of water *per minute*, which amounts to between 0.5 and 0.7 gallons per kilowatt.[31] These are staggering figures. Energy is the most underreported story in the looming water crisis of the Southwest and the nation.

Water needs energy and energy needs water. About half of all the water withdrawn from rivers and lakes in the United States is applied to thermo-electric power generation, which is to say, to producing or cooling steam by one means or another. That number will climb in the years ahead as the demand for electricity grows.[32] Power-related withdrawals are lower in the West, where hydropower contributes heftily to the grid, but anyone watching the declining levels of Lakes Mead and Powell can see that existing sources of hydropower face an uncertain future. And moving water from rivers, like the Colorado, to desert cities, like Phoenix, LA, or even Las Vegas, consumes a lot of juice.

Between 10 and 20 percent of all the electricity produced in the United States is used to move water around, and the schemes of the future promise to be ever more elaborate and costly, both in cash and electricity. Desalinization, the chimerical, sound-bite panacea beloved of many pundits, is notoriously energy-intensive. And if the energy to move or purify water is to come from coal or other petrochemicals, the cycle of greenhouse gas production and climate warming only gets worse. Utah's hoped-for pipeline to pump 100,000 acre-feet annually across the south of the state from Lake Powell to fast-growing St. George would have a carbon footprint equivalent to burning 39 million gallons of gasoline.[33]

As if the noose were not already tight enough, it is worth noting that carbon capture and sequestration technologies that might improve the carbon performance of fossil-fuel power plants are intensely water-consumptive.[34] While wind generation and some (but not all) solar technologies are comparatively thrifty in their use of water, the only sure way out of the energy/water death spiral is to use less, an extremely unlikely outcome. Current forecasts indicate that electrical consumption in the United States will grow 50 percent by 2030, and that consumption in the rest of the world will climb an equally steep curve. Notably, the interdependence of energy and water assures that the droughts of the future will sorely limit the availability not only of water, but also of kilowatts. Future brownouts in people's lawns will likely be mirrored—with far more consequence—in the regional electrical grid.[35]

THE UPPER BASIN has never failed to meet its compact obligations to the Lower Basin, but for the Lower Basin, that's not enough. The metastatic growth of southern California, central Arizona, and southern Nevada has been predicated on the notion that economic expansion will always find the water it needs. It is a gambler's game, and to play it you have to keep raising the stakes. If you try to stop or slow down, you lose. As Mulroy puts it, "You can't take a community as thriving as this one and put a stop sign out there. The train will run right over you."[36]

The thirst of the Lower Basin has addicted it not just to the 7.5 maf it receives as its due from the Upper Basin but to the surplus flows that also regularly came its way—until they abruptly stopped in 2001. These flows covered much of the 1.4 maf that Lower Basin evaporation and system losses consume.[37] The Lower Basin never allocated these costs to its member states, and essentially pretended that they did not exist. Even without revising the flow fiction on which the compact is based (which would mean accepting a share of responsibility for the treaty obligation to Mexico), the Lower Basin is living beyond its means. You can parse the figures any number of ways—Barnett and Pierce argue that Mead has been overdrafted by 1.3 maf since 1999[38]—but it is beyond debate that the current level of water consumption in the Lower Basin cannot be sustained.

Meanwhile, the rapids of climate change lie ahead. Far from calculating its best point of entry, the Lower Basin is headed for the first rocks. True, the Interim Guidelines of 2007 present a schedule for modest consumption cutbacks, but even the most fervent admirers of the guidelines praise only the fact that agreement was reached, not the substance of the agreement itself. The Interim Guidelines are shortage-sharing with training wheels; graduation to an actual bicycle needs to come soon.

As Las Vegas struggles to build its way to water security with a bathtub drain in Lake Mead and a pipeline to the north, it faces yet another set of problems.

Since 1969, metropolitan Las Vegas has grown more than twice as fast as any other major U.S. city.[39] Logically, it has used its growing population to finance the rapid expansion of its infrastructure, including its water system. Special fees for connecting new homes and businesses to SNWA services typically generated 57 percent of the funding for the utility's capital projects. This ordinary bargain turned Faustian with drought: infrastructure expansion and growth were well matched in flush times, but the cost of each successive new gallon in a drying era is higher than the one before, requiring yet more growth to fund the infrastructure to service what is already there. The spiral of need and obligation spins on, seemingly without limit, but a taste of the end came with the bursting of the housing bubble in 2008. Connection fees and other growth-related revenues fell to a trickle at the worst possible time: SNWA needs more than $800 million for Intake 3 and $3.5 billion for the northern pipeline. Las Vegas, a city built on odds-making and risk-taking, is betting the moon on its water projects, even though its billfold is suddenly slim.

BESIDES THE STRAIGHT-UP power play of the northern pipeline and the technical semi-fix of installing a drain in the bottom of Lake Mead, Patricia Mulroy is putting big money into cooperation and mutual assistance. "The only way to get through what climate change is going to bring us," she says, "is to recognize that we are inextricably interdependent and to use that interdependence as a strength rather than an impediment." Nowadays she waxes almost evangelical on the subject of cooperation, arguing for "a fundamental cultural shift, an ethics shift, from competition for water resources for economic gain to sharing of water resources for joint survival." Old hands may marvel that Mulroy, one of the most intimidating and hardnosed competitors of recent decades, should be preaching so much sweetness and light, but the change is less a conversion than a reappraisal of self-interest, which today dictates that the Colorado's users move to "almost a regional management of municipal supplies [and] an improved relationship and synergy between agriculture and urban [interests]." Says Mulroy, "I think you will see the boundary on water issues between the United States and Mexico be so blurred that it's almost invisible. We're going to manage as one system."

An example of this kind of unity may take shape along the following lines: Mulroy's SNWA joins forces with the CAP and the Metropolitan Water Department (MWD, the water utility serving greater Los Angeles) to build and operate a desalination plant in Mexico near the head of the Gulf of California. The water produced might then be piped into Mexico's water infrastructure, both to serve local needs and to feed the existing transmountain pipeline to Tijuana. This water would be swapped for Colorado River water being held in Lake Mead for delivery to Mexico under the 1944 treaty. The swapped water would then be apportioned among MWD, CAP, and SNWA. "The more interagency cooperation you have, the less structures you have to build." In this case, one structure—the desalination plant—would serve three major U.S. utilities, as well as several in Mexico.

But desalination and other high-tech fixes, however helpful they may be, are insufficient to slake the thirst of the overdrawn Lower Basin. With current technology, desalination consumes about $2,000 worth of electricity per acre-foot of water produced. It is generally practical only when the desalt plant and its consumers are close to the source of saline water.[40] The high cost of long pipelines pushes the economics of desalination far beyond the limits of feasibility.

Not even conservation provides much of a silver bullet. In fact, it is a double-edged strategy. As people xeriscape their yards and install low-flow toilets and showerheads and water-saving appliances, they pare down to a level of water use that has no slop in it, no waste. What is used one week must also be used the next, even in times of drought. This is called "demand hardening." When demand hardens past a certain point, a utility cannot simply ban lawn sprinklers and car washing to stretch supplies through a dry time. If the utility had placed the savings from conservation in an emergency reserve, the system might have become more resilient, but this almost never happens. In virtually every community that has pursued it, water conservation has principally resulted in freeing up resources that are subsequently consumed by growth—more hookups, and more hardened demand, producing a system that is less flexible and more fragile than before. Phoenix is in a particularly delicate position. Its wastewater stream cools the Palo Verde Nuclear Generating Station. If the people of Phoenix save too much water, more than the usual bad things can happen.

THERE ARE REALLY only three strategies by which the Lower Basin can hope to rebalance its water budget and prepare for the rapids. The first is to shift large amounts of water from agriculture to municipal and industrial uses. Agriculture consumes by far the largest share of western water. In California, 79 percent of diverted surface water and pumped groundwater goes toward producing crops and livestock. Most other western states devote a similar portion of their water to agriculture, although in Arizona the percentage drops as low as 68 percent, a result both of Arizona's high degree of urbanization and of the thirst of its power plants.[41]

In the march of settlement across the West, farms came before cities, and so water rights for irrigation are generally senior to rights for drinking water. This creates what Brad Udall describes as an "upside-down" water system, and it sets the stage for a collision "between nineteenth-century water law, twentieth-century infrastructure, and twenty-first-century population and climate." If the collision is to be resolved with minimum damage, "we've got to design a system that's fair to ag but also gives certainty to non-ag users that ag [water] will be there at least in some quantity should we ever hit these terrible tree-ring paleodroughts or what we think will happen under climate change."

Being "fair to ag" turns out to be more than a little bit complicated. The quickest path to water savings is to fallow farms—to cease farming them, either temporarily or permanently. Any defensible fallowing program has to compensate the farmers whose fields are idled, paying them for their water. This is already happening in many areas. The Imperial Irrigation District of southeastern California, for instance, sells water to both the Metropolitan Water District of Los Angeles and the San Diego County Water Authority under elaborate agreements that will reap billions of dollars of income for the Irrigation District over the life of the contracts. Fallowing means easy money for landowners, but it also idles the truck drivers, equipment operators, implement and seed salesmen, pickers and packers, and all the rest of the extended family of workers and suppliers who contribute to the productivity of the fields. Not to mention the diners where they eat, the gas stations where they fill their tanks, and the dealerships where they buy their pickups. Or the counties, towns, and school districts that depend on taxes those businesses pay. The farm community rightly views fallowing as an economic wrecking ball, yet more and more land is sure to be idled in the years ahead. Lest fallowing programs produce more lawsuits than water, the trick will be to plan them, as equitably as possible, in advance of the emergencies that make them necessary.

Mulroy isn't so sure that enough water can be—or should be—squeezed out of agriculture to satisfy the cities. If agriculture in the Southwest, or anywhere, gets heavily cut back, she asks, "How are you going to feed 9 billion people? When I started, I was a fierce urban advocate, fiercely believing that agriculture was wasteful, that we had too much agriculture." Now, she says, the "larger water picture" has taught her a fuller appreciation of unintended consequences, and one such consequence concerns the United States' agricultural self-sufficiency. We can't maintain that, she says, "if we fallow all the agriculture."

Her preferred solution to the long-term water woes of the region is the second major strategy. Like generations of water managers before her, she dreams of augmenting the flow of the Colorado River with water from outside the basin. The granddaddy of such plans was the North American Water and Power Alliance—NAWAPA. A 1960s brainchild of the Army Corps of Engineers and Parsons Engineering, NAWAPA would have diverted waters of the Yukon and other far northern rivers, rerouting them southward to deliver 22 maf to southern Canada, 20 maf to Mexico, and 78 maf to the United States, much of it to the Southwest. With NAWAPA, no one in western North America would ever again have been obliged to take a short shower.[42]

Although NAWAPA died a victim of its own grandiosity, the hope for out-of-basin rescue lives on. Mulroy points out that the climate models that predict drying for the Southwest also prophesy wetter times in the upper Midwest. When communities along the Missouri and Mississippi scream for flood relief, she says, aridland interests will have an opportunity. She believes they should come forward with support—and an agenda to capture and store flood flows, and pipe them westward across the Great Plains (an uphill, energy-intensive trajectory, to put it mildly), and deliver them to, say, Denver and the Front Range of Colorado. Newly flush with water, the Front Range would no longer need to divert water, as it now does, from tributaries of the Colorado River. The saved water might be left in the Colorado River watershed to flow to the mainstem, boosting supplies for users downstream.

Mulroy's concept involves not so much transferring water from one basin to another as "wheeling" it to produce leap-frogging swaps. Trouble is, even if the diversions and reservoirs were placed just right to capture flood flows, and if the energy to lift the water across the Great Plains and the billions of dollars to build the system might also be found, and if all of the water now transferred eastward through the Continental Divide were to be saved—a gauntlet of intimidating *ifs*—the total net benefit to the Colorado River system would not exceed 600,000 acre-feet, at present rates of diversion.[43] This is not enough to rescue the Lower Basin from its overtapped predicament.

Notwithstanding the possible realization of far-fetched augmentation schemes, and notwithstanding the eventual transfer of large amounts of water from agriculture, the Southwest and particularly the Lower Basin cannot secure a stable water future without pursuing a third strategy, an inherently painful one. The basin will have to consume less water. Reducing demand is the only sure way of arresting the relentless decline of Lake Mead and of the region's water resources generally.

Cuts can come from increased farm efficiency in the use of water, both by using drip irrigation and other water-saving technologies and by switching from water hogs like alfalfa and cotton to thriftier crops. Much progress has been made in these areas and more can still be achieved, but in terms of scale, the pursuit of efficiency will only dent, not satisfy, the regional need. Farms can be fallowed, at first temporarily, then permanently, with dusty environmental consequences and economically wrenching human costs. Municipal and household conservation can provide further cuts, until demand hardens to an adamantine core, but if continued growth never ceases to add more taps to the system, the benefit will be lost. Growth must be limited; the train must be stopped. But who is willing to stand on the tracks and flag it down? The economy of the region is predicated on continuous growth. Conversion to a new paradigm will require revolutionary economic change. In all likelihood such change won't come until the train runs out of track. And—although saying so causes the metaphor to melt from sight—the track won't end until the water does.

Barnett and Pierce estimate that long-term sustainable deliveries from the Colorado River lie in the range of 11.0 to 13.5 maf per year, a far cry from the 17.0 maf presumed by the Colorado River Compact. Adjusting to that level of supply will require reducing current consumption by as much as 20 percent.[44] And that's just for starters—future depletions in the Upper Basin may complicate that number a good deal.

The problem of scaling back the consumption of water in the Southwest is not far different from the problem of limiting greenhouse gases. People haggle to exhaustion over the need to take action. Then they haggle over inadequate and largely symbolic reductions. For a host of well-considered, eminently understandable, and ultimately erroneous reasons, inaction becomes their main achievement. Members of the water establishment of the Colorado River have known since at least the 1930s that the assumptions regarding river flow that undergird the compact were unsupportable. Since the 1970s, it has been clear that large-scale augmentation—the hydrologic equivalent of the Seventh Cavalry riding to the rescue—would not be forthcoming. The problem of overallocation has been known, ignored, and passed from generation to generation.

Now climate change tells us that the time available for remedy has shortened and that the needed magnitude of response is greater than previously thought. The paleorecord, newly rich with detail, underscores the imperative for remedial action. The message on water consumption is as clear as the relentless upward creep of the Keeling Curve, which begs action on greenhouse gases. Patricia Mulroy and a few others have heard it and are scrambling energetically, at times ruthlessly, to respond. But the efforts of a minority cannot address the entirety of the problem; indeed, no one has suggested that the train of growth must one day be stopped, or even slowed. Without a societal commitment to live within the limits of finite resources, in an environment where climate speaks last and loudest, the ultimate train wreck, the final reckoning with aridity, becomes a certainty. In the meantime, the most onerous consequences of inaction will be pushed off on the poorest classes of society, those least able to bear them, and the responsibility for balancing the water budget will be left to the next generation. They, in turn, will face a still graver situation with fewer alternatives and less of what policymakers call "decision space."

This pattern of behavior may be quintessentially human. If one were to write a survey of all the instances in the history of civilization when societies accepted difficult medicine in order to spare their descendants worse pain in the future, it would make a very short book.

HOWEVER LOUD THE roar of Lava Falls may be, today's Colorado River is a caged, if not thoroughly domesticated, beast. It is an old lion, beset by many ailments. Its volume, carefully controlled by the operators of Glen Canyon Dam, rises and falls with demand for hydroelectric "peaking power" to run the air conditioners of Phoenix and other sun-drenched communities. Here and there along the riverbanks of the Grand Canyon you can still make out the rough line in the vegetation above which mesquite grows, and below which it can't be found. This is the old high-water line of the undammed river. Before Glen Canyon Dam was built, flood flows used to scour away those woody plants that dared to sink roots within their reach. But now the flood zone is transformed. It wears a tutu of fast-growing tamarisk along the river's edge and boutonnières of opportunistic Apache plume and Mormon tea above it. The banks of the river have dressed themselves up, now that the river no longer rips their clothes off. It does, however, gradually wash away the sand bars that the river's silt and bedload used to nourish. Nowadays nearly all the eroded cargo of the river is trapped in Lake Powell. The water the dam releases is cold (drawn from the depths of the lake), clear, and fairly regular in its flow—just the thing for nonnative trout, and bad news for homegrown chubs and squawfish.

From a certain point of view, the growl of Lava Falls is less like a caged lion and more like the rumble of a freight train. It is the sound of property being delivered over a long distance. Every pint in that thundering cascade is promised to someone or something. Even the urine and dishwater contributed to the river by boaters (whose solution to pollution is dilution) enters the system of ownership, with the molecules of any given ounce scattering to myriad destinations: to the taps of Las Vegas, Tijuana, and Phoenix, to the invisible evaporative mist of Lake Mead, to heads of lettuce outside Yuma and date palms in the Coachella Valley, to artichokes and almonds near El Centro, to electrical stations and foundries outside LA, to the shipyards of Long Beach and the backlots of Hollywood, to Dodger Stadium, and to the beaver tank at the Arizona-Sonora Desert Museum. In a good year, a few of those molecules, but only a few, might find their way to the river's yawning mouth at the Gulf of California, which, together with the surrounding delta wetlands, formerly constituted one of the greatest natural areas on the planet.

We ask too much of the Colorado River, as we also do of the Rio Grande, Gila, San Pedro, Pecos, Canadian, Platte, Conchos, Yaqui, and every other river of the aridlands. The Texas historian Walter Prescott Webb once wrote that the American West is "a semi-desert with a desert heart" that necessarily gave birth to an "oasis civilization." Rivers provide the primary oases to which the region owes its life, but one by one they have succumbed to settlement's excessive thirst. The Salt River through Phoenix is no longer a river. Nor is the Santa Cruz though Tucson, the Gila below Coolidge Dam, or the Rio Grande through El Paso and Ciudad Juárez. Today these rivers and many others are little other than large-scale ditches, much like the no-longer-mighty Colorado in its last 300 miles.

In the course of writing this chapter, I e-mailed Tim Barnett at Scripps to clear up some minor questions. He quickly wrote back, inserting answers in the appropriate places in the original message. Because Barnett did not know much about me or my intentions, I included a line to remind him what I was about. I wrote, "I am not trying to write a good guys/bad guys story, just trying to understand how we got into this fix and what our chances are for getting out."

It wasn't a question—I hadn't intended it to be—but in his return e-mail he nevertheless provided an answer: "NIL if climate change scenarios are reasonably correct."

# 6

## THE CANAL AT RIVER'S END:
## THIRSTY ARIZONA

"IF YOU RUN the math," says Brad Udall, "You sort of go, wow, Arizona, they may be totally out of their Central Arizona Project water." Udall is referring to Arizona's unenviable position as California's aquatic whipping boy. The two states have long fought over water, and although Arizona has won a battle or two, it has taken a beating in the war. A key result of their combat has been to make the majority of Arizona's Colorado River water rights expressly junior to California's. This means that during inevitable and possibly imminent periods of shortage, the people of southern California, under a strict interpretation of the law, will be able to wash their cars, water their lawns, and keep their showers streaming while the millions who live in Phoenix, Tucson, and points between watch the flow from their taps slow to a dribble.

Fortunately, events are unlikely to turn out so apocalyptically. When crisis comes, emergency negotiations will produce a less black-and-white outcome, and Arizona's groundwater reserves (some of them recharged in recent years with CAP water) will be tapped to meet priority needs—at least for a time. Nevertheless, the potential for a winner-take-all showdown between large populations highlights the vulnerability of the urban centers of the arid West in an era of climate change.

Fates are hardly fixed. How the cities of the region grow and change in the years ahead will significantly determine their ability to withstand the shocks of a hotter and drier future. How well they respond to the challenges ahead will also determine the future of their states and of the entire West, for in an arid land, a modern society is obliged to be an urban society. The survival of aridland cities and the struggle to preserve their quality of life will become a matter of national concern, even obsession, and the entire world will watch their stories unfold.

Arizona has always been jealous of California's economic power, its political heft in Congress, and its early and abundant claims to Colorado River water. Too often, its strategy for contending with California has been to sulk. Mulishly, Arizona refused to ratify the Colorado River Compact after it was negotiated in 1922. A dozen years later, Arizona's grandstanding governor, Benjamin Moeur, capped obstinance with absurdity when he dispatched a unit of the Arizona National Guard to oppose construction of Parker Dam, a lesser sibling of Hoover Dam. Building Parker was an initial step in the physical implementation of the Boulder Canyon Project Act, a lineal descendant of the 1922 compact that made it possible. Arizona's unlucky soldiers boldly commandeered a ferry, which was immediately dubbed the "Arizona Navy." Their ensuing exploits cost one man's life (pneumonia), delayed work on the dam for a year, and added impressively to the accumulated weirdness of the historical record.[1] But they did not materially change Arizona's underdog position or sense of inferiority.

To be fair, Arizona had reason to be anxious. Because the compact had divided the Colorado River into Upper and Lower Basins and endowed each (theoretically) with equal shares of the river's flow, Arizona was now isolated from its allies to the north. Like Wyoming, Colorado, Utah, and New Mexico, Arizona in the 1920s was almost entirely rural, no match for the economic and political powerhouse of California. But the Upper Basin states had made a separate peace: the compact now shielded them against California's potential aggression. Nevada, with its tiny interest in the river, was too small to be a player. Arizona had to face the California juggernaut alone.

From the outset, Arizona feared that California would get more of the Colorado River than it deserved, and that Arizona would get less. In hindsight, the prophecy appears to have been self-fulfilling. Arizona finally ratified the Colorado River Compact in 1944, so that it might appeal to Congress for the funding it needed to move a copious portion of the river's water to its central farming region in the landlocked heart of the state. Its appeal for federal largesse fell on deaf ears. Some say that Californians subverted the project, arguing that Arizona might not have defensible rights to the water it aspired to move. Others hold that ceaseless squabbling between Arizona and California fatigued congressional appropriators, who resisted the idea of further public works in the Lower Basin until the two states settled their differences.[2]

But the two states were no more likely to come to agreement than Ahab and the white whale. Instead, Arizona filed suit against California in the Supreme Court, seeking a judicial determination of its rights to the river. The legal process thus unleashed proved to be colossally labyrinthine. Four years spun by in pretrial preparation. Four more years were consumed (and 22,500 pages of trial transcripts accumulated) before the court-appointed Special Master, in 1960, filed his 433-page final report. The Supreme Court then required another two and a half years to render judgment. It ultimately issued a decree in March 1964.[3]

At first blush, the outcome of *Arizona v. California* seemed to offer resolution. The Supreme Court ordered that Arizona was entitled to 2.8 million acre-feet from the river, as had been set forth in the compact; that shortages in the Lower Basin should be shared proportionately among the member states; and that Arizona was entitled to use the water of tributary streams that entered the Colorado below Lees Ferry without applying those volumes against its compact limit.[4] The decision not only positioned Arizona where it would have been anyway, had it ratified the compact in the 1920s when the ink was wet, but provided an extra boost: by exempting tributaries from the compact's allocations, it said that the roughly 1 million acre-feet of the Gila River belonged to Arizona alone, and would not be considered in dividing the rest of the river with California.

Arizona's good fortune, however, was fleeting. Even the highest court in the land could not move Arizona's water from the river to its principal cities and most extensive farmlands, more than a hundred miles away. The state needed federal authorization and funding for construction of the Central Arizona Project—the aqueduct that would carry water across the state. Without the CAP, Arizona could not deliver the water where it mattered, and where it might fuel the state's growth. Although the U.S. Senate, where Arizona's presence was equal to California's, had approved predecessor versions of the CAP as early as 1950, Arizona was powerless in the House of Representatives, where California outnumbered its delegation by 38 to 3. For almost two decades congressional authorization of the CAP, not to mention funding, remained a dream. California essentially made a hostage of Arizona's plans for a trans-state aqueduct, and demanded ransom.

The standoff had the potential to last indefinitely, and in the 1960s, as the new phenomenon of Sunbelt growth gathered momentum, Arizonans became impatient. They knew their time for explosive economic expansion had arrived, but their water had not. Something had to be done. The price California ultimately forced Arizona to pay was a heavy one. Arizona had to agree to subordinate the water rights pertaining to the CAP to those of California. This meant that shortages would not be shared proportionately. Instead, California would take its entitlement of 4.4 maf, no matter what. After that, if the river and the reservoirs could provide Arizona its full 2.8 maf, fine; but if not, tough luck: Arizona would have to take the hit, all of it, until the CAP's aqueducts were bone-dry. In theory, proportional shortage-sharing would apply only to Arizona's older Indian and agricultural water rights close to the river. In a showdown between California and the CAP, California would get every pint; the CAP would not get enough to fill a thimble.

This lopsided relationship is incorporated in the Interim Guidelines of 2007, which specify how the Lower Basin states will deal with shortages in Lake Mead. The worst situation contemplated by the guidelines is a drop in the elevation of the lake to 1,025 feet. At this point, Las Vegas will already have lost its Intake No. 1, power generation from Hoover Dam will cease, and the guidelines will impose an overall reduction in downstream deliveries of 6.67 percent. California, how-ever, will be unaffected, because Arizona will absorb California's share of the reduction. While Nevada, logically and appropriately, watches its delivery reduced by 6.67 percent, Arizona's apportionment will be slashed by more than 17 percent—an amount equal to 6.67 percent of both its and California's share of the river.[5] And if conditions worsen while this punitive framework stays in place, they will worsen more for Arizona than anybody else.

"That's a huge problem, and it's going to have to go," says Patricia Mulroy. "We're going to have to finesse that somehow." But the law is the law, and in order to finesse it, California's water establishment will have to be persuaded to draw on reserves of flexibility and generosity that have heretofore escaped public notice. Or else Arizona will have to trade California something the Golden State wants more than water, and it is hard to imagine what that might be. Part of the problem is the intransigence of California's farm bloc, particularly the power centers of the Imperial and Coachella Valleys, which hold rights to more than 3.4 million of California's 4.4 million acre-feet of Colorado River water.

California's urban water rights from the Colorado River, notwithstanding that more than 20 million people depend on them, are junior to the preponder-ance of its agricultural rights. This means that, under the law, cities like Los

Angeles and San Diego would have to bear the burden of any system of cut-backs. As Brad Udall puts it, a system that values the watering of alfalfa and lettuce over the well-being of people is "upside-down." Mulroy adds that it is crazy for one city to have priority over another, for Los Angeles, say, to be superior and more secure in its rights than Phoenix. But that is the way things work under western American water law, which, in the absence of specific exceptions, is governed by the doctrine of Prior Appropriation. Put simply, the doctrine holds that "to be first in time is to be first in right," which means that the claim of the individual or group who first diverts water from a river and puts it to "beneficial" use is thereafter superior to the claims of all who follow. In California, the turn-of-the-century farmers who watered the Colorado Desert from the Colorado River and transformed it into the Imperial Valley earned a priority date of 1901 for the majority of their 3.1 million acre-feet of water rights, and even the subsequent rights they obtained in the 1930s are senior to all but a fraction of those held by southern California's cities.[6] This peculiar sequence of events means that Los Angeles and San Diego are only one step less exposed than the CAP to draconian cutbacks in time of shortage. They are not likely to surrender their defenses easily.

ARIZONA'S POLITICAL LEADERS in the 1960s, who included Brad Udall's father Morris, then a young congressman, and his uncle Stewart, the secretary of the interior, accepted Arizona's second-class water citizenship only because they had to. There was no other way to get CAP legislation out of the House of Representatives. It was a bad deal and they knew it, but they took solace in the expectation that Arizona's subordination to California would never be invoked.

Their confidence in the future was not blind. Although the first paleoreconstruction of Colorado River flows was still more than a decade off, they already understood that the supply of native Colorado River water available to the Lower Basin was insufficient over the long term. In a 1967 speech before skeptical Californians at the Biltmore Hotel in Los Angeles, Morris Udall warned of future Colorado River shortages that would affect both states. He located the day of reckoning in a distant time—the 1990s—and placed his hope in what was effectively the Holy Grail of water technology: *augmentation*. Shortages might be avoided, he said, if "we take steps to make basin augmentation a reality. When that is done, there will be enough water in the river and the question of priorities will be entirely academic."[7]

The idea of massive augmentation, adding water to the Colorado River from sources outside the watershed, not in small amounts, but at the scale of a major tributary, constitutes an important subtext in the Colorado River Basin Project Act, the legislation that finally authorized the CAP in 1968. Three years earlier, representatives of all seven Colorado River Compact states had gathered in Washington, DC, and agreed on a set of joint priorities. One of them was to persuade the federal government to support projects that would import 2.5 maf to the Colorado River system. (The amount was calculated to cover the 1.5 maf treaty obligation to Mexico and roughly 1.0 maf in evaporation and transmission losses.) Upper Basin states endorsed these extreme measures not only because they hoped for a piece of the action but also because they feared the Lower Basin in the same way Arizona feared California. It was the lion more powerful than the rest of them combined. The best way to keep the hungry beast from eating their food was to feed it from somebody else's plate.

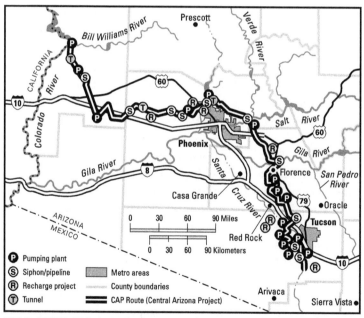

**MAP 7: CENTRAL ARIZONA AND THE CENTRAL ARIZONA PROJECT.** *ADAPTED WITH PERMISSION FROM INFORMATION PRODUCED BY CAP.*

By the time Mo Udall gave his Biltmore Hotel speech, some of the more outlandish augmentation schemes were already off the table. NAWAPA (the North American Water and Power Agreement), which would have spread the waters of the Yukon over the entire West, was remembered like a bad hallucination. Lesser plans, like importing water from the Columbia River, had also succumbed to outbreaks of sobriety, to say nothing of resistance from political and economic interests in the Pacific Northwest. Replumbing the West, at least at a continental scale, had been judged politically impossible. Nevertheless, Udall, a pragmatic politician who occupied the middle ground on such issues as water development, had faith that more conservative strategies, including desalination, weather modification, and agricultural conservation, would meet the goals of augmentation. Even as he spoke, the United States and Mexico were jointly investigating the feasibility of using nuclear power to purify seawater from the Gulf of California for use by both countries. The prospects, said the commission conducting the study, were promising.[8]

The expectation of augmentation made the subordination of Arizona's rights acceptable to Arizona's leaders. It was the tonic with which they washed down a bitter pill, ensuring passage of the Colorado River Basin Project Act and authorization of the CAP. In fact, the Act specifically orders that, once augmentation of 2.5 maf is achieved, Arizona's subordination to California shall cease to exist. The states would thenceforward be on a par. The Act further specifies that the first priority for augmented water will be to satisfy the 1.5 maf treaty obligation of the United States to Mexico.[9]

FOR A HOST of reasons, including cost, environmental impacts, and the variability of political winds, the hoped-for augmentation of the Colorado River never materialized. Moreover, the onset of wet weather and bountiful water supplies in the mid-1970s helped everyone forget about the region's intrinsic thirst. The river remained merely itself, or a little less so, as portions of its tributaries were diverted through the Continental Divide to provide water to cities and farms on the Front Range of Colorado and to augment the Rio Grande in New Mexico (which was that state's way of capturing a share of political pork and moving water to its urban core). Arizona, meanwhile, had to bite its metaphorical lip and be content with a 2.8 maf apportionment and a backseat position to California. The shortages that Mo Udall predicted for the 1990s were scarcely glimpsed until the close of the decade, and people went about their business.

Much of Arizona's Colorado River allocation irrigates crops in the agricultural districts along the river, but more than half—approximately 1.5 maf—is tied to the Central Arizona Project.[10] As of 2009, about a third of this (512,000 acre-feet) flowed to municipal and industrial uses. Another 41 percent was consumed by agriculture, with an additional 11 percent recharging aquifers or otherwise being stored against future needs. (Most of the rest of the CAP's water is delivered to Indian tribes.) Water used for agriculture and recharge approaches 800,000 acre-feet per year. This represents considerable insulation for central Arizona's cities against delivery reductions because, under the rules of the CAP, priorities are "right-side up"—agriculture will suffer cuts before people do. The maximum reduction contemplated by the Interim Guidelines (17.15 percent of Arizona's 2.8 maf) would amount to 480,000 acre-feet—a disaster for Arizona's agricultural economy but well within the cities' current margin for error. Unfortunately, the problem remains that the Interim Guidelines are to a real drought strategy what a movie set is to a real town. They are good for photographs and rehearsals, but their usefulness stops there. When Barnett and Pierce factored the guidelines into their analysis, the reductions they would make were too small to influence outcomes significantly.[11]

THINGS WILL BE fine for the roughly 3.5 million people who drink, cook, and wash with CAP water only for as long as (1) the predictions of the climate change models prove groundless, (2) the kinds of droughts documented by the paleorecord don't recur, and (3) the cities of central Arizona don't grow so much that they consume their agricultural buffer, which is their main protection against the uncertain years ahead. Yet another problem amplifies all of these liabilities. The persistent overdraft of resources in the Lower Basin, if unaddressed, means that increased water consumption in the Upper Basin, which is universally expected, will at some point remove water, on a gallon-to-gallon basis, from the intakes of the state of Arizona. From a Lower Basin perspective, Upper Basin development operates exactly like climate change: it means less water in Lake Mead. So this becomes a fourth condition: that the Upper Basin continue to send downstream its "excess" water, which the Lower Basin will then use to meet its overgrown demand. Believing that this situation is sustainable is like believing in the Tooth Fairy, but the Lower Basin, locked in a vise of its own making, presses on as though tomorrow morning there will be money under its pillow.

If any of these four conditions fail to be met, let alone if two or more fail in concert, people living in what Arizonans proudly call the "Sun Corridor"— the emerging megalopolis anchored by Phoenix and Tucson—will find it difficult to keep their thirsts slaked and their cities running.

THE VULNERABILITY OF aridland communities invites a historical comparison. Following the close of the Civil War, many westering Americans staked their fortunes on the 1862 Homestead Act, which promised title to 160 acres of public domain for any settler or family who could turn those acres into a farm and make a go of living on them. The Homestead Act stood at the center of a national vision of opportunity. It embodied the "American Dream," long before the phrase had currency. It promised that with sweat and industry, people who possessed nothing could become owners of property.

Not so fast, said John Wesley Powell, the one-armed explorer of the Colorado River who became the nation's first great bioregional thinker. After his days probing the river and studying and mapping the Colorado Plateau, Powell embarked on an even more important career as a federal scientist and agency head. For many years he simultaneously directed both the Bureau of American Ethnology and the United States Geological Survey. In the latter capacity he became known as one of the nation's foremost authorities on the character and potential of western lands. He argued that, in the aridlands, the Homestead Act promised not dreams but nightmares. It was a cruel hoax, betraying the hopes of those who would do its bidding. In the humid East, where the family farm had become enshrined as a national ideal, the 160-acre farmstead made sense. With ample rainfall, a family could wrest a decent life from the soil. But in the West, except in irrigable bottomlands, only tragedy, not prosperity, would ensue. Powell set his dividing line at the longitude of the eastern boundary of the Texas Panhandle—the 100th meridian. East of it, the land generally received twenty or more inches of precipitation, enough for raising crops without irrigation. West of it, he told all who would listen, the attempt to subsist from farming 160 acres of dry land would break hearts and ruin lives.[12]

Powell's message was unwelcome and unheeded, but history proved him right, not once but repeatedly—on the Great Plains during the droughts of the

AQUEDUCT OF THE CENTRAL ARIZONA PROJECT, NEAR FLORENCE, ARIZONA.
*AUTHOR PHOTO.*

late 1800s, when the line of advancing settlement actually pulled backward; in the calamity of the 1930s Dust Bowl, which rendered hundreds of thousands of Americans homeless; and in countless other homesteading failures scattered across the West that were no less painful for being local, particular, and limited in area.

In our own time, another dream of opportunity has lured a sweeping migration to the aridlands. The attraction of the Sunbelt, and its life of warmth, health, and clear skies, is founded on an assumption of sufficient water resources, but sufficient for how many, for how long? Powell, if he were alive today, would rail against the renewed courtship of calamity in the aridlands, and he would be not only as unpopular, but also as correct, as ever.

# 7

## HIGHWAY 79 REVISITED: "MEGA"
## TRENDS IN THE SUN CORRIDOR

WHETHER YOU ARE breaking prairie sod in the nineteenth century or raising a family and scrambling to make ends meet in the twenty-first, it is hard to get worked up over abstract possibilities. There is too much that needs doing, right here, right now. Even knowing the odds, people still live in earthquake zones, hurricane alleys, and the unprotected floodplains of mighty rivers. The warm embrace of a thirsty aridland city is not so different. Generally speaking, it is hard for any of us to get seriously concerned about what *might* happen until it *does* happen. That's why the politics of climate change are so difficult. The measurements and observations that convince scientists about the warming of Earth are invisible to the rest of us. We fail to sense them at the scale of our personal lives. And believing in the verdicts of computer models about what might happen twenty or forty years in the future, well, that is tantamount to a leap of faith, and most people don't ordinarily jump that far.

Believing in the growth of cities can be difficult, too. Beginning in 2007, the domino of subprime mortgage defaults knocked over the domino of overleveraged investment banks, which toppled a wobbly world credit system, which upended industries around the globe and ushered in the Great Recession.[1] The home-building industries of growth-crazy cities like Las Vegas and Phoenix collapsed virtually overnight. Suburbs from Florida to California became ghost towns where wind-driven litter piled up in doorways and weeds grew higher than the sills of boarded-up windows. Some analysts predicted the emergence of a new generation of suburban slums and the death of gas-guzzling, car-dependent, long-commute suburban lifestyles.[2] Indeed, in the long run, considering the implications of peak oil and peak water and the likelihood of more severe climate reckonings than we've yet seen, such a demise seems likely—though maybe not quite yet.

In mid-2009 Grady Gammage Jr., a senior fellow at Arizona State University's Morrison Institute of Policy Studies, wrote of the Sun Corridor, "Today, somewhere near the bottom of our chronic 'bust' cycle, it just doesn't seem likely that we'll ever again be in hyper growth mode. We have to remember that myopia is the dominant characteristic of places that live from boom to bust—at the top we believe it will go on forever, at the bottom we think recovery will take decades."[3] Gammage is a close observer of the Arizona real estate scene. As an attorney, he devotes half of his time to the practice of land-use law, serving clients with big stakes in the real estate development industry. The other half of his time he gives to "this think-tanky-type stuff" at the Morrison Institute. If you ask around for people who think hard about Arizona's land-use future, Gammage's name keeps coming up.

The bursting of Arizona's real estate bubble—like the downturns in greater Las Vegas, Florida, California, and every other fast-growth region in the country—affords an opportunity to reevaluate the patterns of its cities and suburbs and redirect them toward a more energy-thrifty, water-resilient, and sustainable future. But the opportunity will not last forever because demographics will ultimately prove Gammage right and growth will resume. The U.S. Bureau of the Census estimates that the nation will add another 100 million people by 2050, approximately the same number it added during the forty years from 1970 to 2010. These new residents will have to make homes somewhere. If the values of the recent past hold true, they'll settle in Georgia, Florida, Texas, and other warm-weather Meccas, and a large proportion of them, shunning the humid oppression of summertime Houston and Atlanta, will opt for the West. They will seek the sunshine and cloudless (if smoggy) skies of western Sunbelt cities that are likely to become even sunnier and drier as the decades roll by. Says Gammage, "Despite the problems we have right now, sunshine is still attractive to people. I don't think people are yet worried about it becoming so hot and dry here that it's not a habitable place."[4]

The new southwesterners will need water and energy, new towns and urban centers, highways and rail lines, houses, jobs, and parks. They'll want to enjoy some semblance of the "American Dream" that previous generations pursued and to varying degrees attained. Whether the desert cities to which they migrate offer them resilient, carbon-smart oases or dusty dystopias of water- and fuel-stressed "slumburbia" will depend in large measure on the patterns that emerge from the Great Recession's suspension of growth.

Maybe those patterns will be shaped by novel forces. Maybe Baby Boomers won't choose to retire the way their parents did, surrendering to "underwater basket-weaving classes and playing golf all day," as Gammage puts it. Maybe they will save more and consume less. Maybe a shaky economy and a shakier Social Security payout, plus their lack of ensured pensions, will make them more cautious, less inclined to uproot themselves for the passcode to a gated desert community. Maybe more of them will stay at home in New York or Milwaukee, possibly working part-time to stretch their 401(k) a little further. Still, even if the migration slows a good deal, people will come to the arid-lands, although no one knows in what numbers, because so many more of them will be looking for a place to roost, and because the sun will still shine on the Sunbelt.

TO GET A sense of the region's urban future, I headed for Tucson and Phoenix, two cities directly in the crosshairs of climate change. Salt Lake City, Denver, Albuquerque, Las Vegas, El Paso, and other urban centers of the "dry Sunbelt," including scores of communities large and small in California, will face similar water-supply, transportation, and land-use challenges, but the Sun Corridor, recession notwithstanding, is poised for especially powerful expansion. Planners view it as encompassing, not just the two big cities at its core, but a broad swath of central Arizona running from Prescott northwest of Phoenix to Nogales and Sierra Vista hard against the border south of Tucson. In 2005 approximately 5 million people lived in the Sun Corridor. Prior to the reces-sion, that number was expected to reach 8 million sometime after 2030, and to double not long after that.[5]

Even if you dial back those numbers substantially, you are still talking about a radical expansion of Arizona's urban core. Gammage, for instance, thinks that the recession, by knocking down real estate prices and making the Sun Corridor more competitive relative to other markets, may have the counterintuitive effect of reigniting Arizona's real estate economy, once the current oversupply of housing is absorbed. Conceiving of the radical growth of cities, in its way, poses as great an imaginative challenge as taking to heart the idea of a much warmer world or the severity of droughts in the paleorecord. This is nothing new. Consider what it was like to stand among the arid expanses of central Arizona in 1950, when saguaro cactuses far outnumbered people and Phoenix and Tucson had a combined population of only 150,000. All the neighboring towns were mere specks on the map, places where you could get water for the radiator of your overheated automobile, and not much else. Imagine saying to the person beside you that sixty years into the future no fewer than 5 million people would inhabit that same slice of desert. Your companion would have questioned your sanity. Probably you would have, too.

LUTHER PROPST DIRECTS the Sonoran Institute, a nonprofit organization heavily engaged in issues touching conservation and land development. Propst is hard not to notice. He has a loud voice that causes everyone within earshot to turn their heads, and he speaks bluntly, directly, and sometimes profusely. The words come out of his mouth with the rounded edges and open vowels of North Carolina, where Propst was raised and educated. Like nearly two-thirds of contemporary Arizonans, he came from somewhere else.[6] Propst is long and lanky, at least half a head taller than any crowd outside a basketball convention. He is a triathlete who likes to test himself in Iron Man competitions, bicycling, swimming, and running grueling distances. He also participates in more narrowly thematic contests, like running from one rim of the Grand Canyon to the other, an extremely precipitous forty-four-mile excursion, which he has completed twice in under seventeen hours. Propst genuinely enjoys such torment, and it seems to suit him. The cultivation of endurance would appear to be good training for anyone who works on growth issues throughout the Rocky Mountain West, as he has for twenty-odd years, and particularly for someone who concentrates his efforts in Arizona, where the going is toughest.

Propst, who holds graduate degrees in both law and regional planning, has kindly agreed to show me the lay of the land in the heart of the Sun Corridor. We start out with a breakfast of empanaditas in one of Tucson's few old neighborhoods—few because, like a lot of western towns, Tucson got caught up in the urban renewal fervor of the 1960s and destroyed much of its architectural and cultural past. Propst suggests we drive up to Phoenix the "back way"—following Highway 79 through Florence—and then circle back by the high-speed corridor of Interstate 10. Propst will drive and narrate the landscape; I will look and listen.

As we leave downtown, the city quickly flattens and opens out. Soon we are hard pressed to spot a building over one story tall, except the sprawling

Tucson Mall, which is the size of a middling university and rises fortress-like from a sea of pavement. Propst points out the bigger, natural islands that ring the city: the looming brown hulk of the Santa Catalina Mountains to the north, and the Rincon and Tucson Mountains on the east and west. "Until the recent bust, almost all the real estate activity and interest was on the edge," he says, meaning the edge of nature, of public lands, of undeveloped tracts of land. While "densification" would make a more vibrant, more efficient city, sprawl is what sells. "The real estate formula here is to put development close to public lands. So you protect land and then you just surround it with golf course–oriented retirement communities," he adds wryly, "which destroys any ecological reasons you might have had for protecting it in the first place."

The idea of living "on the edge" touches a key characteristic of Tucson, which it shares with Phoenix and the other major cities of the intermountain West: unlike population centers farther east, they are *bounded*. Public lands, Indian reservations, steep slopes, and National Forests limit their possibilities for expansion. As a result, the big cities of the Southwest, notwithstanding their considerable sprawl, are actually more compact than counterparts like Atlanta, Charlotte, and Birmingham in the "wet Sunbelt" of the Southeast.[7] Instead of grading outward to ever lower population densities, their built-up areas often end abruptly in a sudden transition to wildlands. For this reason, coupled with a preponderance of public land, the states of the arid West have the paradoxical distinction of being at once the most urban and the most rural region of the country. They are the most urban because no other region concentrates a larger proportion of its population in cities, and the most rural because outside the cities, no other region has a lower population density.[8]

Cities in the arid West are also constrained by their water systems, which are necessarily highly centralized—the CAP is a good example. The Phoenix metro area also draws water from the Salt and Verde Rivers, but the CAP is its lifeline and the key to its future. Tucson, until the CAP arrived, was a groundwater city, rapidly pumping up and consuming its aquifer. After a difficult transition (its water utility bungled the switch to salty, sometimes corrosive Colorado River water), it now relies heavily on the CAP and, like Phoenix and the smaller communities in the 100-mile corridor between them, cannot grow without it.

The expansion of the desert cities tends to follow the Interstate Highways, most of which were built before the cities began their rapid growth. The reverse is true of most eastern cities, which emerged as metropolitan areas before or during the boom in highway building. The special characteristics of their boundaries, water systems, and recent growth are among the factors that cause researchers like Gammage and Mark Muro, of the Metropolitan Policy Program at the Brookings Institution, to argue that a new, "megapolitan" urban form is emerging in the arid West—the "Mountain Mega."

The awkward coinage *megapolitan* is intended to denote the result of multiple metro areas growing together, like the blur of urbanization from Boston nearly to Richmond. The newest class of these *über*-cities consists of sharply bounded, water-precarious youngsters like Salt Lake City and its siblings on the west face of the Wasatch Mountains, and the Front Range Mega of Colorado that includes Denver, Fort Collins, and Colorado Springs. Preeminent is Arizona's megacity, the Sun Corridor.[9]

**MAP 8:  THE "SUN CORRIDOR."**

---

PROPST AND I are cruising along in his Prius on the outskirts of Tucson. We are still in Pima County, of which Tucson is the seat. Pima County is roughly the size of New Jersey and contains some of the most botanically exotic desert in the world. Propst pulls into the driveway of a small nature center and stops the car. Around us stand twenty-foot-tall saguaros, icons of the Sonoran Desert, their upreaching arms supplicating a brilliant sky. Tangled among the saguaros are green-limbed palo verdes and gnarled ironwood trees. They are the size of fruit trees, but they are anorexic in their look, angular and alien. If they were people, they would be as lithe as fashion models, but with their short hair in spikes and safety pins in their lips. The gaps among the trees and saguaros are populated by several varieties of cholla cactus, some that branch like weird inverted candelabras, some compact and furred with bright yellow spines—these last are called teddy bear cholla, so named because of their deceptively fuzzy and inviting appearance.

Propst explains that it was habitat like this that led Pima County, and Tucson along with it, to embark on one of the most comprehensive planning efforts undertaken by any equivalent jurisdiction in the country. A large new high school was proposed for a site nearby. The land was part of the same kind of vegetational wonderworld—an "ironwood forest." It so happens that ironwood forest is the habitat of the endangered cactus ferruginous pygmy owl, a curious grapefruit-sized predator that nests in cavities in saguaros and other tall desert plants. Plans for the high school were put on hold as an all-too-familiar sequence of conflicts arose over the quality of the science detailing the owl's habitat needs and over the formal designation of critical habitat where development would be prohibited. There were the usual complaints about valuing critters over people, the environment over the economy, and vice versa. A number of local leaders had the wisdom to see that the issues led cumulatively toward a central question: what kind of place did Tucson and Pima County want to be, and specifically, what kind of relationship did the community want to have with the desert that enfolded it?

Surprisingly, the community came up with an answer.

"Tucson views itself culturally as a desert city—a city in the desert. The dream in Tucson is to live next to Coronado National Forest," says Propst. By contrast, "Phoenix views itself as an oasis city. The dream there is to live on a green golf course." These are overgeneralizations, of course, but the germ of truth at their core is evident in the cities' contrasting histories. "In Phoenix much of the development has taken place on converted farmland, so it's already denatured," Propst explains. Tucson, lacking a source of water as abundant as the Salt River, "didn't have as much farming, and so development is much more likely to go on native desert." And since Tucson's elevation is higher than Phoenix's, "the desert's a little greener, a little richer." Which makes it easier to love.

Which in turn made it easier to think big when the battle over critical habitat was grinding onward. Pima County is endowed with striking mountains and National Forests. It has units of Saguaro National Park on either side of it. Fifty miles southeast of Tucson the Bureau of Land Management acquired the Empire Ranch, which in 2000 became the Las Cienegas National Conservation Area. "So, you know, every agency's working, but it is all serendipity. There's no plan for it. It's just each agency, the Park Service, Forest Service, and BLM, doing its own thing in its own discrete way." And then came the owl and the obligation to protect the ironwood forest on which it depends. Knowing that there were other rare species that would one day take their turn at center stage, Pima County embarked on a countywide multispecies conservation plan. And then things went further. "What's interesting is that the thing morphed from just species protection to a countywide plan for an integrated comprehensive conservation system that included not just endangered species but wildlife habitat generally, archaeological resources, open space, recreation—kind of everything that you would want. There were six goals and they all got thrown into the plan."

The result was the Sonoran Desert Conservation Plan (SDCP). It distinguished between land that would be set aside for preservation and land that would be developed. It also provided for gradations between those extremes, allowing, for instance, small-footprint development to be mixed with reserved open space. The SDCP effectively completed and hardened the boundaries of Pima County's urban and suburban lands. It called for habitat corridors connecting one protected area to another. The voters of Pima County have backed up the plan with a pair of bond authorizations that together approach $190 million. These funds have fueled an acquisition program aimed at completing protection of lands designated for conservation and filling the gaps that exist around the public lands. According to Propst, "We're maybe one more bond package away from completing the job." He should know. He served on the steering committee for the SDCP, and under his direction the

*A Great Aridness*

Sonoran Institute has been deeply involved at every stage of the plan's development and implementation.

"We've done—and by 'we' I mean the whole community—a really good job in this county on deciding what we want to protect and how we want to go about doing it." The big job ahead, he says, lies outside the conservation land system: "We've got to do a much better job of creating a livable city. We've figured out where we are going to live. Now we've got to figure out how to live there."

As soon as we crossed the Pima County line into Pinal County, Propst steered us off the highway and into SaddleBrooke Resort Community. The affectation of the terminal *"e"* was the first sign that something was amiss. The homes were large, set on uniformly sized, walled lots, and the roads were sized for cars—and only cars. There was not a bicycle in sight. Only after we were parked at a small convenience store did we glimpse a lone golf cart. "If I were king of the world," says Propst, "I'd require developers of places like this to give golf carts away when you buy a house." SaddleBrooke is a retirement community patterned after a country club, a staple of the Sunbelt real estate economy. But it is a country club for old folks. You don't hear the shouts of kids down by the pool. In fact, pretty much all you hear are the gasoline drones of automobiles, lawnmowers, and hedge-trimmers. In Propst's view, Saddlebrooke is a pretty good example of how *not* to live.

County governments love developments like Saddlebrooke because they generate taxes without requiring much in the way of services, especially schools or anything youth-oriented. They have no poor people and little crime or rowdiness. They tend to be self-contained, even isolated. Many of the residents are snowbirds who come to Arizona to escape the winter someplace else, and also to avoid poor people, crime, or anything else that might disrupt the tranquility of their day. Saddlebrooke is not near anything except the Coronado National Forest boundary, which it jams up against, affording splendid views of the brown craggy mass of the Santa Catalina Mountains. All of the people who rake the development's enclosed yards or vacuum the carpets of its air-conditioned houses come from Tucson, twenty-five miles away. Residents, who are no less car-dependent than the mostly Mexican workers (legal and illegal) whom they employ, travel the same distance in the opposite direction for nearly all their shopping and service needs.

Saddlebrooke is a stuccoed maze. Behind its exterior walls stand the walls of individual houses, each with its own walled yard—a cubicled existence at landscape scale. The intent is to hold the rest of the world at bay. Security is assured, the golf course beckons, the television is on, and as months stretch into years, the cocktail hour arrives a little earlier, day by day. Not a bad existence, you might say, if leisure is all you want, but if you factor in the gasoline on which such a lifestyle depends and the disconnection from the rest of society, whose benefits its followers enjoy without contributing much to the costs, you come away with a pretty good idea why "gated community" is so often an oxymoron.

NOT FAR FROM SaddleBrooke we intersect Highway 79, the "Pinal Pioneer Parkway," and head northwest. Soon we are in rangeland, no buildings in sight, and the thornscrub vegetation is low and battered-looking, with scant grass and scattered cholla. We've gained elevation and are now briefly too high for saguaro and the other big succulent cactuses, which cannot grow where frosts are deep and prolonged. If the water they store in their tissues freezes, they burst and die. (This is why water-rich cacti like saguaros are absent from the Chihuahuan desert to the east, which is markedly colder than the Sonoran.)

Highway 79 runs as straight as a surveyor's laser, and the high desert stretches on for miles, lapping the feet of isolated mountains that bound the blue distance. The sweeping vistas suggest the rural, cowboy past of Pinal County. We even pass a memorial to the cowboy's cowboy—Tom Mix, the straight-shootin', bronc-bustin' rescuer of distressed damsels in over 300 silent westerns. In 1940 a flash flood destroyed the bridge over a nearby arroyo, and Mix, roaring up Highway 79 in his Cord Phaeton, saw the warning signs too late. When he crashed, an aluminum case full of cash and checks flew forward and hit him on the back of the head, killing him instantly.

Mix would find little resemblance between Pinal County's past and its future. No fewer than 650,000 "residential units," presently unbuilt, are platted for development—the local term is "entitled"—within the boundaries of the county, and they pose a problem. Arizona's peculiar process of entitlement allows developers to gain government approval for specific land uses, housing densities, and crude road alignments many years in advance of actual building. In fact, calling such operators "developers" is misleading; lacking any intention to commence building, they are speculators, pure and simple. During the go-go years of recent decades, speculators in Pinal County got whatever they asked for, thereby setting patterns for land use far into the future. Although the process of development will whittle down those 650,000 "units" to a much smaller number of actual houses, the patterns they impose on the land—uncoordinated from one area to the next—will endure, and will require of their residents a car-dependent, high-consumption lifestyle. Worse, the county has little power to revise these antique authorizations. In 2006 Arizona voters passed a ballot initiative, Proposition 207, that requires governments to pay landowners for value lost as a result of virtually any government action. Prop 207 was an assertion of property-rights-on-steroids. The specter it raises of costly lawsuits and potentially big payments makes downzoning and master planning where entitlements are in place virtually impossible.[10]

As a result, Pinal County faces an energy- and water-stressed megapolitan future with 650,000 albatrosses already strung around its neck. Worse, there's no method in the madness. The lots, says Propst, "are just helter-skelter all over the place." Maybe the county can nudge patterns by withholding services or road construction from certain areas; maybe it can exert influence through open space designations; but Prop 207, in the words of Grady Gammage, means that "government's kind of out of the business of regulating development in those areas." That leaves "the developer and the marketplace and the lenders to figure out what actually happens."

FLORENCE, ARIZONA, IS the seat of Pinal County. It is also a penal colony, which is not a play on words. The largest and oldest of the state's prison complexes is here, but concertina wire twines around many other establishments as well. I count at least seven on our passage around the edge of town. As Propst points out, "There's kind of one of everything. There's a women's prison, a couple of different state prisons for high security, low security, you name it, a couple of federal prisons, an Immigration and Customs holding facility, and more." *More* includes the Pinal County Justice Complex and the Central Arizona Detention Facility, a private prison run by the Corrections Corporation of America, which, Propst allows, "scares the hell out of me because those guys spend so much money lobbying."

Florence is a focal point in the geography of the region (and not just the cultural geography of crime and punishment). The mighty Gila River used to run through here, and for maybe a few weeks per decade, when thunderstorms rage or snowmelt streams out of New Mexico, it still does. The sandy bed of the river cuts through town, with a concrete batching plant—a sure sign of an Arizona "river"—mining the Gila's sand and gravel a short distance upstream of the highway. Long before it gets to Florence, the natural flow of the river is diverted into irrigation canals, and the small amount of water that seeps past the dams soon sinks into the thirsty ground.

Nevertheless, a lot of actual wet water flows through Florence by other means. On our approach to town, soon after the first trailer homes and low-slung bungalows appeared, we encountered the aqueduct of the Central Arizona Project. It enfolds Florence in a long hairpin curve, on a trajectory incongruously perpendicular to that of the natural watershed. It is a deep trough of concrete, surrounded by cyclone fence, and filled with fast-moving, clear, green water. Carp, facing upstream, graze the algae on the smooth canal walls.

THE DRY BED OF THE GILA RIVER, NEAR FLORENCE, ARIZONA. *AUTHOR PHOTO.*

The CAP arrives in Florence after a lengthy and (from an engineering perspective) eventful journey. The water is pumped from an impoundment of the Colorado River called Lake Havasu, three degrees of longitude to the west on the California border, then disappears into a tunnel that carries it through the Buckskin Mountains. After that, it courses down 150 miles of gently sloping aqueduct, is lifted by four more pumping stations, and travels through an additional tunnel and five siphons—all this just to get to the north side of Phoenix. From there the canal bends southeast, siphons under the Salt River, gets a lift from another pumping plant and, many miles later, fetches up at the dry channel of what used to be the Gila River, on the outskirts of Florence.

Hydraulic manipulation is not new in these parts. This particular reach of the Gila River, along with nearly every other stretch of flowing water in what is now central Arizona, was once home to the Hohokam people. For a thousand years, from roughly A.D. 450 to 1450, the Hohokam dwelled along the Gila, Salt, Verde, Santa Cruz, and San Pedro Rivers. Their society was highly organized, not least because they depended on irrigated agriculture. Building and maintaining their waterworks required armies of workers and prodigious commitments of labor. They battled ceaselessly against floods, washouts, erosion, silted canals, and the natural meandering of the rivers on which they depended. The ancient Egyptians or Sumerians would have fully appreciated their situation. The recruitment of adequate labor and the peaceful apportionment of water from one community to the next required of the Hohokam the utmost sophistication. They must have been experts in negotiation and rulemaking, and experts in engineering as well. So skilled was the design and layout of their irrigation systems that when Anglo farmers reached the Phoenix Basin and began to turn out the water of the Gila and Salt Rivers to irrigate fields of cotton, alfalfa, and citrus, they frequently found themselves reopening the canals the Hohokam had left behind.

One of those canals runs by the south side of Florence, along a contour of the shallow valley and roughly paralleling the course of the Gila. At their peak in the early 1300s the Hohokam may have numbered 40,000, and their imprint on the land was both emphatic and easy to read. By 1450, however, they had become—to use an almost mystical phrase—archaeologically invisible, which means that the last members of their civilization either migrated elsewhere or adopted lifeways that left no identifiable imprint. The various tribes of O'odham people of today (formerly known as Pima and Papago) are among the Hohokam's descendants, but the trail from ancient civilization to modern people jumps large gaps in the documentary record.

The decline and disappearance—or transformation—of the Hohokam has fascinated students of the Southwest nearly as much as the depopulation of the Four Corners region that preceded it by 150 years. Various cataclysmic causes—drought, flood, channel incision, disease, conflict—have been examined and found to be important but, on balance, individually insufficient to explain the phenomenon as a whole. Recent scholarship suggests a more synthetic view. Not all at once, but gradually, large numbers of climate refugees began exiting from the drought-stricken Four Corners region in the late thirteenth century. Their advance progressively forced the Hohokam to withdraw defensively from their scattered ranchos and hamlets into larger, more centralized communities, much as the builders of Sand Canyon Pueblo, faced with an influx of newcomers, had also done.

Under pressure, the increasingly urban Hohokam abandoned the vulnerable periphery of their territory, lands where they had hunted and grown dryland crops. The new population centers to which they retreated were safer but in certain ways less salubrious. Mounting stresses included poorer nutrition and health, a result probably of both congestion and a narrowing diet. In time, these ills produced not a dramatic, identifiable Waterloo, but the brutal arithmetic of population decline: more people died each year than were being born. At some point, Hohokam society passed a threshold: the number of able-bodied workers it could muster was no longer sufficient to meet the challenge of rebuilding dams when they washed out and cleaning canals as they inevitably silted up. Eventually the hydraulic system collapsed, and the society that depended on it could no longer exist.[11] The survivors turned their backs on their cities and scattered into the vastness of the land, doing what they could to survive.

No doubt the foregoing is excessively simplified and leaves out important elements of what actually befell the Hohokam. Perhaps, in time, we will learn what those elements are. But, at a minimum, it seems reasonable to believe that over centuries, the Hohokam scaled up their irrigation system to meet the needs of a large and growing population. They mastered life in an austere land. Yet the system that they built, past a certain point, paradoxically became their nemesis. It could not be scaled down when external circumstances and the capability of their society changed.

These reflections come to mind, not far from the prisons of Florence, as the brimming canal of the Central Arizona Project comes again into view, its waters quietly flowing toward 1 million living residents of greater Tucson and a hefty share of 650,000 undeveloped lots. The waters are approaching the end of a 300-mile journey through mountains, over plateaus, and across deserts. Virtually every drop in the canal's tightly controlled, manmade flood first sojourned in overtapped, imperiled Lake Mead, which is the shining central achievement of a system of water distribution as impressively upscaled as any on the planet.

FLORENCE HAS A lot at stake in the emerging megalopolis of the Sun Corridor. Transportation infrastructure will shape the physical contours of the expanding cityscape more surely than water, and rail is the key. Rail makes possible high-density, compact communities. Density, in turn, allows for downtowns with pedestrian-based retail and cultural activities, a nightlife, and a tangible sense of community identity. Compared to car-dependent bedroom suburbs, development along a high-speed commuter rail line is carbon-thrifty and fuel-efficient. It is also much more cheaply provided with police, fire, and emergency health services, not to mention libraries and clinics. The trouble with rail is that it requires large investments and creates new sets of winners and losers, so that the haggling over whether to do it sometimes outlives the opportunity that makes it possible. One of the big losers to a Sun Corridor rail system would be the Tucson Airport Authority, which would surrender passengers to Phoenix's larger and better-connected Sky Harbor Airport. Still others will include any town the rail line doesn't go through.

One of them might be Florence. Propst explains that a leading option for the rail line, if it gets built, would be to follow Interstate 10, the shortest route between Tucson and Phoenix, possibly with a detour around the Gila River Indian Reservation if gaining right of way through the land of a sovereign tribe proves difficult. But Florence comes out a winner if the choice goes to an alternate and slightly longer route, which Propst, perhaps with too much optimism, says currently enjoys "the most support." For much of its length this route roughly parallels the CAP canal. From south to north, it would follow the interstate out of Tucson as far as Red Rock, then angle toward Florence. From Florence it would approximately follow the path we are about to take: north to Florence Junction, then obliquely westward to approach Phoenix from the east, through Mesa.

This route has the advantage of being almost entirely on state trust land, and cuts through the heart of the greatest prize in that vast inventory: a large state-owned district of more than 175,000 acres called Superstition Vistas. The land is presently undeveloped, but Arizona's master planners and real estate experts say that Superstition Vistas, which adjoins the Phoenix metro area, will one day be home to more than 1 million people. That's where we are headed next.

"STATE LAND REFORM is critical," says Propst, not least because state trust lands constitute 12.7 percent of Arizona. "The state owns something like 60 percent of the total land that's likely to go into development in the next fifty years." Unlike other western states, whose state land is often scattered in small parcels across the landscape—"checkerboarded" is the regional term—Arizona's trust lands are pretty well "blocked up." They have been assembled into large continuous tracts, and they are concentrated in the Sun Corridor, where 80 percent of the state's population resides within 20 percent of the state's land area. The financial return from development of state land within the corridor has the potential to enrich the beneficiaries of trust lands enormously. Those beneficiaries consist mainly of Arizona's schools and universities and, through them, its children. It is no exaggeration to say that the fate of the state trust lands will materially affect the development of Arizona's human and intellectual capital and the future of its economy. It will help determine the kind of place Arizona becomes and the future it offers its young people.

Several requirements for optimizing value are already in place. The lands exist in the scale and location necessary to dominate decisions on water, transportation, and energy. At Superstition Vistas key infrastructure—high-tension power lines and the CAP—is already present. Everything else will follow, says Propst, if only the state land department will lead. So far it won't—and in important ways, it can't.

Currently the state sells a small amount of land at auction each year, a process that lends itself to what Propst calls "hopscotch development." He points out that before the housing crash in 2008, "something like 40 percent of the new houses being built to serve the Tucson market were being built in Pinal, Cochise, or Santa Cruz Counties"—jurisdictions that surround Pima County and lack its restrictions, including the Sonoran Desert Conservation Plan. "When that's the case, you just get leapfrogging." The quest for water adds to the incoherence: within "Active Management Areas" that were established under a powerful groundwater protection law in 1980, a developer has to demonstrate a secure 100-year supply of water before he can build. Unfortunately, the AMAs have not been expanded since then, in spite of a near quintupling of the population of the Sun Corridor, and builders routinely place their developments (and seek state trust land) outside AMA boundaries, where they sink wells irrespective of the long-term viability of the aquifer.

Developing Superstition Vistas, by contrast, would be the opposite of "hopscotch." It is big enough, attractive enough, and sited in a desirable enough location that it will make the rules, not evade them.

When we reach the lonely crossroads of Florence Junction and turn west toward Phoenix, we've already been within the Superstition Vistas Growth Area for ten miles or more, a small fraction of its total area. The land is high, empty desert, scarcely a building in sight. To the north the Superstition Mountains rise in rugged volcanic blocks, and the vegetation of the terrain leading to the mountains grows richer as it rises and captures the rain that the mountains squeeze from the sky. Saguaros are again plentiful, tall ones, as well as palo verde and teddy bear cholla. Most of the mountain range is officially wilderness. The boundary of the Superstition Wilderness of the Tonto National Forest comes all the way to the foot of the mountains, where it adjoins state land. Propst and others are campaigning for preservation of the habitat close to the mountains and the wilderness area. Across the highway to the south and west, the land is flatter, good for trains and clustered communities, and the vegetation is simpler, with much less conservation value. It's only logical, he says, to concentrate development there.

Toward that end, Propst and the Sonoran Institute belong to a group writing a master plan for Superstition Vistas. The effort includes the state land department but is independent of it. Its purpose is to demonstrate the benefits of master planning, if only the state would—and could—engage in it. Both politics and current law stand in the way.

Propst's is the principal conservation voice on the committee writing the plan. It is a diverse assemblage that includes Pinal County, several municipalities, the Salt River Project (the area's dominant utility), Rio Tinto (a mining company with local interests), the land department, and various others. The council has no legal standing, but it possesses the suasion of broad representation and hard-earned consensus. "What we really want to do is influence the state land department so that when they sell the land, they've created a plan for the whole 275 square miles, and then they sell according to the plan." A second goal lies behind the obvious one of producing a good plan. "We are making plans that cannot be implemented under current law," Propst explains. The idea is to use the plan "to build the case that we need to change the law.... We want to show the [financial] numbers for the kind of future the land department can have and to use that to build a case for reforming the tools that the land department is permitted to use. We just want the land department to have the power to act like any rational landowner."

Among the tools the state needs is the authority to determine the location of roads and rail lines. "You need the ability for the state to decide where infrastructure's going to go and what kind of infrastructure, so that the infrastructure is not just left to each individual buyer.... You need the state to change the rules so that you can more easily do mixed-use development. The state generally sells land for residential development and leases it for commercial development. So what do you do with a three-story building that's got retail, office, and condos? That kind of stuff doesn't compute. The system's not designed for it." The system, he says, was designed in 1912, when Arizona gained statehood and most of its people "were still riding mules."

According to Propst, the state needs to be able to participate as a codeveloper in projects, not just as a seller of land, and it needs the authority to plan jointly with cities and counties and to give away rights of way "so that a rail line can go through state trust land and they can make ten times more money." In general, the state needs to lose its passivity. "They more or less just sit back and say to the development community, 'Bring us what you want to do and we'll look at it.' All over the world, that kind of land bank— large parcels in one ownership—creates the opportunity for great planning. And here we've got the big land in one ownership, but we manage it as if that's not the case."

Several years ago Propst and the Sonoran Institute campaigned for a ballot initiative that would have reformed the land department along the lines he describes. The measure lost by less than 3 percentage points. The homebuilding industry, which was "very happy with the model where the land department sells eighty acres at a time," vastly outspent Sonoran and other advocates in the final weeks of the campaign.

Now, however, the housing crash has left the homebuilders struggling, and the advocates of reform see a new opportunity. Propst expects a fresh and stronger citizens' initiative to be on the ballot in 2012. "We expect by then the economy will be back, conservation philanthropy will feel robust again"— robust enough to fund a strong campaign—"but the homebuilders will still be down in the back." The reformers hope to be like jackrabbits after a drought, doing what they please before the coyotes build back their numbers.

We leave the highway and drive to the Peralta Trailhead, almost at the boundary of the Superstition Wilderness. The desert is opulent with saguaros, palo verde, and the twisted forms of other cactuses and shrubs. A likely feature of the 2012 citizens' initiative, besides giving the land department the authority to "act rationally," will be the creation of a system of protected lands within the state trust inventory. If such a system materializes, this portion of Superstition Vistas will be one of its jewels. Only half an hour from the core of the Phoenix metro area, it adjoins a federally designated wilderness area. This is how life in a western Mega can be: several million nearby neighbors, and 160,000 acres of rugged wild lands at the back door.

Propst and his organization work on land-use issues in many areas of the West—on Colorado's West Slope, in towns surrounding Yellowstone National Park, even in the dried-out delta of the Colorado River in northwest Mexico— but "trying to make the Sun Corridor a more sustainable place is probably the most challenging, ambitious work the Sonoran Institute does." Usually, he says, a kind of parity exists between the challenge of what needs to be done and the resources available to take it on. Not so in central Arizona: "There's a huge disconnect between the conservation voices and this leviathan of economic growth, and sometimes you feel like you're just whittling on the very edges."

We are driving into the leviathan now. We've entered Maricopa County. The highway has swollen to eight lanes, and the sky, which once attracted refugees from the tubercular East, is murky with smog and dust.[12] The Mesa-Tempe-Phoenix-Scottsdale-Glendale complex of cities-within-cities spreads to the horizon and beyond. Propst grows reflective. "There is a level of denial here that is so deep. The economy is so dependent on housing. The legislature is so thoroughly debased—this is a hard state. You look at other states and there are challenges, but my goodness, I think Arizona might be the hardest to make any real progress in. And I lay that at the feet of Maricopa

County, where there's just this get-rich-quick mentality and then if you score, move to La Jolla. There's not a widespread sense of stewardship in the state, certainly not in the state legislature, not in the major landowners. It's not like Colorado, where you have a mature society in Denver that thinks about the welfare of the state as a whole. You've got very little money in the state that pushes for a more sustainable future. You've got a few foundations that do great work, but mostly you've just got this go-go mentality: make your money and move on."

Phoenix took root as a farming town in the 1870s, a young brash sibling to older Tucson. Whereas Tucson was still predominantly Catholic, Spanish-speaking, and shaped by traditions from south of the border, Phoenix was Protestant, English-only, and defiantly Anglo. It was the all-American alternative, and projected a self-righteous and mercenary spirit from the start. In 1920, notwithstanding that the federal census counted fewer than 30,000 people in Phoenix, the chamber of commerce proudly declared that the city had grown to 40,000, and according to the chamber's directory, they were "the best kind of people, too. A very small percentage of Mexicans, negroes, or foreigners."[13]

A reminder of the current Maricopa attitude toward ethnic diversity roars past our car. It's a prisoner transport bus with blacked-out windows—"very Gestapo-looking," Propst observes. The words "Office of the Sheriff" are emblazoned in giant letters on the sides and back. The sheriff in question, Joe Arpaio, is both popular and notorious. He is an anti-immigration zealot and a darling of right-wing television and talk radio. "The default vote in Maricopa County is about as strong as it can be towards law and order, towards anti-immigration, and towards this kind of every-man-for-himself worldview."

I don't have hard data on the subject, but I would be willing to bet that the particular vehemence of the pro-Arpaio, anti-Latino majority in Greater Phoenix has much to do with the shallow roots of its angriest members. They, too, are immigrants to the region—from places like Milwaukee, Pittsburgh, Little Rock, and a thousand other towns and cities, where they never heard anything but English spoken and where the only significant racial issue was the ancient tension between black and white. Now having torn themselves from those familiar environments and plunged into the land of their dreams, they find their Eden strangely corrupted. People they deem foreign in language, values, and behavior surround them, numbers swelling. Beyond appreciating the occa-

sional taco or margarita, the newcomers may not have been in Arizona long
enough to absorb much of the Hispanicism in their new environment, nor do
they want to. Instead they soak up the never-extinguished shoot-first ethos of
the western frontier. They take their stand with the kind of defiance that put the
Earps and the Clantons in the O.K. Corral. They vote their frustration with the
petulance that led their adopted state to deploy its "navy" against California.
The sorry outcomes of these events do not register; it is the feeling that counts.

The growth engines of modern Phoenix and Tucson revved high as World
War II approached. The war department needed to train pilots, lots of them,
and the clear dry skies of Arizona were perfect for flying. Airplanes need parts
and electronics, and soon industries grew up to meet those needs. As the
national economy boomed after the war, other industries with similar
technical demands began locating in Arizona, especially Phoenix. They could
do this without displeasing their workers, because the scorching desert climate had been tamed. Design advances and mass production forced down
the cost of air conditioning, and almost overnight the negative of a too-hot
climate became the positive of an escape from cold winters and gloomy skies.
Workers flocked to the jobs, and more new jobs arose simply because workers
wanted to be there and because land and energy were cheap—Glen Canyon's
hydropower came online in the 1960s, and the big coal-fired plants of the
Four Corners were not far behind. In the race for growth, Phoenix quickly
outstripped Tucson. It had more and cheaper land, its banks were more
aggressive, and local government was more compliant and less corrupt.
Phoenix proved that it was decidedly "business-friendly." And water, once the
CAP was authorized, was never considered a limitation.

Propst and I stop for lunch in downtown Tempe, one of the independent
cells in the overcrowded petri dish of Phoenix communities. The fact that
Tempe, amid a landscape of largely undifferentiated sprawl, actually possesses a
functioning and recognizable downtown deserves comment. It has tall buildings, broad, tree-shaded sidewalks, and an atmosphere of bustle and possibility.
It is served by the metro area's new light rail system and is home to gigantic
Arizona State University. Multistory apartment and condominium complexes
ring the concentration of the center. In downtown Tempe, a family can lead a
nearly carless life. One of the magnets for downtown development is a lake,
some dozens of acres in size, crossed by a pair of graceful bridges. It occupies the
bed of the Salt River, which—except for infrequent floods—has otherwise

been dry since the 1940s. Dams made of giant inflated bladders impound the lake, which holds mostly CAP water. Its considerable evaporative losses (more than five acre-feet per day) are replaced by additional purchases. It is "a flamboyant waste of water," Propst acknowledges, but he suggests that the waste may be warranted by the higher densities, the reduced amount of driving people do, and other efficiencies of the downtown. Whatever its justification, the lake is a work in progress. In July 2010, a bladder in one of the dams burst, draining the lake and generating a bounty for a local (captive) alligator, which feasted on stranded fish. The dam has since been repaired and the lake refilled.

THERE IS NO more central fact of life in the emerging megacity of the Sun Corridor than the ubiquity and essentiality of the CAP. It nourishes homes and businesses. It provides amenities like Tempe Town Lake. Without it, substantial growth, including the development of places like Superstition Vistas, would be flatly impossible. And yet, if Lake Mead fails in the years following 2026, when, in Brad Udall's words, the risk of failure "just skyrockets," the CAP will fail too. If that were to happen, Tucson and other communities would be obliged to pump groundwater at a dramatically unsustainable rate, draining anew the aquifers they have recharged against just such a day.

Phoenix and its neighbors, meanwhile, would have to stretch the supplies of the Salt and Verde Rivers to unimaginable thinness. But the supplies would likely be thin to begin with. In any scenario for the failure of Lake Mead, the same triggering forces of climate change or a return to the kind of drought known in paleo times would almost certainly shrink the Salt's and Verde's flows. Disaster will not be felt immediately. Stored supplies of various kinds would be brought into play, but eventually Tempe would drink its Town Lake and the "water farms" where today's surplus of CAP water is injected or allowed to percolate into the ground would be fully "harvested." Then what? No one knows. If the crisis comes in 2015, something like 5 million people in the Sun Corridor will feel the pain. If the crisis comes in 2030 or 2040, perhaps half again as many will be affected, maybe more. And their options will be fewer.

For the sake of argument, let's assume that 7.5 million people will call the Sun Corridor home in 2030, and that these people will consume water at the same rate Arizonans did in 2009, about 204 gallons per person per day.[14] That translates to just under a quarter of an acre-foot (0.2285 af) for each person every year. Total Sun Corridor demand will jump from 1,142,500 acre-feet per year (of which CAP provides about 45 percent) in 2010 to more than 1.7 million acre-feet in 2030. Barring successful augmentation and the provision of new supplies, the additional demand would be met by transferring water out of other uses, principally agriculture and recharge programs. The effect on the statewide water budget—to borrow a term from a previous chapter—would be to *harden demand* immensely. In times of shortage, you can withhold a half million acre-feet of water from alfalfa fields without causing the sky to fall. If you withhold an equivalent amount of water from the people who depend on it, people whose demand has already been hardened by their water conservation efforts, the sky comes crashing down.

Interestingly, no comprehensive and generally accepted study of the hydraulic carrying capacity of the Sun Corridor yet exists.[15] As Propst says, "Arizona is a hard state" in which to raise issues of limits and sustainability. When recession hit the state in 2008, the fledgling Arizona Water Institute, a cooperative venture among three universities and an appropriate source for long-range water planning, was an early casualty of the legislature's efforts to rebalance a depleted state budget.[16] Buoyed by the seeming abundance of the CAP, the state has welcomed all the growth it can attract. Even if CAP deliveries remain as stable as a fixed star, a limit of prudence—the point past which the risk of harm outweighs prospective gain—will one day be reached. But if Lake Mead ultimately succumbs to some combination of drought, climate change, and overuse, no amount of planning will provide a remedy.

Even if Lake Mead remains functional, the CAP remains in peril. When (not *if*) climate change and Upper Basin consumption combine to restrict the Lower Basin to its compact allocation and no more, and when releases from Lake Mead reflect this, the Lower Basin will have to absorb its structural deficit of evaporation and other inadvertent depletions, plus a share of the obligation to Mexico. There exist multiple ways to do the math, but none of the numbers look good for Arizona.[17] Brad Udall thinks the structural deficit may be as large as 1.5 maf.[18] That is roughly equal to all of the water that flows in the CAP, and under a strict interpretation of the law, all of it could be sacrificed to the Lower Basin's shortfall. Things get even worse if climate change dries up the Grand Canyon inflows to the Colorado River, which heretofore have been a kind of compact-free bonus to the Lower Basin. With such prospects even remotely possible, the limit of prudence might be said to have disappeared from society's rearview mirror quite a while ago.

I CALLED TOM McCann to ask his thoughts on these matters. McCann is the CAP's assistant general manager for operations, planning, and engineering, and his schedule is as heavy as his responsibilities. When we finally found a brief window in which to talk, I asked him what he thought of the analysis of Lake Mead's viability by Tim Barnett and David Pierce. He said the paper was "significantly flawed," and that the authors chose their assumptions in order to "make the system fail."[19] McCann's language was more circumspect than that of other members of the water establishment, who had complained to me that the Barnett–Pierce study was unfair and conceptually flawed. Nearly all of them, including McCann, had been unaware of revisions to the original paper that corrected those flaws. Therefore, they were also unaware that the corrections only pushed the day of Lake Mead's reckoning a few years farther into the future, without altering the fundamental conclusion that the lake is at risk. Nevertheless, McCann and the Scripps researchers are not as far apart as might be supposed. McCann volunteered that the Lower Basin is operating at an annual deficit of 1.2 to 1.3 maf, a view that Barnett and Pierce share and that, combined with the prospect of declining river flows, accounts in large measure for the lake's increasing vulnerability.

"All managers recognize there are difficulties ahead," McCann said. "Collectively we need to be augmenting the system." McCann is a lawyer—formerly he served as senior attorney for the Central Arizona Water Conservation District, the parent institution of the CAP. He pointed out that Arizona accepted subordination of its rights to California's "because we were going to do augmentation," and he emphasized that today the Lower Basin still requires augmentation on a scale sufficient, at the least, to offset the 1.5 maf treaty obligation to Mexico. Otherwise, "we're going to have problems."

He quickly ticked through the leading alternatives, which have been the subject of repeated, almost continuous, study by the basin states: weather modification, vegetation management, importation (either directly or by replacement), and desalination. Each of these, as McCann is intensely aware, faces formidable obstacles. Weather modification, which involves wringing moisture from clouds by seeding them with silver nitrate crystals, has many strong advocates, but its efficacy has been hard to demonstrate. Moreover, if cloud seeding were found capable of increasing precipitation on a regional or

even subregional scale, jurisdictions downwind of the cloud seeding would
no doubt sue to have it stopped, and with reason. At best, weather modification does not increase the amount of moisture in the sky, it just squeezes it
out at different places.

Vegetation management, which can take the form of forest thinning, similarly fails to deliver a solution. Even where forests are not on a course toward
replacement by shrubs and grassland (as discussed in Chapter 9), warmer
temperatures and the earlier onset of spring will likely offset any additions to
runoff that thinning might produce. The removal of riverside plants such as
salt cedar produces similarly ambiguous results and is even more laden with
environmental conflicts.

Which leaves importation and desalination. McCann thought Patricia
Mulroy's idea for delivering Mississippi floodwaters to the Front Range of
Colorado might have merit. When I objected that the amount of water to be
saved was relatively small, he replied that the greater benefit might be in fore-
stalling future diversions from the Upper Basin to the Front Range. Point
taken. But the complexity and cost (in both dollars and energy consumption)
of so major a project remain daunting in the extreme.

The same may be said of desalination. Its chief drawbacks are expense,
gluttonous energy consumption, and the environmental impacts of disposing
of the brine it produces. In 2010 the CAP, Metropolitan Water District, and
Southern Nevada Water Authority partnered with the Bureau of Reclamation
to finance a pilot run of the mothballed Yuma Desalting Plant, which was
originally built to provide treaty-acceptable water to Mexico. They expected
to spend $800 per acre-foot to produce 29,000 acre-feet of relatively fresh
water, which would be applied against the 1,500,000 acre-feet annually owed
to Mexico. These costs, however, are strictly operational; they do not include
the capital cost of the desalting plant itself, which was completed in the early
1990s for more than a quarter of a billion dollars.

The Yuma Desalt Plant remains the largest reverse-osmosis desalination
facility in the world. Its designed capacity is about 73,000 acre-feet per year,
although a plague of problems has prevented it from ever reaching that level of
production. If you scale up the capital and operational costs to a level that—
theoretically—would describe a plant offsetting a major portion of the Mexican
treaty obligation, you reach numbers that are truly astronomical. And that's
before adding in the capital cost of the nuclear plant it would take to provide

enough electricity to push all that water through an infinitude of purifying membranes.[20] (For reference: $800 per acre-foot × 1.5 million acre-feet comes to $1.2 billion, in annual operating costs, if everything goes well. Add at least $4 billion to build enough plants to produce 1.5 maf of sweet water and an astronomical investment for the nuclear reactor, and pretty soon, as the saying goes, you're talking real money. Of course, maybe there will be technological breakthroughs or economies of scale that radically reduce these costs. But maybe not.)

The price tag for large-scale desalination, to say nothing of other obstacles, currently appears prohibitive, at least with present technology in the present economy—this is roughly the conclusion reached by the National Research Council of the National Academies of Science in 2007. In the long run, said the NRC, no augmentation strategy offers "a panacea for coping with the reality that water supplies in the Colorado River basin are limited and that demand is inexorably rising."[21]

McCann notes with some anguish the current lack of enthusiasm at the federal level for expensive, large-scale augmentation projects. He appears resigned to the fact that the basin states may have to undertake augmentation on their own. His sense of the unfairness of this outcome is amply documented in a lengthy memorandum he kindly shared with me on the problem of CAP's priority. One of its major headings is titled "The Broken Promise of Augmentation."[22]

WATER SHORTAGE WILL not be the only gift of climate change to the Sun Corridor. As Propst and I cruise the freeways of the Phoenix metro area, we see mile after mile of highways and streetscapes, parking lots and sidewalks, shopping centers and buildings, buildings, buildings. Nothing green, nothing soft. The earth has been reformed from scruffy desert floor to pavement, stucco, and concrete—all of them substances that very effectively absorb heat and hold it. At the close of World War II, before the city's prodigious career of expansion began, overnight low temperatures of 90°F were unheard of.[23] Today a 90-degree summer night in Phoenix warrants no comment, and according to Propst, "it's not a joke but a question about when Phoenix is going to have its first night that it doesn't drop below 100 degrees. Because it's inching closer and closer."

The phenomenon known as the "urban heat island" is hard to quantify, but the heat-storing and warming effect of built-up land appears to be substantial. According to one study, Phoenix's overnight lows in the month of

June rose 2–4°C (3.6–7.2°F) in just fourteen years. Other measurements suggest that daily low temperatures average almost 10°F higher at the Phoenix airport than at an equivalent site beyond the city's edge. Daily high temperatures, meanwhile, remain roughly the same at the two sites, illustrating the fact that the impact of the urban heat island is mainly felt at night, when the hard surfaces of the city release heat stored during the day.[24]

The problems raised by heat islands go beyond comfort. Heat stress can kill people, especially the old and the very young. And it can make everybody miserable, sometimes sick, and as jumpy and quick to anger as tomcats. As with most unpleasant things, the effects of heat islands are felt more strongly in poor neighborhoods, which have higher densities and less mitigating open space and greenery, than in wealthy ones. And poor people, of course, have fewer resources with which to combat the higher temperatures.[25]

The urban heat island effect drives people to use more water, both to cool off and to keep their landscaping alive—even when, from a supply standpoint, they need to use less. For urban planners, heat islands are a major pothole in the road to successful design. The heat island effect rises with density, and density is the key to energy-efficient transport, vibrant downtowns, the prevention of sprawl, and the protection of open space. But downtowns like Tempe's won't survive if the nights are unbearably hot, and the allure of the radiant Sunbelt will tarnish irredeemably if chronic heat and water shortage reduce yards and parks to dustbowls. Perhaps the mavens of design will solve the heat island problem through the use of water-wise greenery and new kinds of shading and building materials. Perhaps through some miracle of public financing such benefits will be equitably shared among all economic classes. In the meantime, one way to understand urban heat islands is that they double the punch of climate change. Life will get hotter everywhere, but the increase will be twice as great in Phoenix.

INTERSTATE 10 STRETCHES roughly 100 miles from Phoenix to Tucson, and there is not much along the way to please the eye. The penumbra of Phoenician smog extends miles beyond the city, imparting a turgid look to the horizon even where it is unframed by buildings. The land itself is the bleakest kind of desert, much of it formerly given to cotton, oranges, and alfalfa, for which the native vegetation was scraped away or plowed under. Now the crops are gone, too, because the water that used to nourish them has since been diverted to the supply mains of subdivisions. Local wisdom in the Sun Corridor holds

that "agriculture is what you do until development arrives," and these abandoned fields, many of them lying within the Gila River Indian Reservation,
languish in the limbo between uses. A common cure for the dust storms they
cough up is to fasten the dust back to the earth by placing a parking lot and a
casino on top of it, like paperweights on a desk.

We cross the Gila River again. It is broader here than in Florence, and
while it is equally dry, it supports a greater number of batching plants, thus
confirming its riverine importance. In our hurried flight down the interstate,
away from the city, Propst and I are reenacting a ritual known to many Phoenix
families, which is called "Drive 'til you qualify." In the recent past when
lenders of subprime mortgages shoveled money into the pockets of the
unwary, a husband and wife needing a third bedroom or one more bath might
set out as we have done, speeding down not just a highway, but a gradient of
declining home prices. Mile by mile, the prices drop. When the husband and
wife reach the point where, with their combined paychecks, they imagine
they can muster the monthly mortgage payment, they take the exit ramp and
begin shopping for a house. Never mind that the costs of commuting, especially in an era of rising fuel costs, may negate most of the savings conferred
by a low monthly house payment; the point is to get the loan and move in.

The interchange where they begin their search might harbor what real
estate professionals call a "power center." It is more than a shopping mall,
more like a conglomeration of malls, a truly *grand mal*, in the language of epi-
lepsy. We are now back in Pinal County, on the indefinite outskirts of Casa
Grande, ancient home of the Hohokam and an hour's drive from Phoenix.
Many of the families who live here—certainly a large majority of the recent
arrivals—devote two hours a day to their commute, and satisfy most of their
consumer needs at the nearby power center. Every chain and franchise you
can think of is represented here: Starbucks, Subway, Cracker Barrel, and the
various burger vendors; Target, JC Penney, and Ross Dress-for-Less; Radio
Shack, Staples, EyeMasters, and, in its own free-standing building, Walgreens,
which Propst mentions always stands separate, pursuant to its inscrutable
corporate plan. He challenges me to identify one business that is not a national
concern. I can't. No local roots here. Propst explains that the development of
a power center is driven by formula: so many rooftops within so many miles
and lickety-split, once the threshold is reached, here come the big-box stores
with the household names.

Except now, the formula no longer works so well because a frighteningly large percentage of those determinative rooftops have no one living beneath them. In August 2009 one out of every ninety-five living units in Pinal County was in foreclosure. The figure includes older homes as well as new ones, but the foreclosed houses are disproportionately new, and in some neighborhoods they are the rule, not the exception. Many more homes are in other forms of distress, unsold, unpaid-for, weighing down the books of some bank that wished it hadn't run with the herd. In 2009 Pinal County's foreclosure rate was the highest in Arizona, and through the recession Arizona has been a national leader in both unemployment and foreclosures (only California and Florida, much larger states, consistently had more).[26] "So this used to be a farming community," says Propst, "and now its last harvest is these homes, but the crop, for the time being, has failed."

The failure represents an opportunity to save the harvests of the future. The power centers and cookie-cutter commuter way stations of the Sun Corridor are continuing expressions of auto-oriented urban design. They will be difficult to adapt to the thrifty requirements of an energy- and water-limited future. Perhaps the real estate crash that closed the first decade of the twenty-first century and stalled the juggernaut of growth will make room for the introduction of new urban "crops": high-speed rail lines, multiple downtowns and employment centers, and mixed-use, pedestrian-friendly residential areas. Advocates of such a vision argue that if the right choices are made soon enough, the Sun Corridor and other Mountain Megas could become to the twenty-first century what the Rust Belt was to the twentieth—a kind of American heartland, a leader in population, economic clout, and (as a consequence of both) political power.[27]

History, unfortunately, contradicts that vision. As Grady Gammage asks, "Can quality ever hope to compete with easy money and fifty years of tradition?"[28] Arizona's real estate habits run deep, and most of the progressive prescriptions for the future require concentrations of planning authority and capital, which the state's freewheeling traditions have rarely supported. When I reached Gammage by telephone, he said he was in his Tempe office looking out the window at two towers of an intended four-tower condo development. One tower has been enclosed; the other is open to the elements. "Both are abandoned, being taken over by lenders. There are two more in downtown Phoenix. There are probably another three in uptown Phoenix." He said that lenders who tried to build for a high-density, truly urban lifestyle got "hugely burned" in the real estate and credit collapse. He expects them to return to the traditional practice of financing single-family homes almost exclusively. "It's a lot safer, a lot cleaner. You can do it one at a time." The prospect does not cheer him. "I don't want us to just go back to the same old cycle of boom and bust based on cheap houses and single-family detached lots on the edge. But we're so good at that.... All of the temptations are setting up for us to go back to our old patterns."

An even bigger problem, however, is the interconnection of water and climate. On paper, it would appear that the Sun Corridor has the most secure water supply of any of the great conurbations of the aridlands. It has the CAP, and the CAP still irrigates a lot of farms. Propst sums it up: "When you're figuring agriculture and the ability of converting agriculture to municipal, it's easy to reach the conclusion that, hey, there's plenty of water." And there is— for the foreseeable future, as long as Lake Mead keeps functioning and the Lower Basin can keep shielding the insolvency of its water budget. "But then," Propst continues, "you talk to another set of people who look at the climate models and think about Arizona being junior, and you think, 'We're in real trouble here.'"

We are approaching Red Rock now—a bedroom community oriented to Tucson. We've left the gravitational field of Phoenix and have passed through the trough of the mortgage-seeker's economic gradient. Now housing costs will begin to climb as Tucson grows nearer. The CAP canal has come back in view. It contours along the foot of a mountain range and then veers toward the interstate. It is an orderly, straight-line feature in an organic and irregular landscape. It draws your eye. You cannot keep from looking at it. Its regularity seems at first comforting, then chilling. I can't help thinking that each of us has just one heart and one aorta, but we each have our own. When it comes to water in the Sun Corridor, things don't work that way. There is just one aorta serving millions, with millions more to come.

# 8

## APACHE PASS: CROSSING THE LINE

THE *MOCHILA* LAY beside the migrant trail, an abandoned black daypack still heavy with goods. The Border Patrol agent carried it to the shelter of a corner of rocks, where no one could spot him while he searched it. He dug through the contents. There was a package of refried beans in gaudy plastic, a bag of instant oatmeal, fruit punch in a bottle too small to slake a serious thirst, and other convenience food. Also a half-pound or more of white grains in a punctured bag; the agent wet a finger and tasted: only sugar. Then he heard voices approaching and scrambled up the slope to hide in the brush.

There were three of them: a rangy young man with a shadowy face in the lead, an older guy in a ball cap, and a pretty young woman with raven hair behind. They were Americans, not migrants or narcos. Their skin, their clothes, even their posture gave them away. They were too relaxed, too careless to be anything else. The agent stepped out from his hiding place. They slowed but did not stop. "You all out hiking?"

"Yep," said the young man with the shadowy face.

"Where you from?"

"Tucson," came the clipped reply.

Then the hikers, unsmiling and eyes straight ahead, passed him at a fast clip, the chill of the encounter resisting the afternoon heat, the desert absorbing the silence. The hikers had come from the direction of the Rat's Nest, a maze of drainages half a dozen miles above the Line, and they disappeared toward Apache Pass—not the famous Apache Pass in the Chiricahua Mountains in eastern Arizona, but a lesser pass on the shoulder of Bartolo Mountain, well south of Tucson and only nine or ten miles north of Mexico. The agent knew they weren't out for a hike. No one comes just to hike in the contorted and contested, bone-dry mountains along this stretch of the border. Everyone has a purpose. They come to smuggle or to be smuggled. They come to scurry in moonlight and to drag themselves under the blaze of the sun across dozens of miles of steep shadeless rock. Or, like the uniformed agent and his partner parked downslope behind a mesquite in a dogcatcher rig, they come to chase the smugglers and the smuggled, the tireless cat after the desperate mouse.

The young man had answered the agent with the terseness of someone accustomed to deflecting the questions of police. If the agent had come close enough to look him over, he'd have seen that the shadows on his cheeks were not cast by whiskery stubble, as anyone might have expected, but were a beard of tattoos, reaching from the neck almost to the eyes. The young man's face was wreathed in blue-ink swirls of dragonheads and filigreed roses. He wore a mask of intricate design, permanent and exclamatory. The silent shout of his tattooed face was his declaration of resistance. It was seconded by the inked skin that showed from the cuffs of his sleeves and from the gaps in his clothing when he leaned over. And it was more loudly reinforced by the cored-out lobes of his ears that hung like limp doughnuts, bereft of ornaments, a personal testament to the unimportance of pain and the mutability of flesh. Even more literally, he had spelled out his message of defiance on the knuckles of one hand: L-E-F-T, a letter for each knuckle, there to read when the hand is a fist coming at you. But it's etched on his right hand, not his left, and that's part of the message. He does things nobody's way but his own, and if you don't like that, screw you.

The illustrated man, Daniel Nelson, is a volunteer with the humanitarian group No More Deaths, whose volunteers otherwise tend to be as outwardly straight and ordinary, and nearly as young, as any group you might round up on the main strip of a college town. Nelson is within one or two years of thirty, but which side I could not say. He is lean and strong, and that day he led a patrol of five people (two remained back at Apache Pass), replenishing water and food drops at various locations and generally scouring the landscape for signs of people in distress. I know this because I was the older guy wearing the ball cap and trailing behind him. The young woman behind me, Jennifer, was an undergrad from the University of Arizona who, like Nelson, scrambled up and down the rubble of those mountains with the ease of a bighorn. Before the day was done, I was fighting cramps in my quadriceps from the effort of keeping up with them.

An uneasy peace exists between the volunteers and the Border Patrol, whose agents the aid workers suspect of knifing the gallon jugs of water they leave out, sometimes fifty or more at a site, and draining the vital contents into the sand. The same may be said of relations between the aid-givers and Immigration and Customs Enforcement (ICE) and the U.S. Fish and Wildlife Service, which manages the Buenos Aires National Wildlife Refuge, a vast former ranch hard by the border that is a frequent migrant corridor. A few of the aid workers have been ticketed, prosecuted, and convicted of littering public land. Their act of littering consists of leaving filled water jugs in places where thirsty people, sometimes at the outer limits of dehydration, can find them. Sentences ranging from a slap on the wrist to bankrupting fines and hundreds of hours of "public service"—picking up litter—have resulted. Other volunteers have run afoul of enforcement authorities by transporting injured or sick migrants to hospitals, acts of mercy that metamorphose into aiding and abetting criminals in the cold-eyed stare of the law. In such ways, the border makes lawbreakers of the innocent along with the guilty. In contemporary America it is one of the few places, perhaps the only place, where rebels young and old can simultaneously indulge both a passion for helping other human beings and a resentment, sometimes a hatred, of authority, of the nation-state, and of bullies with badges and guns.

Of course, not all of the people with badges are bullies. Most serve their country with integrity and diligence, but the border has a way of placing nearly everybody in a position of compromise. When the moral and legal landscape is as twisted as the physical topography of the Rat's Nest, you can't help but be on the wrong side of somebody's law. The moral and legal coordinates of the narcos running drugs and of the thieves and rapists who prey on the *migrantes* are easy to pinpoint, and the same may be said for many of the *coyotes* who guide the migrants and who are sometimes no better than thieves and rapists. The not inconsiderable cadre of bribe-taking badge-wearers on both sides of the line also belongs to the dark side.[1] But the many officers of the Border Patrol and other agencies, who enforce policies that guarantee human suffering, can find themselves, if they care to look, on shaky moral ground. The moral and legal terrain beneath the humanitarians who, by saving lives and mitigating suffering, effectively undermine those policies is also unsteady. It is like a dunefield in the desert, and no one staggers through it without losing his footing at least a time or two. The ranks of these players, however, are dwarfed by the legions of ordinary people—ordinary in their poverty, their usual decency, their struggle to preserve and benefit their families, and their hunger for opportunity—who trudge those shifting dunes and become criminals the instant they cross the international frontier.

CLIMATE CHANGE WILL stress the Southwest in myriad ways. Resource issues concerning water supply, forest management, even ecosystem collapse will be difficult and perhaps intractable, but they will be relatively straightforward: pay for this program, or pay for that; live with this outcome, or go somewhere else. An even greater challenge will be to contend with the consequences of cumulative stresses on human relations. When the "haves" don't feel they have enough and the "have-nots" are left with less than ever, how will they arbitrate their differences? As long as the economic and environmental tide is rising, it is easy to be a good neighbor. But when the tide goes out, fights break out all along the shore.

Climate change is already loosening the bonds of people to place by augmenting downpours and mudslides in El Salvador and bitter drought in Sinaloa. Under the prod of desperation and the pull of hope, people move. Sometimes it is the only strategy. You see it in sub-Saharan Africa. You see it in Bangladesh. You see it wherever people have abandoned played-out rural lands for the dubious attractions of megacities like Jakarta, Mumbai, and São Paulo. Mexico City's population now exceeds 18 million. Its water supply is questionable, its infrastructure brittle to nonexistent. Still people stream in. Others stream north. One recent study projects that declining crop yields, prompted by climate change, will induce 1.4 to 6.7 million Mexicans to emigrate northward by 2080.[2] Such figures, no matter how rigorously determined, rest on speculative assumptions and are open to debate. Nevertheless, it is reasonable to assume that the mounting stresses of the future will be most fiercely felt and most dramatically displayed along the existing fault lines of world society. And, as everybody knows, the mother of all fault lines in the climate-vulnerable Southwest of North America is the international border between the United States and Mexico. It is a veritable Great Rift of politics, economics, and culture. If you want to see how a nation responds to an accumulation of unresolved, complex, long-term, and interconnected problems, all of which climate change will worsen, take a look at what's happening along *la linea.*

THE BORDER HAS always been porous. At the contact point between San Diego and Tijuana in the mid-1990s, it was especially so. Every night as many as a thousand migrants surged across terrain guarded by a decrepit fence and no more than a hundred Border Patrol agents. Most got through. The relative few who got caught were taken back across the line and released, free to try again the next night, and the next, and the next, until they sprinted into something like freedom. Anyone with functional legs and a modicum of audacity could make the crossing. The Clinton administration decided to act. In 1994 it launched "Operation Gatekeeper," under which it hired thousands of additional Border Patrol agents and built a wall along the urban portion of the international boundary. The Gatekeeper strategy was soon applied at other ports of entry: Calexico in California, Nogales and Douglas in Arizona, Laredo and El Paso and other towns in Texas. The idea was to seal off major points of ingress with physical barriers and swarms of agents. With the urban areas bottled up, undocumented border crossers would be forced into the desert, where they would have to make their way across tens of miles of rugged, waterless, and roadless expanse. Word would get back, it was assumed: the desert is too dangerous; you can't make it; don't try.[3]

Part of the assumption was correct: word got back. But it didn't say what the architects of Operation Gatekeeper thought it would say. The word people heard was that the desert is tough but with a good coyote—a human guide and transportation organizer—you can still get to LA, Phoenix, or San Antonio in just a few days. It is hard, expensive, and risky, but it is worth it.

The idea that the desert would be a deterrent was flat-out wrong, and soon everybody on both sides of the line knew it. Nevertheless, the policy did not change. People kept coming across; they just crossed in different, more difficult, more dangerous places. What changed was the power and straight-up cost of the coyotes, the vulnerability of the migrants to robbery, rape, injury, and abuse, and the frequency of death and the certainty of suffering. All of these factors rose faster than a surveillance balloon.

And after the terrorist attacks of September 11, 2001, the idea that more agents and more walls were needed in the name of national security became an easy political sell, so the United States armored its border with more steel and more badges, and the coyotes and their charges were pushed into ever more severe stretches of desert. The logistics of the people-smuggling business became more demanding as the pickup points grew more remote, the coordination of rendezvous more delicate, the dodging past checkpoints and traffic stops more elaborate, the need for safehouses (where the migrants could be held hostage until their relatives paid the balance of their ballooning fees) more urgent. But business thrived and the migrants kept coming, because people who have nothing will do anything to get something—especially when the well-being of their children and families is at stake.

Nothing happens in a vacuum. It doesn't take an MBA to understand that when the customer base of a business is expanding and its logistical and capital requirements are commensurately rising, business operatives of increasing capacity are likely to participate in the boom. Which means the big boys shoulder their way in and capture market share. That's what happened on the border. One unintended consequence of the Gatekeeper strategy was to make the business of smuggling people, which was once a kind of mom-and-pop operation, attractive to the same ruthless and powerful organizations that controlled the movement of drugs from Mexico to the United States. It enabled the *narcotraficantes* to diversify their business plans.

THE NIGHT BRINGS a chill. We are crowded around a campfire stoked with chunks of juniper that one of the volunteers has hauled down from Flagstaff. The wood was buried in the back of her pickup beneath several cubic yards of bundled sweatshirts, jeans, socks, and other used garments. Most migrants need clothes, especially now that it is November. In violation of camp rules, I have uncorked some cheap red wine. The main group of volunteers will arrive toward midnight, in a few hours, and so far there are only a half dozen of us. No one seems to mind. The bottle travels the circle, but Nelson abstains. In fact, he is oblivious to the rest of us as we chatter, tell tales, and shift our camp chairs to avoid the swirling woodsmoke. Nelson has a textbook spread on his knees, and he hunches over it, reading by the light of his headlamp. He's got an exam on Monday.

Nelson is studying for his A.A. at the Pima Community College in Tucson. From there he will go for a B.A., probably at the University of Arizona, and after that his sights are set on earning certification as a Physician's Assistant. He says he is headed toward a career in medicine because of his experience giving aid to migrants in the desert, and giving aid to himself.

He came out to No More Deaths' "Bird Camp" near Arivaca in early 2008. "I was going through this time in my life where I was healing from a lot of stuff, like a lot of garbage, like I'd been pretty screwed-up previous." Nelson told me this the next day as we scrambled up a hillside on our way to Apache Pass. His voice projects none of the toughness of his tattoos. It has a sweet tenor sound, and every word is pronounced with care and fullness. "I'd gotten to this point where I'd realized that I needed to keep healing from whatever was going on with me, that I had to, like, participate in something that's a little bit larger than myself." The outdoor skills weren't a problem. He had jumped and ridden many a freight train, hobo-like, except he did it for the thrill, "and that's kind of like outdoors stuff because you pack your own food and you take care of yourself and you get wet and you deal with the environment." Plus, as a teenager he'd been in a wilderness drug rehab program "where you'd go off by yourself and find your reasons for staying alive and not getting screwed up on drugs."

Nelson arrived in Arivaca in 2008, just a few weeks after four other volunteers from No More Deaths had found the remains of a fourteen-year-old girl. They were taking a shortcut through a nameless canyon to deliver supplies to a water and food drop on a heavily used trail. The girl's name turned out to be Josseline Janiletha Hernandez Quinteros. She was from El Salvador, and she'd been dead about two weeks. Almost two years later, her name came up in virtually every conversation I had with volunteers.

Part of the intensity of Nelson's early experience with No More Deaths grew from treating medical emergencies: "Coming out here as a medically unskilled person was really scary because you run into people who are really fucked up, really frequently. Like you run into people who are dehydrated, going through hypothermia, having heart attacks, going into insulin shock, all kinds of things. It's scary to not know how to deal with them." So he first got certified as a Wilderness First Responder—a WFR or "woofer." "Then I realized how much I totally love knowing this stuff." Notwithstanding the long path before him, he dreams of earning PA certification and plying his skills not in well-ordered hospitals but in aid stations and field clinics in crisis zones.

Another big part of the intensity Nelson feels about his work grew from watching his friends recover from the trauma of finding Josseline. It was not a quick process. Her death became a touchstone. You could not avoid thinking about the despair and suffering of the two weeks that elapsed between the time her group abandoned her—because she was vomiting and too weak to go on—and the time of her death. You could not avoid imagining your sister, your daughter, or your friend's sister or daughter, in Josseline's green tennis shoes, starving and dehydrating in the canyon. Utterly alone. People talked it out, and the more they talked, the more they reassured each other that saving lives in the desert was the most important thing any of them had ever done.

On the last day of Nelson's initial stint with No More Deaths (he's since been back for over a year, spending roughly half of every week in the desert), he humped a sack of cement to the spot where Josseline spent her last bewildered days. Many others also hiked into the canyon that day, including Josseline's documented U.S. relatives: her grandmother, an aunt, a couple of uncles, and various cousins, some of whom were born in the United States and were therefore American citizens. But Josseline's ten-year-old brother, who had traveled with her up from El Salvador, surviving the trip, and her mother, whom they'd been trying to reach in California, could not attend, lest they be detained and deported by the Border Patrol. A priest said Mass, and

SHRINE AT THE DEATH SITE OF JOSSELINE JANILETHA HERNANDEZ QUINTEROS. *AUTHOR PHOTO.*

everyone piled stones into a yard-high monument, and they set a white cross with Josseline's name on it in front of the stones, in the cement Nelson had brought. Throughout the ceremony, everyone kept looking at a framed picture of Josseline that the family had set before a kind of makeshift altar. It showed a pretty adolescent girl with wide dark eyes and a hopeful smile.

IN GRITTY DOWNTOWN Nogales, where the Mexican city presses up against the Border Wall, there is a second cross dedicated to Josseline. From the U.S. side, you come through the old port of entry at the bottom of the canyon that the city fills, a canyon once graced by a stream lined with walnut trees—*nogales*. You turn right as you exit the Mexican customs facility, and on the far side of the bus stop, the art begins—hung, wired, welded, and painted onto the Wall. Some of it is merely average graffiti—white-painted slogans like *fronteras: cicatrizes en la tierra* (borders: scars on the earth) and *las paredes vueltas de lado son puentes* (walls turned on their sides are bridges). But much of the art, spread along nearly half a mile of wall, would stop you in your tracks if you saw it in a museum. Here, displayed in the open air above a line of parked cars, it seems otherworldly, a wild and colorful visitation. It is more polished and inventive than the garish painting that used to decorate the free side of the Berlin Wall. On the East German side, of course, there was no art, and there is also none on the U.S. side of the Border Wall.

The wall through downtown Nogales hails from the oldest generation of border-barrier architecture. It is made of steel panels, obtained from military surplus, which are hooked together and set on end. In World War II they served as portable landing strips on Pacific atolls. The steel is rusted and weathered, the color of defunct tanks and jeeps, and the panels make a brawny, industrial backdrop for the artwork they bear.

There are giant *milagros* (charms against illness and misfortune) in silvery stamped tin, depicting coyotes and dollar signs, trucks hauling cargoes of skulls, border patrolmen with baseball bats shouting imprecations in Latin. There are paintings, poems, prayers, and giant op-art photomurals made up of thousands of snapshots of people who have crossed the line—or died trying. There are tableaux of confrontations, curses, and indictments, sheet metal banshees and ghosts with riveted joints, and brightly painted human figures on fields awash with Aztec and desert Indian symbols. There are also crosses, hundreds of them, stick-thin and painted white, not two feet tall, attached to the crossbars of the Wall with the kind of one-way plastic fastener that riot police use for disposable handcuffs. Some of the crosses, including one for Josseline, bear a name, but most are as anonymous as grains of sand. As numerous as they are, however, the crosses cuffed to the Nogales Border Wall fail to represent the actual toll of the border.

Reliable data are hard to come by. Since 1998 the Border Patrol has kept count of the fatalities that its agents report, but their tally does not include deaths—like Josseline's—for which local sheriffs and medical personnel were the first official responders. The Border Patrol's tally through the first half of 2009 was 3,861 deaths among migrants attempting to enter the United States from Mexico. A more detailed analysis, cobbled from additional sources, including Mexican consular offices, academic studies, and other materials, estimates a total of 5,607 deaths for the period beginning in 1994, when Operation Gatekeeper commenced, and ending in 2008, the last year for which Mexican data are available. If you add the Border Patrol's partial-year count of 304 for 2009, you get an approximate figure of 5,911 fatalities during the Gatekeeper era.[4] (By comparison, the total number of deaths of

PHOTO MURAL ON THE MEXICAN SIDE OF THE BORDER WALL, NOGALES, SONORA, BY TALLER YONKE (GUADALUPE SERRANO, DIEGO TADDEI, AND ALBERTO MORACKIS), WITH THE COLLABORATION OF SOPHIA JANANTY AND SAMUEL RUIZ. *AUTHOR PHOTO.*

DETAIL OF MURAL. THE IMAGE IS COMPOSED OF THOUSANDS OF SNAPSHOTS OF INDIVIDUALS.
*AUTHOR PHOTO.*

U.S. military servicemen and -women in the course of both the Afghanistan and Iraq wars, begun in 2001 and 2003 respectively, stood at 5,309 in the first days of 2010.)[5] Deaths along the border have clipped along at an average of more than one per day since 1998. It is not a war—it's hard to say what it is— but it approximates a war's cruelty and loss.

One thing about which all observers of the border can agree is that the various official death counts, based as they are on actual corpses and recovered remains, are low. A lot of people die and don't get found. Partly this is a function of the ruggedness of the terrain; partly it is an expression of the biology of death. A well-trained woofer, for instance, will tell you that in the late stages of dehydration, people do strange things. They scramble, slide, or tumble down into canyon bottoms that not even a raven would visit. They curl up in the shade of a mesquite or a leaning boulder, where the blistering sun can't see them—and nobody else can either. To find them, you have to be extraordinarily lucky, if luck is the right word. Or you have to look everywhere, including the places where no human should ever have gone.

Dan Millis was one of the volunteers from No More Deaths who found the remains of Josseline. He's a soft-spoken former prep school teacher, now a border activist, who was drawn to humanitarian work for personal and ethical reasons. On February 20, 2008, he was leading a patrol of three other volunteers to cache food and clothing at a remote drop. Remoteness is a good quality in a drop site, he explains, because it means the goods cached there are less likely to be vandalized and destroyed. The shortcut he intended to take looked good on the map, although he knew of no one who had been that way before. Millis and the others bushwhacked over a ridge and down a prickly slope. At the bottom, the canyon walls were close on either side. They picked their way along a jumbled arroyo, heading "upstream," where there is almost never a stream. There wasn't much of what you would call a trail, either, and they struggled under awkward loads, especially a large box that would serve as a storage locker for the drop. Then they turned a corner. Millis was in the lead, and he thinks he was carrying the box, but he is not sure. What happened next erased part of his memory. "I was walking in front. First I saw some sort of bright green tennis shoes that someone had taken off." He called out to reassure any migrants who might be hiding nearby, "¡Hola hermanos, tenemos agua, tenemos comida!" He would have continued: *Don't be afraid. We come from the church.* "I started to yell out but I didn't get very far before I saw there she was. I told my friends to stop, you know, but they already saw her, too, by then. It was just a horrible moment."

MILLIS TELLS ALSO of a woman named Lucrecia who disappeared in 2005. Her son, who was crossing with her, made it to safety and was able to describe the general area in which she succumbed to exhaustion and probably dehydration, unable to continue with the others. Lucrecia's father obtained permission and came up from Mexico to look for her. Aid workers organized a search to assist him and stayed at it for several days. They found no sign of Lucrecia. But the father would not quit. He kept searching for weeks more, combing canyon after canyon, wash after wash. Ultimately he found his daughter, but before he found her, he also found the remains of several other victims of thirst and injury. Millis sums it up: "A man who spends a month walking up and down the washes finds three or four dead people. A [separate] group that decides to take a shortcut off the beaten path comes across a dead little girl.

And I think that speaks for itself in terms of how many dead people there are out there, how many people have perished and never have been found, and never will be found."

THE BREAKFAST LINE at the Centro de Apoyo al Migrante Deportado in Nogales forms early. Deportees stand quietly along the sidewalk, women in front, men behind, as Sister María Engracia and other Missionarias de la Eucaristía and several parish matrons bustle through the chain-link gate with food. The aid station is little more than a sheet metal roof shoved into a rocky hillside, with a cement floor underneath and chain-link fencing for walls. It is wedged so tightly between the street and the hill that the arrow of a traffic sign directing motorists to the United States actually projects inside the eating area, useless to motorists, who can't see it, and weirdly ironic to the deportees, who can.

A kitchen fills one corner of the *comedor*, and two bathrooms another. Unfortunately the needs of the deportees overwhelmed the bathrooms soon after they were installed, and they were shut down. They now serve as closets for spare clothing, medical supplies, and the audio equipment Padre Martín uses to address the crowd. No one seems to know where the overwhelming needs are now met. Martín, dark-eyed and serious, arrives with his fellow Jesuit, American-born Father Donald—*Pato Donaldo*, Donald Duck, he likes to call himself. He's well past seventy and has ministered to the poor in Latin America for decades. He seems irrepressibly merry.

The gate opens and the deportees file in. Some of them know the ropes— you can get two free meals a day at the *comedor* for up to two weeks after deportation, so quite a few stick around. But each morning usually brings a fair number of others who were hauled from a U.S. detention center the night before and deposited in Mexico by a Wackenhut bus. Most of them spend the night on the street, lucky if they can find or buy water, and when they first come to the *comedor* they are only beginning to get their bearings. Martín, speaking into the microphone, tells them where to sit, and when to get up and serve themselves. The fare is simple: chicken broth with potatoes, beans, rice, and tortillas. Crammed to its pipe-frame rafters, the *comedor* might accommodate 130 people. It holds almost that today. The men, who are more numerous than the women, sit at the tables on the noisy side of the room, toward the street. They crowd around long trestles and sit on benches and stackable plastic chairs. They talk quietly, listening to each other with a measure of animation. Their interaction fills the space with a hivelike murmur.

On the opposite side of the *comedor* toward the nonfunctioning toilets, the women, unlike the men, are utterly silent. They sit together but they are not together. Nearly all seem to be in their twenties and thirties. One holds a five-year-old girl on her lap. No one speaks. Their faces are solemn masks, and their vacant eyes are fixed in the same thousand-yard stares you see on soldiers returning from combat. Padre Martín leads the group in the Lord's prayer. Pato Donaldo follows with a blessing in heavily accented Spanish in which he says, "Remember, despite the horrible experience you have just had, you still have life, and so let us give thanks *al Señor* for the gift of life and for the dreams for a better life that remain in our hearts and will never be extinguished." Then the meal begins. The women, who barely glanced at the priests as they were speaking, eat as though the food has no taste.

All migrations attract predators. Pacific salmon run a gauntlet of sharks, orcas, seals, sea lions, otters, herons, bears, and fishermen. Human migrants seeking entry to the United States face an analogous bestiary. There are the *pandillas*, gangs like the feared MS-13, that prey on migrants from El Salvador to Sonora, climbing aboard northbound trains to rob the migrants who cling to the roofs of the freight cars. This is what happened to Mario, with whom I spoke at the *comedor*. He was there for the meal and also to seek help for a broken (and probably abscessed) tooth. He rode the roofs of boxcars from his native Honduras into Mexico. At one stop, on a rainy, moonless night, gang members climbed atop the train and ordered him off at gunpoint. They held him under guard in the glare of a strong flashlight, while they gathered other victims. When the flashlight wavered, Mario bolted for the trees. The pistol shots missed him. He doesn't know how he broke this tooth. Maybe just clenching his teeth. After hiding out a day or so, he hopped a train that rumbled slowly by, and continued his journey north.

There are also the *bajadores*, the bad guys who will drop you, take you down, take you *bajo*, lower you. And, of course, the *narcos* whom you just want to stay away from, and Los Zetas, a cartel that includes many ex-commandos, known to kidnap *migrantes* for ransom, and other *secuestradores* who not only kidnap for ransom but sometimes sell women into the sex trade. You also try to stay away from the dirty cops who shake migrants down for bribes, threatening detention if they don't get their money. With intense apprehension, you eventually have to deal with the *enganchadores*, "hookmen" who line up recruits for the coyotes, and then with one of the actual coyotes, who often nowadays prefer the term *pollero,* which loosely translates as "chicken wrangler" and evokes an image of herding a flock of human poultry through the desert. Even having run the gauntlet of all those predators, the migrant still hasn't crossed the border.

An important, unintended consequence of pushing migrant passage into increasingly hostile environments, besides killing more people than a medium-sized war, has been to reduce markedly what a sociologist might call the "circularity" of migration. Some aspects of border enforcement are undoubtedly less effective in keeping migrants *out of* the United States than in keeping them *in.* Before the Operation Gatekeeper era, nearly all the migrants were male, and they typically went home most years to see their wives, children, parents, and other kin, and to enjoy the texture of life in their home village or town. As the difficulty and expense of crossing the border have increased (a coyote fee of nearly $3,000 is no longer uncommon), the old practices have become impossible. Instead, an increasing number of successful migrants send for their families to come and join them in the United States, as Josseline's mother did. A consequence of this shift is that more and more women, like those in the aid station, and children, like Josseline and her brother, have attempted the crossing.

Migrants as a class are vulnerable. They are marks for thieves; and they are often friendless, adrift in unfamiliar territory, and forced to trust untrustworthy people. But clearly women and children, especially girls, are vulnerable in ways men are not. Perhaps the most chilling thing I heard at the aid station in Nogales was that rape is so common in the migration north that many women, anticipating the worst, start taking birth control pills before their journeys begin. And if not rape, then the obligation to give sex in exchange for the protection of a coyote or some other male figure, although the difference between that form of coercive sex and outright rape can be pretty slim.[6] Nevertheless, knowing all they know, tens of thousands of women, many with children in tow, continue to undertake the journey every year.

THE GLASSY STARES of the women in the *comedor* may have been the product of nothing worse than exhaustion and disappointment. One may hope. All of them, however, now faced a crushing decision: whether to go home in defeat—to a shack in Oaxaca, a farmstead in Guatemala, or any of a hundred thousand places—and to forfeit the money and tears they'd so far devoted to their dream of opportunity and (in all likelihood) family reunification, or to attempt again the trial that left them in their present depleted and disconsolate condition. Most coyotes, for their basic fee, allow multiple attempts at passage: if you are captured and deported in your first try, you are allowed a second, even a third. Some of the women in the *comedor* and many of the men would be on the trail again soon; indeed, later that afternoon, I spoke briefly with a small group of their compatriots, three women and two men, who waited close by the port of entry for their coyote, their loaded *mochilas* beside them. They had a long night ahead of them. Perhaps they would be crammed into impossible positions inside a car or truck, unable to move or to relieve themselves and scarcely able to breathe for many hours while their driver tried to pass the port of entry without getting searched. Perhaps they would be ferried out into the desert, there to start in darkness, scaling the Border Wall or scurrying through vehicle barriers, dodging the headlights and night-vision scopes of La Migra, plunging into an immensity of rock and thorn they cannot even see.

The trick in the desert is to move fast and keep moving. In the vicinity of Apache Pass migrants must cover at least thirty miles, as the crow flies, often more. Josseline, who may have used the trail through the pass, survived for about fourteen of those miles. If all goes well, the trip might take three or four days. Extreme thirst and dehydration are practically guaranteed: no one can physically carry enough water to maintain him- or herself for that amount of time over so great a distance, and especially not when spring-to-mid-autumn temperatures routinely climb above 100 °F. Yet still they come.

They come hustling across Apache Pass and other mountain gaps, down washes and up ridges in feeble moonlight, across a volcanic landscape littered with loose rhyolitic stones, the small ones like ball bearings sliding underfoot, the big ones easy to crack a shin against. Spraining a knee or turning an ankle is a constant hazard, and a turned ankle might lead to death: if you can't keep up with the rest of your group, you get left behind. In terms of equipment, your defenses are few. If you are a migrant, you are not wearing a sturdy hiking boot with ankle support. Even the least experienced tracker can tell the print of a boot from that of a tennis shoe. And *boot* means Border Patrol or hunter or humanitarian. *Tennis shoe*—low-cut, archless, easily pierced by cactus spines—

means migrant. Eventually, if you are a migrant, you will run out of water and the sun will boil your brains. Then you will come to a stock tank. It may hold only cockleburs and dried mud cracked into jagged plates. Or it may hold a few inches of tepid manure-flavored green slime that, in the moment, looks better to you than the coldest cerveza you ever sipped. But watch out: one swallow can bring on the shits, the cramps, and the heaves all at once. This may be what happened to Josseline, although she didn't have to drink slime to get as sick as she did. Dehydration alone, paradoxically, can bring on vomiting. It doesn't seem fair: you get dried out, then you sicken, losing the fluids that remain in your gut, and you dry out even more. Then you get left behind. Constant movement is the only path to success; you only shade up and rest when it is too dangerous to be in the open. This goes on for days. If you have to stop when the group needs to move, *buena suerte, compadre, y adios.*

THE CENTRO DE Apoyo is a joint venture. The sisters, the Jesuits, and local church members put on the meals. Volunteers from No More Deaths help stock the larder, bring clothes, and provide medical assistance, as well as other services. After breakfast is completed, Christa Sadler, a certified woofer and a No More Deaths regular, kneels on the floor by the toilets amid a clot of anxious people. She probes a needle under a deportee's toenail in order to drain a painful edema. The cure is also decidedly painful, but somehow the tension of merely physical pain (someone else's, to be sure) helps to banish the psychic gloom that hung over the meal. Sadler has already wrapped an ACE bandage around a badly sprained wrist (sustained in a fall on a desert trail), tweezered out a few cactus spines, moleskinned some blisters, and doled out all the *pastillas de dolor*—painkillers, in this case acetaminophen—in her kit. She promises to bring more to the afternoon meal.

I am talking to José, a thirty-something furniture finisher from Phoenix. He came north in 1998 and settled in the San Fernando Valley, where he learned his trade. He says he can match the color for any kind of wood, touch up antiques, prep whole kitchens. The burgundy splotches on one leg of his jeans look like a spatter of stain. In 2006 he and his wife, whom he met in California, and their three young children moved to Phoenix to be closer to his dad (his mother remains in Jalisco). There was more work in Phoenix, and better paid. They bought a condo. Their dreams seemed to be coming true. Then the recession hit. José lost his job. They fell behind on their (no doubt subprime) mortgage. They hung on, but barely—until José was pulled over by a traffic cop. He couldn't produce a U.S. ID or a visa or a green card, so he was sent to detention in what he described as a warehouse. While José was locked up, his mortgage holder foreclosed on the condo. His wife and kids moved in with his dad. After sixty days of detention, he was deported. He doesn't know when he will see his wife or kids again. He doesn't want to try crossing the desert. He says he will go to Tijuana and apply for a visa. He knows he won't get one. He looks at me as though searching my face for something he doesn't find. I offer the cheap, inflated currency of sympathy. He smiles wanly.

Then a commotion breaks out—a babble of urgent voices, benches and chairs scraping across concrete. A small riot seems under way in front of the toilets. I stand as tall as I can to look over the scrum of people pressing forward toward Sadler, who holds a bulging garbage bag and loudly protests in Spanish that she only has two hands. Finished with medical matters, she has been distributing clothing. Sweatshirts and jeans were handed out to whomever wanted them. Then she opened the garbage bag full of socks. People went nuts. They clamored for socks. She could have been doling out money, and they would not have clamored less. These are people whose feet hurt.

Cecilia sits calmly outside the storm of sock-seeking. She tells a story similar to José's, for she is another of the several hundred thousand "aliens removed from the United States" in 2009.[7] Her parents brought her to the United States as a child, nearly twenty years ago. For whatever reason, she was never "naturalized." Now she is grown and has four children of her own. All the family members she knows live in Arizona or Texas. Her Spanish is, at best, rudimentary. English is truly her first language, but now she's been sent "home," where she knows no one, can barely make herself understood, and is an utter foreigner. She aches for her kids. She doesn't know what to do or where to go. People tell her, "Trust in God." She says, "I was already doing that, just living my life." She wears the same kinds of clothes you see on middle-class moms in the mall in Tempe. By the look of her auburn hair, you know that not too long ago she visited the hairdresser. Her hands are soft; in spite of her time in detention, her nails look nice. The only world she knows is the one she has just been ripped out of.

Social theoreticians offer the term "structural violence" to describe the effect of a social institution that prevents people from meeting their basic needs. The simplest way to measure structural violence is to document the increased rates of death and disability suffered by people at the lowest rungs of society. Lack of medical care (when others have it), forced exposure to pollution, the involuntary obligation to put life and limb at risk or to work in systematically dangerous conditions—these kinds of burdens are expressions of structural violence. A border enforcement policy that subjects migrants to high rates of death and suffering is another example, which becomes especially notable when it persists long after its failure to deter migration has become clear. Agricultural and trade policies that prompt such migrations are another example of structural violence, and such policies are embedded in both NAFTA (the North American Free Trade Agreement) and the water and land "reforms" of recent decades that have favored commercial production over traditional subsistence farming in Mexico. By undermining the capacity of the rural poor to support themselves, these policies have in turn stimulated additional migration northward.[8]

The border itself, meanwhile, is an inherently violent "structure." It marks one of the steepest sets of social and economic gradients on the planet: material wealth and a white-dominated society on one side, poverty and a brown-skinned world on the other. The differences are as powerful as the opposing electrical charges within a thunderstorm. As long as they exist, there will always be a current between them, and people will move along that current, irrespective of the risks.

It has been said that for every complex problem there exists a simple solution, which is completely wrong. The simple solution to the United States' perceived border woes is to build a wall between it and Mexico. So far, over 600 miles of wall and vehicle barriers have been constructed along the 1,950-mile international border. The Border Wall (called a fence by its advocates) is a strange-looking thing: in many locations it is a rust-red curtain taller than a house running arrow-straight, down into canyons and up ridges, over mountains and across plains from horizon to horizon. At first glance, it looks like an overbuilt art installation—something Christo might erect to dramatize some esoteric concept. But the Wall is a curtain of steel, not fabric, and it is not made to come down. It is the ultimate refutation of the Statue of Liberty's welcome to the "huddled masses" of Europe, and it puts the United States, as a wall-building nation, in company that includes communist East Germany at the height of the Cold War, Rome under Emperor Hadrian, the Han Dynasty of China, and contemporary Israel.

Dan Millis took me to see the Wall southeast of Fort Huachuca, where the San Pedro River flows north from Mexico across the border. It is one of the prettiest corners of Arizona—or was, before Sierra Vista began overflowing with growth from the Fort. Millis wanted me to meet Bill Odle, who had come to the area a decade earlier and bought a fifty-acre parcel of land wedged up against the border and the San Pedro. With his wife, Odle built a solar house, off the grid, that faces south across a tawny plain toward his nearest neighbor, who lives in Mexico. At least the house used to look into Mexico. Now it looks at the Wall.

Odle meets us at his gate. He wears shorts, sandals, and a blue work shirt. Hair the color of gunmetal flows from under a faded ball cap of indeterminate original color, and his beard is more salt than pepper. There is a Marine Corps pin on one side of the cap, an Eagle Scout pin on the other. Odle is neither tall nor big, but his voice comes out of a thick round chest, and today, if the wind stays down, it might be overheard in Hermosilla. We drive in his truck to the Wall. A pistol in a ballistic nylon holster is Velcroed to the console between the front seats. "I am not anti-gun," he explains. "I'm anti-shithead, which is why I carry a gun around." Three decals decorate the back window of the cab: "Freedom Isn't Free," "Marine Corps," and "NRA." Fencing tools and a nearly exhausted spool of barbed wire clatter in the bed.

Odle served twenty-six months in Vietnam during the war. I ask him where. He rattles off "Dong Ha, Phu Bai, the Rockpile, all over I Corps." Nam stories, though, are old news. He jumps back to the present describing his present location as "twenty miles south of a Disneyland for Japanese, Germans, and Californians called Tombstone" and much too close to "the yuppie taco-deco ranchitos and golf courses of Sierra Vista."

The design of the Wall varies by location. In some areas in and around San Diego it consists of two or even three successive barriers separated by roadways along which Border Patrol vehicles ceaselessly ply back and forth. At Sasabe, Arizona, it is made from fourteen-foot steel pipes, each about four inches in diameter, set vertically in a concrete footing and topped with a horizontal beam. The pipes are spaced wide enough apart that you can get your hands around them, which is exactly what migrants do: you see handprints on the steel where people have climbed it; hands on one pipe, feet pressed against pipes on either side, they inch their way up and over in a basic climbing move any agile eight-year-old can figure out. Here at Odle's place, the engineers opted for a design based on mesh. A twelve-foot-high frame of

I-beams and box beams holds the mesh, which is a grid of steel rods an eighth of an inch in diameter. The rods are set too tightly together for a climber to get his fingers in the gaps.

Soon after this portion of the Wall went up, the Border Patrol began finding panels where the mesh had been cut and pulled back in sections as big as a garage door. The narcos had wasted no time. They would drive their trucks through, sweep away the tracks, and restore the mesh to its original position. Such breaches brought the contractor back to install a second barrier of I-beam posts and rails, immediately behind the mesh, to deter vehicles. This in turn has led the *narcotraficantes* to combine their mesh-cutting strategy with the use of vehicle ramps. And so the escalation of tactics continues.

For migrants, however, no escalation is necessary. They simply climb over, using hooks to defeat the tight weave of the mesh, or, more simply, ladders, or by climbing at corners or angles, where the Wall juts in or out a foot or two. "For a hungry twenty-year-old who wants to work," says Odle, "it's no more than a couple of minutes." And maybe a minute more for the older men, women, and even four- and five-year-old children who seem to cross the Wall and his land "apparently without much struggle." But not without the occasional splintered tibia. The orthopedic service at the Copper Queen Community Hospital in nearby Bisbee has seen cases triple since the Wall went up. The result has been more misery for the unlucky migrants, a crippling financial burden for the hospital, and another "unfunded federal mandate" that Odle resents.

He says that apprehensions by the Border Patrol in the Naco district where he lives totaled roughly 35,000 in the previous year, 2008. The portion of border patrolled from Naco is thirty-six and a half miles long, and the Wall was in place for the entire year. The figure of 35,000 reflects only the people who got caught—almost 1,000 per mile. If you stood them up along the Wall from one end of the district to the other, they would nearly be able to touch hands.

Odle, Millis, and Sadler, who was also there, were talking about Secretary of State Hillary Clinton's visit to Berlin, only the day before, to celebrate the twentieth anniversary of the fall of that city's wall, in 1989. Secretary Clinton had said many a glittering thing and made reference to unnamed other walls and barriers throughout the world that needed to come down. "There are no walls that cannot be torn down when people stand up and work together," she proclaimed to the listening world.[9] Odle was not amused by the irony. He shook his head and gestured toward the Wall. "It's a sad bastard," he said. "It doesn't work, and it causes damage to the image of what this country stands for."

Estimates of the Wall's cost vary—and continue to climb. Early sections were a relative bargain, averaging a merely exorbitant $2.8 million per mile. More recently, according to the Government Accountability Office, the cost of building 140 miles of Border Wall in the thirteen months ending in October 2008 jumped another 36 percent, attaining the more mythic cost of $3.8 million per mile.[10] An outlay of that magnitude equals $2,160 per yard, or $720 per foot, roughly the wholesale cost of a new refrigerator. If you extend that for 140 miles, you have the rough equivalent of a refrigerator every foot of the way from Washington, D. C., to Philadelphia, except that full-size refrigerators are three feet wide, so you would have to stack them three high—about the height of the Border Wall. The result would be, lo and behold, a wall of refrigerators lining I-95 through Maryland, Delaware, and a corner of Pennsylvania, an analogue to the work of American tax dollars in the desert.

DAN MILLIS, BILL ODLE, AND CHRISTA SADLER AT THE SAN PEDRO RIVER, NOVEMBER 2009. *AUTHOR PHOTO.*

During the same period the Department of Homeland Security also constructed seventy-five miles of "Normandy-style" vehicle barriers along the border. Their design is similar to the jetty jacks with which the Germans defended D-Day beaches. They consist of crisscrossed railroad rails or steel beams, arranged so that not even a bulldozer can push them down, let alone get over them. They were a relative bargain at only $1 million per mile, or $568 per yard—think high-end dishwashers as far as the eye can see. (Incidentally, the irony of the terminology seems to go unnoticed: Homeland Security freely uses the term "Normandy-style," perhaps not remembering that the original "Normandy-style" barriers were built by Nazis.)[11]

And then there is maintenance. The Congressional Budget Office estimates the cost of maintaining the physical infrastructure of the Border Wall at 15 percent of construction costs per annum, which means that within seven years maintenance will have drained more dollars from the Treasury than construction did. Given the haste with which the Wall was built and the inattention in its design to drainage and topographical issues—a summer thunderstorm in 2008 put downtown Nogales, Sonora, under several feet of water because the Wall impounded surface flows—the CBO's figure is undoubtedly low.

Odle and Millis are talking about these matters as we stand next to the Wall, toeing the dirt, nodding occasionally to a passing Border Patrol truck, when a light breeze begins to stir. An eerie moan rises like heat from the desert. It waxes and wanes, like voices singing a lamentation. I cannot identify the sound, and my bewilderment amuses Odle, who breaks out in a grin. Finally I get it: the breeze through the grid of steel rods makes the Wall moan, like a child blowing over the top of a Coke bottle. Only this bottle is scores of miles long, and its initial mournful seething promises to become a wail. It seems to come from everywhere and nowhere at once. I think of the mythic La Llorona wailing for the children she drowned. Odle says, "You should hear it when the wind really kicks up."

As we walk back toward the truck at the end of our visit, we see a doe mule deer trot down a low barranca and cross the Border Patrol road to the Wall. It noses around, exploring the structure, evidently searching for passage through it. The doe alternately peers through the mesh and turns to survey the road warily in both directions. Then she looks back over her shoulder toward the barranca. We follow her gaze and spy two yearlings watching her from the

THE BORDER WALL, NACO DISTRICT. *AUTHOR PHOTO.*

brush. After several minutes at the Wall, much too exposed and finding no passage, the doe turns around, recrosses the road, and begins picking her way along the draw at the foot of the barranca. Now the yearlings come down the slope—they flow through the thornscrub, tails flicking—and pick up the trail of the doe. We watch as the three formerly southbound deer make their way northward and melt into the scrublands.

With so many problems attending the Wall's cost and effectiveness and amid the fevered rhetoric that infects most discussions about "securing our borders," the environmental impacts of the Wall receive little notice. But they are substantial. In south Texas, the Wall fatally fragments the habitat of the last remaining U.S. population of endangered ocelots, despite tens of millions of dollars of federal and nonprofit expenditures in previous decades to secure it. In the bootheel of New Mexico, a line of "Normandy-style" vehicle barriers blocks the movement of the southernmost free-roaming bison herd in North America, for whom the border was the center, not the edge, of the world. The Wall and associated roads and barriers also negatively affect jaguars, pygmy owls (which tend to fly lower than the Wall is tall), pronghorn antelope, desert bighorn sheep, and scores of other animals, like the three mule deer we observed, all of which are accustomed to moving freely north and south through the borderlands.[12] Odle gave me pictures of mountain lion, porcupine, javelina, and more deer jammed up against the Wall. His voice boomed: "People say, 'Who cares about the damn deer?' Well, I do. I'd rather see deer than I would you."

By now we were in sight of his house, a low structure as brown as the landscape. Two flagpoles frame the front yard. One flies the Stars and Stripes, the other the banner of the Marine Corps. I ask Odle how he feels about living next to the Wall, how he likes having it as a neighbor. He hems and haws a bit, searching for the right words. Finally he says, "It's gotten to me. *Depression* is the easy word to throw out." Then gesturing broadly to take in the horizon-to-horizon span of the Wall, he deflects the conversation toward one of his practiced one-liners. "Look at it," he said. "It's an ugly damn thing, it makes a lot of noise, it does a lot of damage, it costs way too much, and it doesn't work—that's how you know it's a federal project!"

THE BORDER PATROL has become the federal government's largest law enforcement agency, far exceeding the FBI in the number of agents it places in the field. It now deploys more than 17,000 officers along the Mexican border (up from about 4,000 at the initiation of Operation Gatekeeper). "Apprehensions" in the Southwest reached a mid-decade peak of 1,189,018 in 2005 (almost half were in Arizona), and then began a precipitous decline—a measure, said the Department of Homeland Security, of fewer people getting through. By 2008 the number had fallen to 705,022, an overall drop of 41 percent, which DHS and congressional advocates of a muscular border policy have attributed to the efficacy of the Wall and beefed-up patrolling. They also point with pride to the steady increase in the number of criminals who are identified among those they apprehend.[13] The program to secure the southwestern border is working, they say.

But critics answer, "Not so fast." The growing proportion of criminals among apprehended migrants is particularly easy to explain. Beginning in 2005, pursuant to legislative instruction from Congress, DHS began requiring the criminal prosecution of all undocumented border-crossers, a matter previously handled under civil law. Most of these prosecutions have been accomplished by *en masse* court appearances of scores of migrants at a time, which combine arraignment, pleading, and sentencing into a single proceeding.

Virtually all migrants plead guilty and are thereupon convicted and deported. Besides swamping the courts and violating most jurists' idea of due process (the Ninth Circuit Court of Appeals has found that Tucson's mass hearings violate federal law), "Operation Streamline," as it is called, guarantees that more and more "criminals" will enter the United States because the migrants thus convicted will return to attempt entry again. As of 2008, nearly a fifth of "criminal aliens removed" by the Border Patrol owed their criminal status not to drug trafficking or crimes against persons, but to violations of immigration law. They were ordinary people looking for work or trying to reach their loved ones. The longer Operation Streamline continues, the more the proportion of returning immigration-law "criminals" will rise. It doesn't take much of a sleuth to see that the agency's increasing statistical success is self-reinforcing and that the numbers game reflects no real progress in crime prevention. Meanwhile, with legal resources monopolized by large-scale petty actions against people who pose no criminal threat, drug prosecutions have declined and prosecutions for human trafficking and other serious border crimes have remained essentially flat.[14]

The overall decline in apprehensions also deserves close analysis. Numbers began falling precipitously in 2007, as the U.S. economy teetered into recession.[15] It wasn't a coincidence. Apprehensions had also declined sharply, from an even higher level, following the dot-com bust at the start of the decade. According to Wayne Cornelius, the founding director of the Center for Comparative Immigration Studies at the University of California–San Diego, new migration from Mexico pretty closely tracks the availability of jobs in the U.S. economy. He, his students, and colleagues have conducted extensive interviews with migrants in both Mexico and the United States and have concluded that migrants are more worried about not finding work than they are about the difficulty of evading the Border Patrol. The decline in apprehensions in the late 2000s, he says, likely reflects a postponement, not an abandonment, of migration as a survival strategy, pending economic recovery in the United States.

Meanwhile, the U.S. recession has failed to produce a large-scale reverse migration, as might have been expected, as a result of migrants' losing their jobs. Cornelius attributes this to the "caging effect" of higher coyote fees and physical risk—direct effects of stepped-up border enforcement—that tend to keep migrants in the United States once they have arrived. Other factors diminishing the amount of return migration include Mexico's deplorable economic situation and the roots many migrants have sunk in U.S. society.[16]

Lamentably, the decline in apprehensions and presumed crossings has not been accompanied by a parallel decline in the number of migrants dying in the desert. The U.S. policy of pushing migrants into increasingly extreme environments appears to be maintaining the death rate among border-crossers at a robust level.[17]

ALMOST NOTHING ON the border turns out to be quite as it first appears. Things typically prove more layered, more contingent, more determined by context, more intertwined with everything else. But while the shapes of things change in the flickering light, at least two things remain clear and unyielding. The first is that as long as the economic gradient between the United States and Mexico (and other nations farther south) remains steep—relative prosperity on one side of the line, destitution on the other—large numbers of migrants will continue to travel north, irrespective of walls, patrols, and dangers. Desperate people will take desperate risks to provide for themselves and their families. A border and immigration policy that acknowledges this reality would presumably provide appropriate legal avenues for poor people from south of the border to seek the work they desire, and for employers in the United States to hire the labor they need, subject to enforceable restrictions. No set of arrangements can be perfect, but an effort toward improvement is decades overdue. Unfortunately, the volatility of immigration issues and the deepening divisions of the American political environment show no sign of moderating, and prospects for meaningful reform remain poor.

The second enduring reality of the border is that the most socially destructive features of the border world are shaped by a seemingly insatiable hunger for drugs within the United States. Notwithstanding the enormous sums of money and the incalculable hours of labor that have been expended in the name of the War on Drugs (a term first used by President Nixon in 1969), the market for illegal drugs in the United States is as robust and lucrative today as it has ever been. The transborder effects of that commerce are devastating. Essentially, from a macro perspective, the United States imports drugs and exports anarchy. The exports take two forms. The first is a flood of money, said to range from $18 to $35 billion a year, which is more than sufficient to change the course of governments. The second consists of thousands of lethal military-grade weapons sold across the border by American gun dealers on an annual basis.[18]

The exchange of drugs for anarchy is nothing new. Observers of not only Mexico but also Colombia, Peru, and other countries point this out with regularity, and there are plenty of statistics to back it up. The Government Accountability Office calculates that drug-related murders in Mexican border cities soared to 6,200 in 2008, an average of seventeen people per day, every day of the year.[19] Besides filling morgues, drug trafficking undermines civil authority throughout the northern Mexican states, where the capacity of all levels of government to maintain control is under constant challenge.

Rarely, but significantly, the violence has leaked north of the line. In March 2010, rancher Robert Krentz was shot dead while inspecting a remote part of his spread in far southeastern Arizona. Footprints of his assailant led toward Mexico along a known smuggler's route. A year later the investigation of the Krentz murder had yet to produce an arrest, but evidence suggested that a known narco committed the crime. The Arizona legislature, meanwhile, wasted no time before launching into a righteous frenzy. It hastily passed, and, before Krentz had been dead a month, the governor signed, SB 1070, a measure that obligates state and local police to determine the immigration status of anyone suspected of being an "alien." Ultimately the courts will decide if SB 1070 unlawfully intrudes on federal prerogatives or violates constitutional protections against unreasonable search and seizure, but no matter how one parses it, SB 1070 marked a new high in militancy on border issues. It was both the agent and the expression of heightened fears and hardened positions, further dividing white from brown, rich from poor, conservative from liberal.

Meanwhile, Krentz's neighbors and friends, the progressive ranchers of adjacent corners of Arizona and New Mexico, who nineteen years earlier had banded together as the Malpai Borderlands Group, "changed the way that we carry on our lives." The logo of the organization shows two stylized cowboys on skinny horses rising from their saddles to shake hands, a symbol of the openness of spirit that has guided the Malpai ranchers' efforts to protect the open spaces where they live and work. The death of Rob Krentz placed that spirit under attack. Writes Warner Glenn, a founding member of the group, "We no longer can afford to view the strangers that we see crossing our land with the trust that we once had. We now lock up our homes and vehicles even when we are close by. We view most unknown human activity with suspicion."[20] Thus does the acid of the drug trade corrode everything in its path.

AS LONG AS demand for drugs remains high, and as long as Americans have the means to buy what they want, the movement of drugs across the nation's borders will continue at full throttle. Efforts to stem the traffic have not slowed it so much as forced a mutual escalation of tactics and strategy. They've had the paradoxical effect of causing the traffickers to consolidate their efforts and to become increasingly centralized, powerful, and violent. A much-debated solution would be for the United States to legalize drugs—at least marijuana—and to weaken the *narcotraficantes* by appropriating their market. The strategy has sparked intense debate (advocates on both sides cite a 2010 Rand Corporation study as supporting their case),[21] but in the meantime, decisive action remains remote, and the carnage continues.

If the complexity and difficulty of border issues might be seen as proxy for the equally resistant problems that climate change is likely to bring, then it is hard indeed to view the future with optimism. In both cases, the underlying causes—drug demand, the economic gradient, a warmer, drier, more erratic climate—are large, impersonal forces. They possess an implacable momentum that no amount of barrier building or declarations of "war" can reverse. One might as well, like King Canute, attempt to command the tides.

Adapting to climate change will require hard choices in the allocation of water and the design of cities. It will force reevaluations of land use and forest management. It will prompt the movement of large numbers of people all over the world, and the largest movements in the climate-driven future of North America will be, as they are now, from south to north. The stresses of climate change will be most acutely felt along the existing fracture lines that run through continental society. Where water is concerned, the division between rural and urban interests—essentially between agriculture and the cities—is one of these. But the deepest and widest fault line fracturing continental society is the U.S.–Mexico border, and the potential for seismic activity, as the strains of climate change mount, is greater along that arbitrary but rigid line than anywhere else.

It may be that the history of no other region of the Americas is as violent and militaristic as this border's. To the continuous warfare of the colonial era one may add the Texas War of Independence (1835–1836), the U.S.–Mexican War (1846–1848), the Apache wars (at their peak in the 1870s and '80s), the Mexican Revolution (1910–1920), and the current war of drug cartels fighting multiple law enforcement agencies and each other. Add the occasional filibuster like William Walker, who tried to seize Sonora as his personal colony in 1853; scalp-hunters like John Glanton (who provided real-life inspiration for Cormac McCarthy's *Blood Meridian*); and a continuous parade of border bandits dodging back and forth between jurisdictions, and nearly every historical period is filled in. Depending on your point of view, you might also include a few score of quick-to-shoot Texas Rangers and several brigades of Mexican Federales, the whole saga leading finally to the present Maginot Line reality of a long wall defended by over 17,000 armed police, a force the size of an army corps.

This is a history in which ethnic prejudice and casual brutality are constant themes. Volunteers at the aid stations in Nogales and Naco have interviewed scores of deportees about their experience in Border Patrol and ICE custody. Based on these interviews and their own observations, they have documented hundreds of cases of physical and mental abuse.[22] The offenses range from cursing and bullying to withholding food, water, and medical attention. One aid worker told me of a fourteen-year-old boy who fell into a cactus attempting to flee the Border Patrol. His injuries received no attention during several days in custody, and when he appeared at the aid station in Nogales the spines had begun to fester. She and her coworkers set about removing them. With their work only partly done, they stopped counting at 700 spines.

The litany of inflicted miseries, small and large, is a long one: family groups split up, diabetes or blood pressure medication confiscated and never returned, cells too crowded to lie down in, sexual taunting, no access to toilets, buses with the air conditioning cranked to wintry levels—and all jackets (except the driver's and guard's) taken away and locked in baggage bins. But no one should be surprised. Not every Border Patrol officer is a sadist, but a job that involves wielding power over powerless people is the kind of job that nurtures the trait, which becomes amplified by inevitable frustrations. Your task is to chase people—to hunt them, really. They give you the slip. You get pissed. You get tired. You work in the heat and in the dark. You are a little bit scared. You never know when you might come across a narco with a gun. And you're as thirsty as anybody else; your knees ache; your nerves are frayed. And then your partner, or the bus driver or the guard at the detention facility decides to give them an extra reason not to come back across the border uninvited again. "A lesson they'll remember," he says. What are you going to do? Do you say anything? And after you've played this interminable, self-repeating cat-and-mouse game for two or three or four years, are you going to keep on doing it without ever taking your hands out of your pockets?

Under pressure, people lash out; they take shortcuts, loosen their grip on self-control. So do nations. One of the most disturbing features of the U.S. mania for wall-building is the willingness of Congress and government administrators to suspend the rule of law in order to erect the barrier (and spend the appropriated funds) as quickly as possible. In 2005, Congress passed the Real ID Act, as a rider attached to an appropriations bill. It was a potpourri of measures having to do with driver licenses, visas, and security issues.[23] One of its provisions granted the Secretary of Homeland Security authority to waive any and all state, federal, or local laws if, in his judgment, adherence to the law might delay construction of the Wall and its roads.[24]

On multiple occasions in the ensuing years, Secretary of Homeland Security Michael Chertoff invoked this authority to speed construction of the Wall and related infrastructure in various locations.[25] He suspended application of about three dozen federal laws as well as, "in their entirety, all federal, state, or other laws, regulations and legal requirements" that might be linked to them. All in all, the suspensions constituted an unprecedented vacation from the rule of law.[26] The laws that were shelved were intended to protect endangered species, water and air quality, archaeological resources and historic structures, Native American interests, wilderness areas, wildlife refuges, religious freedom, coastal zone management, and the public's right to question and appeal federal actions. The suspension of these laws implicitly argued that no amount of environmental, cultural, or procedural injury could outweigh the immediate benefit of building the Wall—an ambitious claim for a porous and ineffective "billion-dollar speed bump," as some critics have called it.

As the world warms, flood, storm, and drought will drive populations into motion, challenging borders around the globe. Already India worries about Bangladesh, Spain frets about Morocco, Morocco mistrusts the other nations of the western Sahara, and Egypt eyes its sub-Saharan neighbors with concern—the list could go on at length, and it will grow longer as competition heightens for safety, shelter, and ever-scarcer resources.[27] Faced with swelling ranks of displaced people, more and more communities, from the national to the local, will debate the questions that were undoubtedly argued at Sand Canyon Pueblo seven and a half centuries ago: do we find ways to accommodate the new waves of homeless, or do we fend them off? Lately the United States has pursued the latter strategy at exorbitant cost while distorting its legal traditions to an unprecedented degree. If the test of character for individuals is to remain true to their ethics and ideals even in times of trial, it doesn't seem much of a stretch to suggest that the same may be true for a nation.

# 9

## MOGOLLON PLATEAU: FIRES PRESENT
## AND FUTURE

EARLY ON JUNE 19, 2002, Paul Garcia looked off the rim of the Mogollon Plateau and did not like what he saw. Down toward Cibecue, the capital of the Fort Apache Reservation, home of the White Mountain Apaches, dark smoke boiled into the Arizona sky. The wind was pushing it in Garcia's direction, toward the rim, as the prevailing southwest wind always pushed fires that start down on the Rez. The churning smoke—dark-tinged because of solid materials that volatilized without burning—told Garcia that the fire was gaining energy, building strength. He was the fire management officer of the Lakeside Ranger District, a unit of the Sitgreaves National Forest. His boss, a couple of steps up the chain of command, was Forest Supervisor John Bedell, who remembers getting a call from Garcia: "He said, 'You know, this thing has some potential.... If they don't catch it today, it's going to get pretty big.'"[1]

The firefighters on the reservation didn't catch it. The Rodeo Fire, which began as an act of arson near the Cibecue rodeo grounds, grew from a size of 1,000 acres on June 18 to 55,000 acres the next day. Garcia, Bedell, and a burgeoning army of Forest Service firefighters scrambled to meet the fire atop the rim, hoping to hold it at the rim road that marked the boundary between the reservation and the National Forest. They did not succeed. By mid-afternoon the fire had developed multiple towering plumes of smoke and ash. Its front advanced at an average rate of four miles an hour. Whole stands of eighty-foot trees ignited in an instant, shooting flames 400 feet high and lofting aerial firebrands half a mile downwind. By 4:00 p.m., some of those firebrands were spotting across the rim road.[2]

The Mogollon Rim is one of the most pronounced topographic features of the Southwest. (It is also one of the most mispronounced; in local usage people say *muggy*-yon, a far cry from how the landform's namesake, Spanish colonial governor Juan Ignacio Flores Mogollón, would have introduced himself.) To gaze off the rim, as Paul Garcia did, is like gazing into an abyss a half-mile deep, except that there is no far side to the abyss. The land drops away and does not climb again. The bottom country recedes into the unlimited distance, rough and broken, colors bluing, shapes and details dissolving into the dry haze of faraway desert. The rim is the southwest edge of the Colorado Plateau. Massive faults along the edges of the plateau, over millions of years, caused lands to the south and west to fall away, leaving behind a vast tableland that tilts gently northward. Since the close of the Pleistocene, that tableland has supported the largest ponderosa pine forest on the planet.

The forest sprawls across the southern end of the plateau and spills off the rim, feathering out among the canyons of the Apache reservation. In the past, fires that ignited in the low country typically burned toward the rim in long narrow strips. They were driven by the wind, steered by the canyons they followed, and energized by the steep rise in elevation. (A fire climbing a slope heats the fuels ahead of it and moves much faster than it would on flat terrain.) On top of the rim the fires tended to keep their linear shape, advancing more slowly down the tilted gradient, until they petered out in decreasing densities of fuel or because firefighters, taking advantage of their relative somnolence at night, succeeded in containing them. That was how it used to be.

But the two previous winters in central Arizona had been almost bereft of snow, and the other seasons in those years were close to rainless—this was the same "global-change-type drought" that led to forest and woodland die-offs elsewhere in the region. As a result, the moisture content of potential fuels, from living trees, blowdowns, and accumulated logging slash to needle litter, grasses and shrubs, was perilously low, so low that a post-fire report by the Forest Service characterized the moisture levels as *unprecedented.* Even the heaviest fuels, the decaying stumps and thick logs on the forest floor, which normally hold moisture in their moldering cores, were tinder-dry and ready to burn. And the dehydrated foliage of the trees, rich in volatile terpenes, was a kind of botanical gunpowder, only a spark away from exploding.

A second factor was also unprecedented. Besides being desiccated, the fuels were more abundant than ever before in the history of the forest. This glut of combustible material was a human artifact, the result of extensive grazing and nearly a century of aggressive fire suppression. Grazing removed fine fuels like grasses from the forest floor and thereby prevented natural "light" fires that would have kept the forest thinned. Fire suppression finished the job and guaranteed the ceaseless accumulation of material. In essence, the forest had been converted from an open, savannah-like environment, where in the late 1800s one might have encountered 50 to 100 trees per acre, to a piney jungle in which counts of 1,500 or even 3,000 stems per acre were common. The highest densities consisted of thickets of "doghair"—spindly, scrawny ponderosas only a few inches in diameter (although often more than a half-century old), growing in impenetrable clumps. Many of these patches were an inheritance from extensive logging—places where taking out the big trees had left a lot of sunny, fertile ground for seedlings to take root. When touched by flame, a dried-out doghair thicket could go off like a Molotov cocktail. Many old-style logging operations had also failed to dispose of their slash, leaving tops, branches, and other detritus jumbled where they fell, as though somebody had started to build a bonfire but quit. Not all the rim country was in bad shape, but enough was that if you were looking for an ideal playground for holocaust-level fire, you'd say to yourself, "This is the place."

INITIALLY WIND-DRIVEN, THE Rodeo Fire gained strength as it grew and began to generate its own weather. It drove dense columns of hot, turbulent gases into the sky. The smoke plumes spiraled, tornado-like, whipping up winds many times more forceful than anything outside the fire. For as long as the fuels beneath a given plume lasted, the column kept rising—up tens of thousands of feet, to altitudes where commercial jets fly. But as the furiously burning fuel on the ground approached exhaustion, the upthrust of heat and energy slackened, and the plume began to collapse. Cold air rushed down the outside of the weakening column. The downdrafts grew into violent cataracts of air that crashed into the fire still burning on the ground, driving burning material outward in all directions. Some embers traveled a mile or more on such winds, and when they landed, say, in a patch of brittle doghair, the flames flared upward in long tongues, igniting the canopies of the overstory trees. Pushed by the gale of the falling plume, flames spread rapidly from treetop to treetop, birthing a new crown fire, which soon burned together with other, similar ignitions, their effects cascading, so that from the turbulent collapse of one plume, another soon roared into the air.

A plume-dominated fire is the most dangerous kind to fight, for it spreads suddenly and unpredictably. Firefighters cannot approach it lest they be engulfed in the chaos of a plume's collapse. Repeated plumes can develop, one after another, several times a day. Usually, however, cooler temperatures arrive with nightfall, and the fires calm down.

On June 20, driven by renewed wind and rising temperatures, the Rodeo fire surpassed its fury of the day before, and consumed a staggering 70,800 acres—more than 110 square miles, as plume after plume erupted and collapsed, spreading storms of fire in all directions. Previously the largest documented fire in Arizona history had topped out near 70,000 acres. The Rodeo Fire equaled that in *a single day*.

And the bad news kept coming. Two days earlier, Valinda Jo Elliott had run out of gas while driving in the Fort Apache backcountry. Increasingly distressed, she scrambled from one patch of high ground to another seeking cell phone reception. On June 20 she was on the slope of Chediski Peak, northwest of Cibecue, when she spotted a news helicopter on its way to report on the Rodeo Fire. Ms. Elliott resolved to attract the chopper's attention. What to do? She hastily built a signal fire.

Her signal attracted plenty of attention. By the end of the day, the fire she started—the Chediski Fire—had burned 10,800 acres and exhibited the same extreme, plume-dominated behavior as the Rodeo Fire, fifteen miles to the east. Bedell and other experts soon concluded that it would be impossible to contain the two fires separately, for there was no safe way to place crews between them. They would inevitably, unpreventably burn together and become a single conflagration, which indeed they did on June 23, creating a fire front of burned and burning landscape more than thirty miles across.

Bedell's recollection of the fire's eruption is essentially a litany of successive calls for more firefighters and equipment, from the initial Type 3 Incident Management Team to a Type 2, a Type 1, then a second Type 1, and a third, and ultimately a fourth. Each team was a self-contained operational unit, like an infantry battalion. Meanwhile, orders went out to evacuate nearby towns, like Heber and Overgaard, and even Show Low on June 22, putting thousands of people on the move. Quite a few of them came from tucked-away, ex-urban subdivisions that snuggled into the forest along Highway 260. Some of these encroachments into the woods were trailer parks; some were high-end getaways, like Bison Ranch, "your western resort hometown." One cluster of quaint but hardly rustic log homes boasted its own airstrip and catered to owners of private airplanes. Its residents may have flown to safety before the

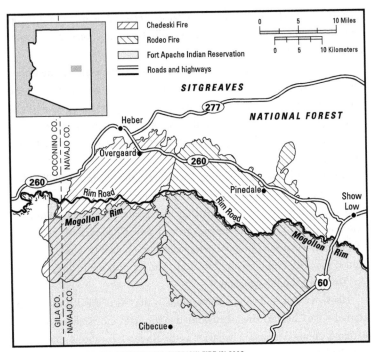

**MAP 9:** THE AREA AFFECTED BY THE RODEO-CHEDISKI FIRE IN 2002.

fire shut down their runway, but most families, rich and poor, piled their photo albums, bags of clothes, tax records, kids and pets, and grandma's silver hairbrush into the family car and crept away in grim lines down congested highways, head-lights on in the daytime, the air dark with ash and bitter to taste.

As the cars streamed out, the politicians streamed in. Every elected official in Arizona and some from Washington, D.C., including the president, uselessly arrived at the command center to show their public concern. They received briefings from sleep-deprived men and women who otherwise struggled to meet the logistical needs and coordinate the efforts of the still-arriving armies of firefighters. Water, food, bedding, and latrines had to be provided, commu-nications and mess tents set up, vehicles gassed, medical stations established, plans developed, maps distributed, priorities assigned, and on and on.

At the peak of the fire, some 4,447 personnel were deployed, not so much to contain the fire—which was essentially impossible—but to protect homes and businesses. Eleven air tankers bombed the fire with fire-retardant slurry. Twenty-three helicopters dumped water and performed reconnaissance; 245 fire engines rumbled into the fray, plus 89 bulldozers and 95 water trucks. Miraculously, although there were the usual injuries, no one was killed. Still the Rodeo and Chediski fires raged. Together they raced through more than 100,000 acres during the single twenty-four-hour span of June 22.[3] That rate of combustion works out pretty closely to the area of a football field going up in smoke every second, sixty football fields every minute, all day and all night long. Think of a steel mill on round-the-clock shifts, except that the blast furnaces have no walls and randomly consume the countryside. Bedell is a veteran of forty years in the Forest Service and throughout his career has battled scores of fires, large and small. As he recalls, "I never saw anything like it.... I'd never witnessed fire behavior like this where we had multiple plumes. I would say there were times when 50,000 acres would just burst into flame. I mean, it was unbelievable."

At the apex of the fire, night provided no relief. "The biggest problem was the nighttime temperatures stayed up. I think we were averaging 85 to 90 degrees in that late June period. We might have hit 100..., and the humidity was single digits and at night it never recovered." And the plumes didn't quit even when the sun went down. "There wasn't anyone that had witnessed these multiple-plume fires at 10 and 11 o'clock at night... but this thing didn't behave like normal fires do.... This thing burned all night long in all directions, multiple plumes."

Against the night sky, the plumes were an eerie, angry echo of the Northern Lights, an incandescent column of red reaching into the heavens. "I can remember when they evacuated Show Low, we went out there, and I mean, those plumes, Bill, were just glowing. The plumes were happening in the middle of the night."

A shift in the wind ultimately saved Show Low, but the fire's appetite raged on: 39,000 acres on June 23; 43,600 on June 24; 35,000 on June 25—by which time the onslaught was eight days old and had burned more than 400,000 acres. If you took that area and packed it into a neat square, it would be exactly twenty-five miles on a side, enough space to hold the city of New York—twice. The fire still had much farther to go. It would top out at just under 470,000 acres and would not be considered "contained" for another week and a half—but after June 25, the daily quantity of landscape it consumed tapered off from apocalyptic to merely catastrophic. According to Bedell, "suppression action accounted for about 40 or 50 percent" of overcoming the fire, "and the weather was the rest, just like it always is when a fire's that big." (The Rodeo Chediski fire's record-setting 470,000 acres seemed unsurpassable until the Wallow Fire of 2011 came along and scorched nearly 540,000 acres.)

After the land cooled down enough for damage to be assessed, the tax assessor for Navajo County tallied 259 trailer homes, 206 nonmovable residences, 6 businesses, and 20 barns or garages destroyed, for a total of 491 structures. The toll was severe, but might have been much worse. The thousands of firefighters deployed to the effort are credited with saving over 2,000 homes. Their efforts cost the government somewhere between $43 and $50 million, but these expenditures represent a relatively small portion of the fire's ultimate cost. When you include property losses, rehabilitation work—from seeding grass and stabilizing soils to rebuilding fences, even rescuing downstream fish—and then add in forgone sales taxes, the cost of job losses (two sawmills on the Apache reservation shut down), public health assistance, and still more damages, you pretty soon top $308 million.[4] And those are the costs that can be quantified. Other costs are intangible, but no less hard to bear.

FIRE IS THE apotheosis of contingency. Nothing better embodies the chance interactions of multiple variables—wind direction and speed, temperature, topography, time of day, fuel abundance, fuel moisture, and fuel type. All of these factors conspire in ways that are highly specific to time and place. Fuel structure is important, too. If "understory" trees are present, ground flames can ignite their lowest branches, and fire can "ladder up" through their tops to reach the canopies of the taller, "overstory" trees. In this way, "crown fires" are set in motion. Change any of these factors, sometimes by only small degrees, and the result of a given ignition can vary from broadly beneficial to ecologically and socially calamitous.

Something caught my eye as I drove along the east edge of the Rodeo Fire in March 2010. I pulled over to the edge of the highway and waited. The movement I'd seen turned out to be a turkey, picking its way through tall grass. In moments, thirteen more birds emerged. They stalked the meadow, probing the ground, snapping at the seedheads of the bunchgrasses. It was almost eight years after the fire. The trunks of the tall trees were still charred, to a level past head height, but their foliage was luxuriant. Here and there stood a dead snag, and a few black timbers lay crisscrossed on the ground, half swallowed by grasses and shrubs. Little time passed before a band of four mule deer stepped gingerly into the meadow. They were trailing the turkeys, using them as sentinels. They drifted along the turkeys' path, silent, ears twitching, browsing bushes of what looked to be mountain mahogany. Here, in this small patch, the fire had clearly reset the dynamics of the ecosystem in a positive way.

But that was the exception, not the rule. The Rodeo-Chediski fire destroyed the forest on more than half of the land it touched.[5] Up on the north side of the fire, along the highway from Show Low to Overgaard, the legacy of the fire is unremittingly negative. For mile after mile, and to the top of distant ridges as far as the eye can see, the forest is simply gone, vanished. The rolling landscape feels more naked than open, its unwanted vistas whiskered by the black husks of former trees and by thousands upon thousands of knee-high junipers, which sprouted after the fire. What was once a great and sprawling ponderosa pine forest is now a kind of failed prairie, a patchwork of

CROWN FIRES IN PONDEROSA PINE FORESTS CAN PRODUCE ECOLOGICAL "HOLES" TOO BIG FOR RECOLONIZATION BY PONDEROSA. THE RODEO-CHEDISKI FIRE OF 2002 PRODUCED SUCH A LANDSCAPE ON AN IMMENSE SCALE. VIEW SOUTHWEST TOWARD MOGOLLON RIM FROM NEAR HIGHWAY 260, MARCH 2010. *AUTHOR PHOTO.*

bare soils and struggling grasses, soon to succeed to scrub fields. Only the elk are pleased; their sign is everywhere.

THE RODEO-CHEDISKI FIRE was by far the largest incendiary disaster of the Southwest's turn-of-the-century drought, but it was by no means the only one. Ten days before flames licked out from the hills behind the Cibecue rodeo grounds, a distraught Forest Service employee in Colorado lit some paper in a campground that had been closed because of drought and fire danger. She claimed to have been burning a letter from her estranged husband. In later legal proceedings, investigators would argue that she started the fire so that she could put it out by herself and appear a hero.[6] Whatever the truth may have been, Terry Barton failed to extinguish what she started, nor was anyone else able to do so for the next two weeks, not until the Hayman Fire had become the largest in Colorado history, burning 137,759 acres of the Front Range southwest of Denver. The bill for this fire included $42.3 million in suppression costs, the loss of 132 incinerated homes, and six deaths. (Five firefighters perished when their vehicle crashed, and an asthma victim was felled by the smoke.)[7]

Two years earlier in New Mexico, the Cerro Grande Fire had seemed gargantuan, utterly calamitous, capturing headlines and dominating news programs throughout the country, but the Rodeo-Chediski and Hayman fires dwarfed it. Still, the Cerro Grande scorched 43,000 acres and triggered the evacuation of 18,000 people from Los Alamos—both the city and the national nuclear laboratory. It also left 400 families homeless. Damage costs were exceedingly well documented and approached $1 billion.[8] While the Cerro Grande Fire burned mainly on National Forest lands, its proximate cause was a prescribed fire on a nearby unit of the National Park System. The prescribed fire's bitterly ironic purpose was to *reduce* fuels and to help safeguard Los Alamos and the National Laboratory from wildfire.[9] The irony does not stop there. Interestingly, the flames that raced across the eastern front of the Jemez Mountains and reduced part of Los Alamos to ashes were not descended from the original prescribed fire. Their genealogy traced to a backfire set by a Forest Service crew to contain the prescribed burn after it got out of hand. And still more: when the fire broke out, a highly relevant Forest Service document had just gone to the printer. It was the environmental analysis for an aggressive program to reduce fuels in the forests above Los Alamos. For obvious reasons, the report was never distributed.[10]

These three notorious fires and many less famous ones share important characteristics, which together define the contemporary wildfire environment in much of the American West. It is an environment unlike any in the past.

First, the fires were started by humans. *Homo sapiens* are fire-wielding creatures, and in today's West, people are everywhere, seemingly all the time, whether they are supposed to be or not. Second, unnaturally high fuel loads and overdense forest structure, principally in ponderosa pine and mixed-conifer forests, were essential to the fires' virulence, and those conditions were the result, not of nature, but of human activity, of culture. Third, before the burned lands were visited by uncontrollable fires, they were first afflicted by extreme drought, which is an expression of climate.

LET'S PUT ASIDE for the moment the issue of human ignition. People do stupid things, and the more people there are, the more stupid things get done. It is an independent variable, not closely linked to the other two.

On the other hand, any attempt to understand the relative contributions of climate and past cultural practices leads to an abundance of thorny and, from the standpoint of public policy, important scientific questions. A principal tool in answering them has been the study of tree rings—the field of dendrochronology, which is said (at least by southwesterners) to be the only science native to the Southwest. A. E. Douglass and other pioneers, working in the early decades of the twentieth century, developed it as a way of dating archaeological sites.

PLUME OF THE CERRO GRANDE FIRE, JEMEZ MOUNTAINS, NEW MEXICO, MAY 10, 2000, OVERLOOKING FRIJOLES CANYON, BANDELIER NATIONAL MONUMENT. *COURTESY OF CRAIG D. ALLEN, USGS.*

An unlikely location for much of what has been accomplished in the name of dendrochronology is the giant football stadium at the University of Arizona in Tucson. Tom Swetnam, for many years the director of the university's Laboratory of Tree-Ring Research (LTRR), likes to point out that the massive concrete pillars that support the stadium's upper deck are roughly the diameter and height of the giant sequoias on which he did some of his early fire-related research. He also enjoys mentioning that the lab's cavernous offices, which fill a wedge of space under the stadium's west side, can boast of "restroom facilities for 50,000."

Swetnam, bearded, burly, and (when he's not fending off draconian budget cuts) unruffled, occupies a windowless office deep under the stadium. If he ever minded having a hundred thousand tons of concrete for a roof, he's gotten over it. In the shadowy parking area outside the Tree-Ring Lab's main door, you might hear strange, wind-warped chords from the university's marching band practicing on a nearby field, but inside Swetnam's office there is the silence of a mausoleum.

We talk for a while and then go upstairs to one of the labs, a crowded room layered with sawdust and redolent of pitch and polyurethane. Swetnam picks up a small cross-section of tree trunk—a demonstration sample from Connie Woodhouse's reconstruction of Colorado River flows—and wipes away the dust. Some 800 tree rings comprise those nine or ten inches of wood, many of them lines as fine as hair. The tree lay dead a long time before Woodhouse found it and stood as a dead snag long before that. Perhaps a century of rings had weathered away. The datable record ended in the seventeenth century and reached back almost to A.D. 800. A small white arrow indicated the Great Drought of the late thirteenth century. There the tree rings were so thin as to disappear entirely before the naked eye; only a microscope would reveal them.

Tree-ring research is fundamentally obsessed with climate. It is based on the fact that a tree, as a long-lived organism with a durable structure, records the changing conditions of its life. Wet years and dry years, and in some cases variations in temperature, cause trees to grow faster or slower, and the amount of growth is recorded in their annual rings. Thick rings reflect favorable conditions, thinner ones stress. Sometimes a tree's growth can be traced to the period of the year in which it occurred—whether it put on wood early or late in the growing season. By matching patterns of thick and thin rings in multiple trees, researchers from A. E. Douglass onward have assembled chronologies of tree rings stretching over millennia. The chronologies, which exist because tree rings distinguish good years from bad, are essentially proxies for climate.

Douglass, the founder of the LTRR, started out as an astronomer. He studied tree rings hoping to detect a record of sunspots and other solar activity. After an ugly break with his mentor, Percival Lowell (over Lowell's embarrassing insistence that Mars was laced with gigantic canals), Douglass directed his attention to the roof beams and lintels of southwestern archaeological sites. By 1929, he had compiled a chronology reaching back to A.D. 700, and unlocked the means to date precisely the ancient occupations of Chaco Canyon, Mesa Verde, and virtually every other site in the region where wood might be recovered. In a sense, he both discovered and deciphered a Rosetta Stone for the ancient Southwest.

Archaeology remains a major preoccupation of the LTRR, which receives more than 3,000 samples for dating each year. (In the archaeology lab, boxes of them are stacked to the twenty-foot-high ceiling, labeled in big black letters: Chetro Ketl, Zuni, Red Hill, etc.) Some of the old hands know the sequences so well they can date a piece of wood just from glancing at it; they are proud to point out that humans are superior to computers at pattern recognition because they can better exclude anomalous information—a missing ring, for instance.

Swetnam, by contrast, has mainly focused his research on ecological reconstructions. Using tree rings, he and his students can estimate the density of trees that stood on a given site at a given time. They can detect the signal of past outbreaks of bark beetles and other insects. In particular, they can tell when fire visited a forest, and how often it came back.

Another kind of Rosetta Stone produced by the LTRR is the master fire chronology that Swetnam and his colleagues have compiled since the 1970s. A fire of moderate intensity burning on a mountain slope will sometimes set burning logs rolling. If such a log lodges against the uphill side of a tree, it will likely burn a triangular scar, called a "cat face," into the base of the tree. Successive future fires, even mild grass fires, will produce a new scar around the healed but resinous edges of the cat face. Some older trees bear the evidence of a dozen or more fires in the scars at their base, each one registered in a specific annular ring, which dendrochronologists can peg to the exact year in which the ring formed.

Imagine now the big picture assembled at the LTRR, as well as other collaborating research centers: 500 or so sites across the West, each site with approximately twenty fire-scarred trees (totaling more than 12,500 trees), each tree with multiple fire scars, and each fire scar a separate datum.[11] The work that began in the Southwest has spread to the Sierra Nevada, the Blue Mountains of Oregon, the Black Hills of South Dakota, and the Sierra Madre. Collectively, tree-ring researchers have compiled a continental biography of wildfire.

Two things jump out from the picture that emerged for the Southwest.

First, at nearly all of the sites the rhythm of fire recurrence abruptly stops in the 1890s, give or take a decade. Up until then, and for as far back as the record shows, which is usually several centuries, relatively hot sites (lower elevation, south- or southwest facing) experienced recurrent fire as often as every five or six years, and cooler sites (higher, greater exposure to the north) every twenty to thirty years. The record also shows every shade of variation between those two extremes, and even then there are many exceptions: Swetnam studied a site near Flagstaff that for decades at a time burned almost every other year.[12] The same is true for portions of the Mogollon Rim. Outside of the Sierra Madre, almost all of this incendiary activity ceases in the late nineteenth century with the abruptness of a switch being thrown. The cause is not hard to trace. It was the arrival of large herds of sheep, goats, cattle, and horses. The herds ate the fuels that made the fires go.

Different kinds of forests experience and respond to fire in different ways. Big stands of lodgepole pine tend to burn all at once in dramatic crown fires every two, three, or four centuries. The Yellowstone fires of 1988 are a good example. Lodgepoles need intense heat to regenerate; their cones are dense and tightly glued, and it takes the high temperature of a powerful fire to open them and release their seeds. The forests of the Yellowstone country soon rose from their own ashes, and today, in sharp contrast to the pineless "balds" of the Rodeo-Chediski burn, the Yellowstone lands that were forested before the big fires raged are mostly forested again.

The spruce and fir forests of high elevations have a somewhat different fire regime, but they, too, burn at frequencies measured in centuries, if they burn at all.

Not so the pines and Douglas-firs lower down. The bark of the mature trees of those species is a giveaway. On the older ponderosas and to some degree on Douglas-firs it grows in thick plates, the organic equivalent of an asbestos vest. Low-intensity fires generally do not trouble the big trees—in fact, the forest giants need the so-called "light" fires to thin out the young upstarts who would compete with them for water and nutrients—and endanger them with fire of higher intensity.

Light fires depend on light fuels, chiefly grass, to carry them across the land. When the big herds arrived, sustained by railroads that carried their wool, hides, and the animals themselves to distant, hungry markets, the grass was soon exhausted. Roads and trails, which proliferated as settlement expanded, also helped snuff out light fires, for they constituted de facto fire breaks. Beginning early in the twentieth century, systematic fire suppression, led by the Forest Service, delivered the coup de grace.

The fact that grazing was the original trigger is proved as much by exception as by the rule. One researcher, Melissa Savage, discovered that fire recurrence in the Chuska Mountains ceased in the 1830s. Such data seemed anomalous until she and Swetnam realized that the 1830s were the period when the Navajo first acquired large herds of sheep.[13] "Sure enough," says Swetnam, "the last widespread date of fires in example after example is within a year or a few years of the original introduction of significant numbers of animals."

The second thing that jumps out from the big picture of the master fire chronology is the synchronicity of fire events. In some years flames seemed to rove across the entire region: "1748, for example. Most of the mountains were burning in 1748; also 1851, 1879. There was a series of classic years and they just keep showing up." As the data came in and the picture took form, Swetnam began telling himself, "It's got to be climate. It's got to be climate." But what was the driver? Was it simple drought? Was it a wet year, good for growing light fuels, followed by drought? And what drove the drought?

Science advances on the strength of good ideas, and good ideas occur to scientists, as they do to anyone, at odd times and in unexpected situations. In the late 1980s Swetnam and his family lived on the north side of Tucson, within easy reach of the cabin of his close friend, Julio Betancourt, a scientist with the U.S. Geological Survey. They hung out a lot. Their wives liked each other. Their children played together. The porch of Betancourt's cabin was a good place to work through a six-pack on a weekend afternoon and gaze out on the oak woodland where it graded into chaparral. On one such occasion, Swetnam mused aloud about the Southwest's years of epidemic fire. As Betancourt recalls, he also noted a few years of opposite effect, when there was no record of anything burning, anywhere, "and he mentioned the years." Betancourt at the time was immersed in climate work, and he immediately recognized the specific years, like faces in a crowd. "Tom," he said, "that's like a who's who of El Niño years in the late 1880s." The moment was almost

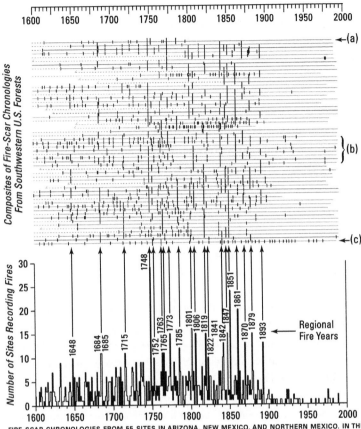

FIRE SCAR CHRONOLOGIES FROM 55 SITES IN ARIZONA, NEW MEXICO, AND NORTHERN MEXICO. IN THE
UPPER CHART, EACH HORIZONTAL LINE REPRESENTS A DIFFERENT SITE, AND THE TICK MARKS REPRESENT
RECORDED FIRES. MOST OF THE SITES WERE IN PONDEROSA PINE OR MIXED-CONIFER FORESTS, AND THE
SAMPLED AREAS TYPICALLY RANGED FROM 25 TO 250 ACRES, WITH AN AVERAGE OF TWENTY TREES SAM-
PLED IN EACH. NOTE THE GENERAL CESSATION OF FIRE BY 1900. SITES DENOTED (A), (B), AND (C) ARE
EXCEPTIONS THAT PROVE THE RULE: (A) RECORDS A SITE IN TRADITIONAL NAVAJO PASTURELANDS WHERE
INTENSIVE SHEEP GRAZING COMMENCED NEAR THE START OF THE NINETEENTH CENTURY; (B) IDENTIFIES
CHRONOLOGIES FROM UNGRAZED SITES WITHIN LAVA FLOWS AT EL MALPAIS NATIONAL MONUMENT,
WHERE FIRE SUPPRESSION DEPENDED ON AIRBORNE EQUIPMENT NOT AVAILABLE UNTIL THE MID-TWEN-
TIETH CENTURY; (C) IS A REMOTE SITE IN SONORA, ISOLATED FROM BOTH LIVESTOCK GRAZING AND
EFFECTIVE FIRE SUPPRESSION. THE LINE GRAPH ON THE BOTTOM SHOWS THE TOTAL NUMBER OF SITES
RECORDING FIRE DATES EACH YEAR. THE LABELED YEARS WITH ARROWS HAD REGIONAL FIRE SIGNIFI-
CANCE, WHEN FIRES OCCURRED IN AT LEAST TEN SITES ACROSS THE NETWORK. *FROM THOMAS
W. SWETNAM, ET AL., "APPLIED HISTORICAL ECOLOGY: USING THE PAST TO MANAGE FOR THE FUTURE,"
ECOLOGICAL APPLICATIONS 9/4 (1999): 1189–1206. © ECOLOGICAL SOCIETY OF AMERICA. USED BY
PERMISSION.*

cartoonish: a lightbulb seemed to go off. They went back and forth, matching years of high and low fire occurrence with, respectively, La Niña, which brings dry winters to the Southwest, and El Niño, which brings abundant rain and snow.[14]

Swetnam recalls another "aha!" moment. He'd been working alone in the crypt he calls his office, plotting big and small fire years against a master list of El Niños and La Niñas based on Peruvian historical documents. The graph surpassed his dreams: it was a complete corroboration of the hypothesis he and Betancourt had hatched. He was so excited he ran down the corridor to tell someone. The first person he encountered was a crusty, skeptical senior colleague, a top scientist "who rarely acknowledged my existence." Swetnam showed him what he had. The older fellow pondered a moment, and then, in language that for him verged on effusion, said, "Yep, you've got something there."

Swetnam and Betancourt produced a paper, published in *Science* in 1990, that established a link between regional fire and macroclimatic processes.[15] It was the first of many journeys Swetnam would make into that subject matter. The potential for connection did not end with the Southwest. By then people were also noticing that the forests of Borneo burned when El Niño was strongest. There was a lot of what meteorologists were calling *teleconnection* between phenomena separated by half the globe. Powerful forces that inhered in the interplay of atmospheric and oceanic currents were shaping events tens of thousands of kilometers away, and for a scientist like Swetnam, the implication was clear that if the patterns of these forces might be deduced, then maybe, just maybe, drought, fire, and other expressions of climatic variation might become predictable.

ONE OBSTACLE TO tracking the interaction of fire and climate, says Swetnam, "is that the historic fire record from the western U.S. is extremely fragmented. It's a sad commentary actually on the lousy record-keeping of the agencies." Because of this, he was especially cheered to encounter the work of a young professor from the Merced campus of the University of California. Tony Westerling had minutely examined the spotty records for fire occurrence in the West and "battled" to fill in the gaps, the worst being the decades from the 1940s through the '60s, when the Forest Service "just kind of started throwing stuff out."

"And then," says Swetnam, Westerling "started doing some really interesting and smart, sophisticated analyses of how rainfall and moisture deficits and snowpack and river runoff might be related to the fire occurrence records." Swetnam and Westerling connected, plugging the tree-ring record into the matrix of relationships Westerling had uncovered. Climate change was much on their minds, and so they looked for evidence that a warming climate was influencing fire activity in the West. They found it.

For years it had seemed that western forest fires were getting bigger, more frequent, and more destructive. In a 2006 paper, Westerling, Swetnam, and two coauthors showed that the change was not imaginary.[16] Comparing the period 1970 to 1986 with subsequent years (1987–2003), they documented a quadrupling of the occurrence of fires consuming 10,000 acres or more. Not only did large fires occur more often, but the more recent large fires also dwarfed their predecessors—on average they were six and a half times bigger, measured by area burned. They also raged through a greater portion of the year. The team found that the fire season, from the onset of firefighting in the spring to the close of activity in the fall, had increased by a whopping seventy-eight days—more than two additional months, with the season lengthening at both ends. The researchers correlated these changes with spring and summer temperatures that were on average 0.87 °C (1.57 °F) hotter than in the baseline period. In sum, they discovered that spring came earlier, warm weather lasted longer in the fall, and summers were hotter—all in all, a pretty good description of a warmer climate.

One of the team's key observations was that large fires showed a parallel increase in the Northern Rockies and the Southwest, although the forests of the two regions are fundamentally different. Fuel buildup affects southwestern forests more than their northern counterparts, and drought less so. Aridity is endemic to the Southwest, but when the normally moist lodgepole pine and spruce-fir forests up north dry out, the consequences are powerful. Meanwhile, many of the northern forests never experienced much light fire, and so their characteristically high fuel loads are essentially what they were in the distant past. There are nuances to the way the forests of the two regions respond to the shifting climate, but significantly, in both regions, big fires have gotten a lot bigger, occur more often, and start earlier and end later over the course of a measurably warmer year.

Are anthropogenic greenhouse gases the underlying cause of these changes? The team cautiously said that such a determination was beyond the scope of their work. They pointed out, however, that virtually all climate-model projections forecast further warming and that "this will accentuate conditions favorable to the occurrence of large wildfires, amplifying the vulnerability the region has experienced since the mid-1980s."

They also note that the forests of the western United States account for somewhere in the range of 20 to 40 percent of total U.S. carbon sequestration. If present trends continue, "even under a relatively modest temperature-increase scenario," they warn that "the forests of the western United States may become a source of increased atmospheric carbon dioxide rather than a sink," thereby exacerbating the buildup of greenhouse gases and the warming of global climate.[17]

No matter how you cut it, says Swetnam, prospects for the pine forests of the Southwest are poor. Ultimately the cycle of fire and destruction will be self-limiting, as the scars of multiple big fires fragment the landscape and bar the spread of future fires. But along the way, as a result of soil loss, the creation of spatial gaps too big for pines to recolonize, and other changes favoring establishment of new species, perhaps half of existing forests may be lost over the decades to come—at least that is what Swetnam told an interviewer for the CBS news program "60 Minutes" in 2007.[18] He immediately regretted the statement—it wasn't scientific; he couldn't back it up; it was a shot from the hip, a WAG, a wild-ass guess.

Some of his subsequent research, however, has tended to buttress that WAG. In a study that appeared late in 2010, Swetnam and a group of colleagues quantified the loss of southwestern forests to insects and fire for the period 1984–2008. It was a hefty 18 percent. They speculate that "with only two more recurrences of droughts and dieoffs similar or worse than the recent events," total forest loss could exceed 50 percent.[19] It was still a guess, but not as wild-ass as before.

In his office in the catacombs of the football stadium, Swetnam laughs off the awkwardness he felt about the "60 Minutes" interview. Standing in front of a camera is not his favorite place to be. But he turns serious as he sums up the changes he has seen in the land. "The Southwest is actually one of those places where it's the perfect firestorm.... It's not all climate change; it's not all warming; it's not all fuels changes; it's not just the problem of people moving into these landscapes. It is all of the above."

LEST ONE THINK that climate-induced forest woes are somehow restricted to western North America, Swetnam's friend and frequent collaborator Craig Allen recruited no fewer than nineteen coauthors to help him assess "recent tree mortality due to drought and heat stress" on a global scale. For the years since 1970, they found eighty-eight instances of such mortality spread from Zimbabwe to Alaska and Australia to Spain. Significantly, the rate of occurrence of such events increased markedly after 1998 and continued to accelerate through the 2000s. The authors concede that the trends they observe could be a function of greater scientific attention being paid to ecological heat stress, but the reported events also "mirror warming global temperatures."[20]

As with North American forests, the implications of the team's findings are disturbing. Two in particular stand out: first, that "no forest or climate zone is invulnerable to anthropogenic climate change, even in environments not normally considered water-limited"; second, that the prospect of forests' becoming a net source of atmospheric carbon as they burn and die, rather than one of the planet's principal means of carbon storage, is a global, not merely a regional problem. It is one of the many feedbacks, like dust and albedo interactions or the release of methane from thawing tundra, that current climate change models do not address, and that have the potential to draw the noose of warming measurably tighter.

FIRE, OF COURSE, has always been with us, and even the honor roll of American literature includes some surprising arsonists. In 1844 Henry David Thoreau, intending to cook some freshly caught fish, managed to set fire to about 300 acres of his cherished Concord woods, an event that prompted considerable moral gymnastics as he worked to clear his conscience and "settle it" with himself. The opprobrium of his neighbors, to say nothing of lingering guilt, probably contributed to his decision the following year to withdraw somewhat from human society and build a cabin beside Walden Pond in an attempt "to live deliberately."[21]

Sam Clemens had a similar but less consequential flirtation with fire. In the early 1860s, well before he'd created the alter ego of Mark Twain, he and a companion camped on the shore of Lake Tahoe—they were "roughing it," as Clemens tells their tale in his book so titled, which appeared a decade later.[22] It seems they kindled their cooking fire on a bed of pine needles, which quickly burst into flame. Moments later a patch of manzanita behind camp was "roaring and popping and crackling," and the two men fled to their boat for safety. "There we remained, spellbound," he wrote, as they watched the fire gallop up one ridge, disappear into the canyon beyond, and reemerge roaring up the next ridge, redder and hotter than before. Then on it went to another ridge, and the next one after that. While Clemens and his friend sat slack-jawed in their skiff, the fire raced far into the distance and spewed, in his telling of the tale, a shower of both embers and exclamation points.

Of all the celebrity arsonists in the American past, however, none offers more insight into the phenomenon of western wildfire than John Wesley Powell, who gained fame from his 1869 descent of the Colorado River. In later years Powell was fond of recounting his experience, one chilly evening in the mountains of western Colorado, when he set fire to a tree for warmth, or just to see it burn—the story comes down to us both ways. "In a few minutes the great forest pine was all one pyramid of flame.... Sparks and flakes of fire were borne by the wind to other trees, and the forest was ablaze.... On it swept for miles and scores of miles, from day to day, until more timber was destroyed than has been used by the people of Colorado for the last ten years."[23] Powell later infuriated Bernard Fernow and Gifford Pinchot, two early champions of forest preservation, by droning on about his act of arson in the presence of Secretary of the Interior John Noble. Fernow and Pinchot were offended not only by Powell's insouciance toward forest destruction, but by his selfishness in boastfully consuming so much of their precious audience with the secretary.

Powell also recounted his feat of fire-setting in an 1890 magazine article, where he went on to say that he'd witnessed more than a dozen fires like the one he had set, each equally large and destructive. He described a trip he made by train through the Dakotas, Montana, Washington, Oregon, and down through Idaho. Over all those hundreds of miles for more than a week, the smoke of forest fires continuously filled the skies with a gray opaqueness "through which it was as vain to peer as through a fog at sea." The forests of the northern Rockies were burning ubiquitously that summer, and their smoke was so dense that Powell and his party were "ever in a mountain land, and never a mountain in sight." Not until they crossed southward into Utah and a rain shower rinsed the air did the Wasatch Range at last come into view.

Years earlier, in his famous *Report on the Lands of the Arid Region*, Powell had blamed the prevalence of western wildfire on Indians. While many of his contemporaries ascribed it to "a wanton desire on the part of the Indians to destroy that which is of value to the white man," Powell was somewhat more charitable. He believed the natives principally used fire to drive game and that as the press of settlement made their situation more desperate, they resorted to game drives—and fire—with increasing frequency. But by 1890, Powell had come around to a different view. Already he had noted the cessation of light fire in many areas and the buildup of fuel that inevitably resulted. He asserted that fires, as a consequence of fuel accumulation, were becoming more destructive. "It is thus that, under conditions of civilization, the great forests of the arid lands are being swept from the mountains and plateaus."[24]

As Powell rightly observed, fires attended every aspect of western settlement. They were ignited by accident, by intention, and even (as Powell well knew) for the hell of it. Sheepmen set fires to clear more land for grazing. Miners sometimes burned off the vegetation of a mountainside, the better to see the underlying geology. Loggers left behind not only untended campfires, but also masses of dried-out slash that were quick to ignite. Travelers on virtually every road and trail built fires to cook their food and to keep themselves warm. Lacking the cozy sleeping bags in use today, they built their fires of logs, not sticks, sometimes making several fires that they might sleep between. Such fires could smolder for days, harboring coals that a rising wind could fan into flame.

Increasingly, as the nineteenth century wore on, many in the West deemed the pervasiveness of western fires to be both a waste of timber and a direct threat to the water supply of downstream towns and irrigators. Progressive thinkers argued that fire was the enemy of a sane and durable civilization. To stamp it out was to eliminate a kind of barbarity, and to suppress fire was therefore as much an act of taming the western wilds as overcoming Indians had been. Campaigns for forest protection, with fire control at their forefront, grew in strength and became a national cause. As parks and forest reserves were set aside, at first tentatively, then in grand waves of executive and congressional action, the ethic of fire suppression became invisibly embedded in the redrawn map of the public domain.

By 1905, 56 million acres of western lands had been set aside as forest reserves. That year, with the urging of President Teddy Roosevelt, Congress transferred administration of the reserves to the Department of Agriculture's Bureau of Forestry, headed by Roosevelt's friend and advisor Gifford Pinchot.

The bureau was renamed the Forest Service, and two years later the Forest Reserves were recast as National Forests. The fledgling agency and the National Forest System continued to grow, and the faster they did, the more opposition swelled against them. Many westerners, accustomed to free use of the public domain, objected to Roosevelt's assertion of control over whole mountain ranges. The moneyed interests behind the West's railroad, mining, and timber companies objected even more, arguing that Roosevelt was tying up resources the country needed to develop. Roosevelt and Pinchot shot back that conservation was the preeminent moral issue of the age, and they kept on reserving forests at a steady pace.

In 1907 Roosevelt's enemies in Congress resolved to clip his wings. They attached a rider to a spending bill he would have to sign, terminating the president's authority to create National Forests in six western states. Future withdrawals would require congressional approval, an unlikely prospect. It looked like Roosevelt was cornered.

But not quite. While the bill awaited signature on the president's desk, Roosevelt and Pinchot summoned sheaves of maps to the White House. They unfurled them on the floors and pored over them, sketching lines and drafting notes into the small hours. Within days, Roosevelt issued executive orders establishing 16 million more acres of National Forests, an area larger than West Virginia. Then TR signed the bill that forbade what he'd just done.

All told, Roosevelt added more than 100 million acres to the National Forest System, and by the end of his presidency in 1908, the Forest Service employed 1,500 men. His achievement, however, was hardly safe. His adversaries in Congress starved the agency for funds, and his handpicked successor, William Howard Taft, proved flaccid and inconstant, ultimately firing Pinchot as chief forester. With Roosevelt out of the country on extended safari and Pinchot banished, the central conservation achievements of his presidency had no protectors. It seemed that the Forest Service and the National Forests might waste away into insignificance or, worse, be disassembled.

Fire changed all that. In the winter of 1909–1910, snowpack was good across the northern Rockies, but spring came without rain and, as Timothy Egan writes, "the snow melted early, all of it. Creeks withered and stilled, and the forest leathered and baked."[25] When summer came, high temperatures and hundreds of lightning strikes, railroad sparks, and other ignitions set the region ablaze. Every able-bodied man who could be induced to help was put to work on fire lines. Under pressure, President Taft called out units of the army to assist. Matters were merely desperate until August 20, when hurricane winds pushed into the region. Like the bellows of a furnace, the winds fanned the blazes to new intensity. Fires spawned plumes. They merged into multiple equivalents of Rodeo–Chediski, spread across states. It was a holocaust.

Flame engulfed whole towns. More than 3 million acres burned in two days. Thousands fled their homes; somewhere between 100 and 200 people died, including many firefighters. The most notable field officers in the battle to contain the fire and to save both lives and property were the rangers of the Forest Service, who emerged as a singed and sooty corps of heroes, both in the public mind and in their own. Preeminent among them was Edward Pulaski, who, surrounded by flames, herded forty-five of his men into an abandoned mine, and held them there under gunpoint, lest they panic and run for it, and thus succumb to certain death. Ranger Pulaski battled to keep his men in line and to keep the timbers at the mouth of the tunnel from burning and collapsing, until he lost consciousness. Five of the men in the tunnel suffocated; Pulaski and the rest barely survived.[26]

Newspapers and popular journals throughout the country told Pulaski's story again and again. He and other rangers who battled the Big Burn became national heroes. Until the fire, efforts to stifle forest protection had been strengthening, but the combined trauma and heroism of 1910 turned the tide. Firefighting was now a national cause, and the men of the Forest Service were its champions.

If there exists a neurosis, directly opposite to pyromania, which consists of a pathological aversion to fire, the Big Burn implanted it in the Forest Service. Anti-fire crusading became not just policy but ideology, and the high priests of this managerial religion tolerated no dissent. The notion that benefits might accrue from the kind of "light fire" that sustained sequoias in California and loblollies in Carolina was viewed as apostasy. Even Aldo Leopold got in the act. In 1920 he wrote an essay titled "'Piute Fire' vs. Forest Fire Prevention," in which he argued that "light fire propagandists" were simply greedy land-owners. They thought that by using light fire to limit the supply of seedlings, they might trigger an increase in demand—and price—for the mature trees they already owned. The effects of the burning they advocated, said Leopold, were entirely destructive, because they inhibited forest regeneration and the production of needed timber. (The term *Piute fire* probably derived from John Wesley Powell's public approval, decades earlier, of Indian fire practices, and its use as a pejorative may be rooted in the meeting with Secretary of the Interior Noble at which Powell so annoyed Fernow and Pinchot.)[27]

Leopold's opposition to light burning is notable because by the early 1920s he was on the cusp of an insight that would find expression in one of his most widely appreciated essays, "Grass, Brush, Timber, and Fire in Southern Arizona," which appeared in 1924. In it he noted that, in the absence of fire, many of the savannah grasslands of the Southwest were changing rapidly, and not for the better. In a word, the country was growing brushier. Practicing a bit of dendrochonology, Leopold examined the fire scars on some of the older trees he encountered, and used them to deduce, rightly, that the landscapes the early pioneers encountered owed their openness to the recurrent fires of the past. But even having made this realization, he was still not ready in 1924 to endorse the intentional use of fire to improve ecosystems. Only after his transformative trip to the Río Gavilán in Sonora, in 1936 would he at last acknowledge that fire might play a beneficial and even necessary ecological role.[28]

The forest service took even longer to accept what Leopold called "the plain story written on the face of Nature." Ultimately, the impetus for change came from a distant region. In the 1940s, faced with the opportunity to extend the National Forest System into the notoriously fire-dependent piney woods of the South, the Forest Service realized it had to adopt fire as a management tool if it was to win local support.[29] This essentially political consideration cracked open the policy door, and new thinking about fire gradually seeped to other regions. Translating thinking into action, however, was hardly automatic, and decades more would pass before the agency's theoretical flexibility began to influence land management practices on a significant scale.

THE INSISTENCE THAT fire was an unnecessary, if not evil, force had roots deeper than the trauma of the Big Burn. It reached into the spirit of the age. Scientific forestry and the idea of land management, born at the high tide of the Industrial Revolution, developed from a view of the world that was essentially mechanical. A factory was a big machine. A forest was a bigger one. The same scientific principles that rendered the factory floor more productive would also increase the efficiency of nature. As Frederick Taylor taught in his analysis of industrial processes, the first thing to do was to eliminate waste and superfluous movement by removing unneeded parts. Among the things people identified as "unneeded" in nature were floods in river basins and freshwater flowing to the sea. The water of a river like the Colorado, all of it, was better used for irrigation. Predators were equally unneeded, for they consumed livestock and desirable wild game like deer. To exterminate them was to eliminate waste. Ditto for prairie dogs and other varmints, even porcupines—each was a threat to something valued by society. Best to get rid of them. Get rid of fire, especially, because it is disorderly and kills trees, which are the desired output of the forests. And get rid of outmoded ways of thinking, like folk and Indian practices that favor light burning, which is just another form of woods-burning, no better than the vandalism Thoreau, Clemens, and Powell bragged about.[30] Granted, a lot of other cultural imperatives were entwined with the urge to simplify—taming a wild land was certainly one of them—but the quest for efficiency remained a unifying thread.

The old mechanical model did not stand alone. It existed in tension with its opposite, which was an essentially Edenic view of nature, a myth of the pristine. This rival view held that the affairs of the natural world were in a state of near perfection prior to settlement, and that the best way to preserve what was left and regain as much as possible of what had been lost was to cease human interference in the dynamics of the land.

Never mind that such a view oversimplifies the ecological history of the continent (notably the role of Native Americans), and that it fares even worse when exported to other regions of the globe. The rightness or wrongness of the Edenic view is immaterial in a post-Edenic world. In 2004, federal agencies estimated that the vegetation on 183 million acres across eleven western states was "highly altered from historical conditions and would experience severe effects from wildland fire."[31] No Eden there, and anyone who believes that matters might correct themselves if only humans would cease their manipulations will find his or her argument burdened by a problem that did not weigh upon early formulations of the Edenic view. That problem, of course, is climate change. Anthropogenic alteration of the climate is the mother of all manipulations. Now Eden is warming, like everywhere else.

The middle ground between the two positions constitutes a large space, broad enough to accommodate wide differences of opinion, but the areas of agreement are substantial. They center on two ideas. First, people need to intervene in forest pyrotechnics by reducing fuels where possible. This means thinning trees. The undertaking may be couched in terms of *restoration*, which begs the endlessly arguable question "restoration to what?," or it may be presented simply in terms of fuel management. Certainly some fuel reduction treatments are better conceived and executed than others, but generally speaking, in ecosystems characteristically structured by light fire, any treatment is better than none at all.[32]

A second key idea is that the job that needs doing is too big for the resources available to do it. Therefore land managers must practice a kind of triage, directing their limited dollars to high-priority ecological areas and to zones of contact between wildlands and human habitation. In the years leading up to the Rodeo-Chediski Fire, no one knew better than John Bedell that his forest was badly overgrown and primed to blow up. There were plenty of places so thickly overgrown "you couldn't even walk through them." Yet in the years leading up to the fire, his funding allowed him to thin no more than "twelve to fourteen hundred acres a year." At that rate, it would take more than three centuries to thin the area burned by the fire.

YOU DON'T HAVE to invoke climate change to explain the Rodeo-Chediski Fire. Human activities on the Mogollon Rim, beginning with grazing and followed by nearly a century of fire suppression, produced the makings of a bomb. In the early 2000s a prolonged drought, a common phenomenon in the Southwest, provided the fuse. Two people with lamentable judgment (still more common) struck their matches, and—boom!

Climate change takes things a step further, altering the bomb-making recipe by adding a yeast of higher mean temperatures and goosing it with upward temperature spikes that will be higher than anything previously known. Amped-up evaporation will enhance the explosive mix. The result will be a supercharged concoction. It is as though in cooking up our forest bombs, of which an incalculable number lie latent across the land, we've swapped our old fertilizer-and-diesel formula for munitions-quality TNT.

Tom Swetnam is frequently called upon to talk about what he calls the "double whammy" of climate change and fuel buildup. He lectures undergraduates, makes professional presentations, and gives talks for a wide range of organizations. To the usual catalogue of threats to western forests, he usually adds insect outbreaks, another of his areas of expertise. Sometimes he also discusses invasive grasses, especially hot-burning, human-introduced species, like buffelgrass in the Sonoran desert and cheatgrass in the sagebrush plains of the Great Basin, which disrupt native fire regimes and threaten the wholesale alteration of ecosystems. As he moves through the list of woes, he says, "I see the glaze come over people."

"I am not the most exciting speaker," he says, "but it's also something else." Helplessness, he believes, is at the root. "People don't have a sense of what we can do about these problems." Their reaction seems to be, "Okay, there's nothing more. I'm just kind of shutting off."

"I've been struggling myself," he admits. "My talks still don't include but the last few minutes of talking about what we can do about it—because, partly, I don't know." The question is, "What can be done to move our forests to a state of greater resilience and ability to withstand both climate change and fire?" So that when drought comes and fire follows, "it burns as a surface fire, which is primarily what those forest types and landscapes sustained for centuries, without blowing 2,000- to 3,000-acre canopy holes and losing the soil in an arroyo."

The answer, he says, is not singular but plural, and part of it inheres in the problem itself. The big stand-destroying burns leave opportunities in their wake. The vast balds—the canopy holes thousands of acres in size—that the big fires create are unlikely to regenerate in a form similar to the forests they replaced; ponderosa pine, for instance, with its heavy seed and tricky requirements for successful seedling establishment, is ill equipped to recolonize such large areas, even over a period of centuries.[33] The opportunity exists to use those balds as firebreaks and to allow them to help determine where limited funds for thinning should be spent for maximum effect.

Strategy is the main thing, in Swetnam's view. "There's been modeling to suggest that you don't necessarily have to treat the whole landscape, but if you treat particular places you're more likely to interrupt a running crown fire. You're more likely to have a chance to catch it. So some smart strategic placement of treatments could have an effect."

A successful strategy, however, depends on execution, and the Forest Service and other agencies will have to be vigilant in maintaining the conditions that allow the strategy to work. For starters, many of those balds will grow up in highly flammable shrubs. Keeping them bald enough to serve as firebreaks will require the use of recurrent, prescribed fire. The strategic stands that receive thinning treatments will also need periodic burning, lest they grow thick and explosive again. And prescribed burning, unfortunately, is a smoky and difficult practice.

In my own corner of northern New Mexico, the Carson National Forest wisely enlisted local people to reduce fuel loads. Nearly everyone hereabouts heats their home with wood, and every able-bodied male and quite a few women are competent in the woods with a chainsaw. So the Forest Service organized the local firewood harvest to thin certain strategic patches of forest. The woods on both sides of the dirt road linking my village to the paved highway—a dangerous zone for ignitions—was one of those patches. With strong popular support, hundreds of cords of wood were removed from the corridor and the slash neatly cleaned up. The mature piñon and ponderosa were left in place, an eye-pleasing savannah.

Two or three years later, the Forest Service prepped it for a prescribed burn. It was imperative to check the reestablishment of seedling trees and prevent the forest from reverting to its prior condition. District staff cleared fire lines and laboriously obtained the necessary clearances and approvals. At last October arrived, bringing weeks of warm, dry days—perfect burning conditions. But the match was never struck. It turns out that a similar burn a few miles up the highway had prompted complaints about smoke. Or someone was transferred. Or the budget ran out. Or all of the above, and more besides. The Forest Service abandoned the project. Years later, in 2010, they finally came back and lit it off. But by then a lot of juniper and pine seedlings had grown too big for a light fire to kill, and they delayed their ignition until Halloween, when the weather had turned cool. In a few places the fire did some good, but for the most part it just turned the ground black, if it burned at all. Someone will no doubt record a substantial number of "acres treated" on his or her list of accomplishments, but the forest is little changed. The most lasting ecological impact of the prescribed fire will likely be the fire line plowed around it, which future visitors to the area will misidentify as an illogical trail. One of "the plain stories written on the face of Nature" around here is that the chain of necessary actions in public land management includes many weak links.

ANYONE ALIVE TODAY who likes watching landscapes change can congratulate himself or herself for having been born into a good time. Plenty of forests in the years ahead will likely undergo "state changes." Driven by climate change, insects, and fire, they will transition from one kind of ecosystem to another. They will "flip" from trees to shrubs, like scrub oak and locust. Or from trees to grass. Or from one kind of tree to another—pines, say, to juniper. A comparable period for observing landscape change may not have been seen for thousands of years, perhaps since the end of the Pleistocene, the last great shift in Earth's climate (caused by variations in the planet's slightly irregular orbit around the sun).[34]

If, on the other hand, you are the kind of person who becomes bonded to specific places, the changes of the coming years may reward your attachments with woe. People develop affinities for all kinds of locations: favorite glades, fishing holes, hilltop views, trails, a jutting rock. From prairies to parklands and mesas to mountains, there is scarcely a creek or canyon that lacks an affectionate human constituency. These connections are nearly universal, at least among people who enjoy the outdoors, and they often possess enormous emotional and spiritual significance. Even so, they cannot compare to the feeling for landscape possessed by people for whom the natural world is both text and repository for much of their cultural memory. Imagine if the land around you were not only a source of refuge, inspiration, and pleasure, but were also the physical embodiment of your scriptures, your literature, your moral code, history, and philosophy. If that were the case, the broad-scale destruction of such landscape would come at terrifying individual and collective cost. And if that were the case, you would be in the same predicament as the White Mountain Apaches.

A long tradition in American literature sentimentalizes the relationship of Native Americans to the environment and portrays them as "first ecologists" who lived in perpetual harmony with nature. That line of thinking tends to be more wishful than factual, and I do not mean to veer in its direction. On the other hand, it is an extreme understatement to say that the White Mountain Apaches have traditionally possessed a finely tuned sensitivity to their land and that they rely on it, spiritually and morally, in profound ways. In this respect, they are far from alone among the tribes of the continent, but they stand out, from an outsider's point of view, because outsiders have been afforded an exceptionally detailed glimpse of what their natural surroundings mean to them.

The source of this insight is an extraordinary book called *Wisdom Sits in Places*. Its author, Keith Basso, first came to Cibecue in 1959 while still an undergraduate at Harvard. He never stopped coming back. Basso went on to become a respected anthropologist with a particular interest in linguistics. Over a period of decades he conducted countless interviews with scores of Western Apaches, probing questions of language and worldview. Basso gradually mastered the dialect of the White Mountain tribe, at least as far as a non-native speaker might, and published his work in prestigious journals and taught at top universities. His was a typical academic career, you might think. But not so typical were the long days on horseback, cowboying with the Apache men who became close friends and in some cases mentors. Ultimately Basso bought a little ranch near Heber and took a plunge into the cattle business himself.

In the 1990s, the tribal chairman suggested Basso make maps for the tribe. "Not whitemen's maps, we've got plenty of them, but Apache maps with Apache places and names. We could use them. Find out something about how we know our country." And he chided Basso, the outsider, in a familiar way: "You should have done this before."[35]

The project required that Basso travel to hundreds of specific places within the Apache homeland, often by horseback, in the company of tribal members who knew not merely the names of the places, but the stories those names invoked. Basso's job was to record all of it, and getting the stories right often led to long, follow-up visits at the homes of his "consultants." The actual maps that resulted are proprietary to the tribe, but the more general lessons Basso absorbed along the way and the understanding he distilled about the Apache relationship to land are the stuff of *Wisdom Sits in Places*.

The stories of places were sometimes historical, relating how the ancestors came to a certain spot and what they did there; or they were mythological, as when Owl Man, old and horny, became bedazzled by a pair of beautiful sisters, who taught him how lust could turn a man's mind to mush; or they included cautionary tales, like the story of a child gathering firewood who becomes inattentive, loses the path on a steep slope, and suffers a rattlesnake bite. Every story, which is to say, *every place*, provided an instructive lesson, and cumulatively the stories of those places comprised a complete body of moral guidance. They were a kind of physical, geographical literature that expressed the history, values, and cosmology of the tribe. It was more than a Bible written on the land: the Apache landscape included a library's equivalent of fables, histories, and handbooks, all the way down to the kind of guidance you might get from a newspaper advice column.

Many of these place-based stories were known well and widely enough that people could use them to communicate in a shorthand they called "speaking with names." By simply quoting the name of certain place, a speaker might induce a kinsman or friend to visualize the place and recall its story. This was a gentle way of suggesting that the hearer should take the lesson of that place to heart and thus be guided or corrected, without the unpleasantness of direct chiding or confrontation. In such ways, wisdom was deemed to "sit" in places, and the knowledgeable individual, whether seeking insight, guidance, or reassurance, might go to a place and "drink" from that wisdom and be helped. In an emergency, it was said, he or she might gulp more than sip. The right place, at the right time, could be a powerful ally.

But then came Rodeo–Chediski. The fire blazed across the places where wisdom sat. In certain instances, it burned away overgrowth and rendered locations more recognizable. More often, though, it obliterated cherished locations, or scorched and scarred them. According to Basso, a number of important places have been so altered that people refuse to go back, even to inspect the damage. The trauma is too deep. There is a sense that the fire did not so much destroy the wisdom that sat in revered places as it impeached it, rendering the wisdom problematic and questionable. The fire seemed to be an indictment of how people were living their lives, and it left them wondering where they might now turn for guidance.

One of the northernmost fingers of flame stopped 150 feet from Basso's barn. For the first three or four years afterward, the vastness of the burned lands seemed a wasteland. Then the elk began to come back, then antelope and other species. He even saw a rare Coue's deer in 2009. But eight years on, the White Mountain Apaches still struggle to recover. Like a victim of severe physical burns, they have endured a long, painful, and partial journey of recuperation. "They weep," says Basso, "those that have gone back, they cry, they sing, they pray. It is a profound loss."[36]

Basso believes it is the meaning of the fire, the meaning of the destruction and loss, that they struggle with the most. The fire has bequeathed them not just a sense of physical disorientation—in a landscape that no longer looks or functions as it did before—but moral disorientation as well. At one level or another, people search for the lesson of the fire. Was it a punishment, or a random turn of fate? Will the old wisdom regain its strength, or does the fire signal its senescence and obsolescence?

One of the legacies of so massive a change in the composition of the world, says Basso, is that it leaves behind a *demoralized* landscape, a world drained of meaning. For a people who have relied on the strength of the land to guide and buttress their own resolve, this is a grievous blow.

Different communities, like ecosystems, respond to traumatic events in different ways. Basso's non-Indian neighbors in Heber and Overgaard may not have felt the kind of metaphysical pain the Apaches suffered, but the fire nevertheless prompted grief and sorrow. An evacuee told me that when she came back after the fire, the sight of the blackened land was "heart-rending, almost heart-stopping." Two things in particular helped her stick with the job of cleaning up and rebuilding. One was the arrival of hundreds of volunteers who streamed in to lend a hand, some from as far away as Europe. "It helped a lot to know that people cared that much." The other was the yellow straw that the volunteers spread on the burned ground. "Just covering up all that black had a healing effect. It made it easier to be here and carry on with all that needed to be done."

There will be a lot to carry on with in the aftermath of the fires of the future. Reconstruction of homes and restoration of watershed stability will top the list of priorities. Right next to them, *re-moralization* of the landscape will deserve its own place, as people struggle, not always as dramatically as the White Mountain Apaches, to find meaning in the events that changed the land and also in the character of the new shrublands and grasslands that will have been created. These ecological communities will develop in a new climatic environment, and we will be fortunate indeed if we discern that any kind of wisdom resides in them.

===== **10** =====

## MT. GRAHAM: THE BIOPOLITICS
## OF CHANGE

THE SITE MANAGER for the Mt. Graham International Observatory (MGIO) met us at the locked gate to the Telescope Road. He was not there to greet us. John Ratje had driven out from Tucson, starting before dawn and traveling not less than two and a half hours, to demand that we surrender to him a key that would have opened the gate. A member of his staff had loaned it to us the afternoon before, but Ratje swept aside that troublesome fact. Ratje also demanded the two-way radio we'd been issued to assure safety on the narrow road to the mountaintop. "That's University of Arizona property," he said. "It should not have been given to you." Our shock at his demands no doubt showed in our faces. Ratje wanted to forestall argument, so he added ominously, "University police are on their way."

I'd been about to unlock the gate when he stepped from the trees. His truck was parked nearby, and he'd no doubt been waiting quite a while. He was a big man, but uncomfortable. His voice had a quaver of anxiety. At first I thought he merely wanted to inspect our one-day permit for entering the Red Squirrel Refugium, the restricted area that surrounds the observatory, at the top of the mountain. I'd obtained the permit from the Forest Service the previous day in Safford; I pulled it now from my shirt pocket, unfolded it, and extended it to him, but he ignored it. We did not have permission to go in, he repeated: "This area is ours."

I stammered a question or two, but the answer did not vary: our entry was barred, and police would soon arrive to ensure our compliance. Then my companion, Peter Warshall, a conservation biologist whose name was recorded on the permit along with mine, asked, "Is this about me? Am I the reason you won't let us in?"

"Your name was part of the discussion," Ratje said. A discussion? I had thought our visit was routine. Evidently we'd been the subject of telephone calls the evening before among administrators of the UA's astrophysics program, and probably others. At some point Ratje had volunteered, or was ordered, to interrupt his summer vacation, make the long drive out from Tucson at an exceedingly inconvenient hour, and execute the unpleasant task of telling us to turn around and go away. He knew that we also had made long journeys to get to Mt. Graham (we had camped on the mountain the previous night) and that we were invested in our visit. We had followed the rules and obeyed procedures. He knew this, too. Now he was playing bouncer for his observatory. He didn't like it, and he could look at us and see that we didn't like it, but he still didn't know how we were going to respond or how mad we were going to get. So he mentioned the police again, and watched us nervously.

Warshall was a veteran of the long battle of the 1980s and early '90s over whether an astronomical observatory—fourteen telescopes were initially proposed—should be sited along the crest of the mountain, amid the dwindling high-altitude habitat of the Mt. Graham red squirrel, an endangered species. He had not been back to the mountain in nearly a decade. Ratje also remembered that battle, albeit from the opposite side, and his presence at the locked gate early on an August morning indicated that the terms of the not-quite-truce that had suspended the battle did not extend to allowing retired combatants access to the contested terrain. In conversations leading up to our visit, Warshall had warned that Mt. Graham attracted the lightning of acrimony as abundantly as it drew the natural stuff. It was, he said, "the most biopolitical mountain in the Southwest, maybe on the continent." He wasn't kidding.

MOUNT GRAHAM IS a mountain of superlatives. Not only is it the most "biopolitical," it is also the biggest, tallest, and steepest Sky Island in the Madrean Archipelago. A Sky Island is a solitary mountain surrounded by ecological barriers that separate its plants and animals from others of their kind. In the case of Mt. Graham, the barrier is the desert from which the mountain precipitously rises. Most of the mountain's creatures are incapable of traversing the desert and, together with their progeny, are thus confined to the mountain for all time, just as if they occupied a solitary atoll in the Pacific. The majority of the mountain's plants are similarly stranded. Their progenitors, like the ancestors of the animals, were marooned on the mountain at the close of the Pleistocene, as a warming and drying climate caused surrounding lands to succeed to desert. Nevertheless, at certain times and for key species, the mountain has served as a kind of stepping stone, together with other mountains to the south, spanning the divide between great biomes. Mt. Graham stands at the northern terminus of the Madrean Archipelago, a chain of Sky Islands that links the biota of the Sierra Madre Occidental in Sonora and Chihuahua to the flora and fauna of the Rocky Mountains. Biologically, the Rockies reach their southern limit at the Mogollon Rim, which faces Mt. Graham across the valley of the Gila River.[1]

Mt. Graham's biogeographical position accounts for still more superlatives: it marks the southernmost occurrence of a number of northern species (including the water ouzel and mountain chickadee), and the northernmost occurrence of several others (the ridge-nosed rattlesnake, sulphur-bellied flycatcher, and Mexican long-tongued bat). But that's not all: because of its island character and isolation from other gene pools, Mt. Graham is a self-contained evolutionary laboratory. It is home to eighteen varieties of plants and animals that occur nowhere else. The most famous of these endemics is the frisky and diminutive subspecies of red squirrel, *Tamiasciurus hudsonicus grahamensis*, that caused Mt. Graham to become so fiercely biopolitical.[2]

When people say "Mt. Graham," they are often referring to the entirety of the Pinaleños Mountains. Mt. Graham might properly be considered the cluster of peaks at the crest of the range, where the fight over the telescopes was focused. Mt. Graham is the tip of the Sky Island; the Pinaleños constitute the sea mount on which it rests. That the Pinaleños constitute the biggest of the Sky Islands (another superlative) is confirmed by their Apache name, Dzil nchaa si'an, which translates as "Big-Footed Mountain." The poets who birthed that name appreciated the mountain's ample girth: it took a long time to go around. That Mt. Graham is the tallest of the Sky Islands can be discerned both directly and indirectly: altimeters show that it crests at 10,720 feet, a record in the Madrean Archipelago, and nearly a record for Arizona. The mountain's height is also attested by its ecology. It is the only Sky Island that supports a Rocky Mountain spruce–fir forest. In the Chiricahua Mountains one finds the same spruce tree (*Picea engelmannii*), and in the Catalina Mountains the same corkbark fir (*Abies lasiocarpa*), but only on Mt. Graham do the two species occur in the defining combination that is the essence of High Country and that characterizes the subalpine zone of the Rocky Mountains all the way from the Southwest to British Columbia (with a little genetic drift in the fir).

That Mt. Graham is the steepest of the Sky Islands is a function of its height and abruptness. Its granitic bulk rises as though punched from the desert floor. The vertical span from Chihuahuan desert base to spruce-fir tip is nearly 7,000 feet. At roughly a mile and a third, this is a longer spread than that of any other Arizona mountain. If you could somehow upend Mt. Graham and thrust it spearlike into the Grand Canyon, a thousand feet of its base would be left sticking out, looming above the Colorado Plateau. With so much elevation, Mt. Graham is also superlative in the diversity of its habitats. It is wealthier in niches and life zones than any other comparable mountain, although exactly how many it has depends on how the enumerator lumps or splits them.

Finally: of all the mountains, plains, valleys, mesas, hills, and canyons in the aridlands of the West, Mt. Graham provides a superlative vantage from which to view the effect of climate change on ecosystems and wildlife. At least three facts support this assertion: (1) the Mt. Graham red squirrel depends fundamentally on habitat that a warming climate will destroy, leaving the squirrel, absent rapid adaptation or human intervention, with no place to go, on a mountain no longer hospitable to it; (2) the effects of climate change are already evident within the squirrel's primary habitat, the spruce-fir forest, which fire and insects in recent decades have gone far toward destroying; and (3) the squirrel's prospects are tied intimately to human activities and agreements, past and future. How well the stewards of its fate keep the bargains they make on the squirrel's behalf will likely determine whether the squirrel is snatched from the edge of extinction or nudged into the abyss. Either way, the saga of wildlife protection on Mt. Graham helps tell us what to expect when other species, beset by diminishing habitat and changing climate, hover on the brink.

THE AFTERNOON BEFORE our scheduled visit to the refugium, Warshall and I took a short walk on Grant Hill, a shoulder of the mountain clothed in mixed conifer forest, the ecological community immediately below the spruce-fir zone. Warshall wore a ball cap, work shirt, and pants that hung on him like a tent. A recent painful illness had made him gaunt. His gray hair is gathered in back, not quite making a ponytail. He has the deep-set eyes of a brooder, and his low-pitched voice accentuates the seriousness of his manner. Decades ago he earned a doctorate in anthropology and biology from

Harvard, but declined to follow a conventional academic path. For a long time he edited the *Whole Earth Catalog* and the *Co-Evolution Quarterly*. In the early 1980s, he worked for the Office of Arid Land Studies at the University of Arizona. When the idea of placing telescopes on Mt. Graham first surfaced, the job of assembling the ecological information on which an environmental impact statement might be based fell to him. He had not intended to become a squirrel biologist, but once he embarked on the project, the squirrel seduced him. He said he found the eight-ounce bundle of hyperactivity "completely interesting."

Warshall pauses on the trail, leans over, and picks up the fallen cone of a Douglas-fir. "Red squirrel," he says, handing it to me. The cone has been gnawed to its woody core, each scale precisely pared away to free the seed it guarded. Three dozen paces farther, Warshall picks up another cone. This one is shaggy, the remnants of its scales frayed and feathery. "Abert's," he says. The Abert's or tassel-eared squirrel (*Sciurus aberti*) is three times the size of the red squirrel. Oddly, it is not native to Mt. Graham but was introduced in the 1940s, presumably to please hunters. Warshall explains that the Abert's lacks its cousin's hummingbird intensity. It chews cones more casually. "Sloppy eater," he smiles.

THE MT. GRAHAM RED SQUIRREL, EIGHT-OUNCE PROTAGONIST OF THE MT. GRAHAM CONTROVERSY.
*PHOTO COURTESY OF CLAIRE ZUGMEYER.*

Conifer seeds are the principal food of both species, with the red squirrel specializing in the small, dense cones of spruces and the Abert's in the larger, woodier cones of ponderosa pines. They overlap in their use of the mixed conifer zone, where Douglas-fir and southwestern white pine predominate. The trick of survival for both species is twofold: to procure sufficient food to carry them through the lean months of winter and to avoid predation. The red squirrel requires mature, dense forest to survive those challenges. Warshall leads me to a clump of venerable Douglas-firs, their trunks hoary with a hair of lichens. Several immense logs lie jackstrawed among their bases. The ground is spongy with cone scales. In places the litter is a foot and a half deep. This is what a red squirrel needs, Warshall says: four or five good cone-producing trees, a mostly closed tree canopy, and a jumble of logs affording good places to stash cones for winter storage. The logs also provide escape routes, allowing a squirrel to dash the length of a log in the blink of a bobcat's eye.

The squirrel's food cache is called a *midden*, and active, established middens like this one, he explains, show a steady accumulation of cone scales, thicker than a mattress, in which the red squirrel buries fresh cones for winter retrieval. In the lingo of squirrel biology, the red squirrel is a "larder hoarder"—it establishes and defends a well-defined midden, one per squirrel. The Abert's squirrel, by contrast, is a "scatter hoarder"—it stashes cones in multiple sites, or none at all. The Abert's squirrel is less dependent on stored food because it can survive on tree cambium for months at a time, a diet that would kill its red cousins.

It is vital that a red squirrel's midden remain deeply cold, for if it warms, the cones will open, allowing the released seeds to rot or be poached by other species. For this reason the tree canopy above a red squirrel midden must be dense enough to keep out the winter sun. A dense canopy also enables a squirrel to travel from tree to tree without descending to the ground, and it inhibits the maneuvering of the squirrel's worst enemies, hawks and owls. Few, if any, squirrels die of old age, or even see it. Less than half live to see their second birthday. Probably a majority come to the end of their short allotment of days being picked apart by a hooked beak.[3]

On the way back to the car, we spot a squirrel scrambling up a tree with a leaf in its mouth. It is a red squirrel. It perches on a high branch, nibbles furiously at the leaf, and stares at us intently. I am pretty familiar with red squirrels in the southern Rockies, for they have frequently reminded me that, in their eyes, I am a trespasser of the vilest sort. Nearly all squirrels will scold, but red squirrels vituperate. They far surpass their equivalents among human critics in their consistency and passion. The sound they make, loud and prolonged, is like the alarm of a cheap, wind-up clock, and if you could translate it to English, most of the words would be four letters long and acutely familiar. In terms of language, red squirrels are the sailors of the animal kingdom, vocal and blue-tongued. Strangely, though, the red squirrel in the tree on Grant Hill did not cuss us out. We stood below it, pilloried by its glare, but it did not utter a sound.

Warshall explained that the near voicelessness of the Mt. Graham red squirrel is one of its distinguishing characteristics. Genetic studies have shown that of the twenty-five subspecies of red squirrels, the Mt. Graham red squirrel differs to the greatest degree from the rest of the group.[4] Its silence is an expression of that difference. Living in isolation from its relatives for 10,000 years, either it forgot the old way of talking, or the rest of the family learned a new one.

We've stayed too long. Our presence has upset the squirrel. It drops its leaf (the basal leaf of a wild sunflower) and begins to hop between two positions. First, it glares at us sidelong, then, faster than the eye can record, spins round to face us. Then sidelong again. Each movement is a convulsion, a violent twitch: facing, sidelong, facing, sidelong. Twitch-twitch-twitch. Each twitch wrings out a high-pitched squeak, like the sound of an infant's squeeze toy. The criticism is mild, hardly red-squirrel caliber. Warshall notices teats on the squirrel's belly. It is a lactating female, a cheering sign. Her nest and young must be nearby. We move away to leave her in peace, and drift down the trail.

IF THE CONFRONTATION at the gate to the telescope road offered a glimpse of the anxiety that prompted it, the underlying motive remained opaque. Warshall speculated that the directors of the observatory did not want him to see how bad conditions in the refugium had become. Or that they had something to hide that we had not begun to divine. Or, alternatively, that they acted more from vindictiveness than fear. We had no way of knowing. Even more surprising than the denial of access, however, was Ratje's reversal of position. Ultimately, the site manager allowed us to pass. Warshall, his demeanor calm—but, as he later told me, boiling inside—had suggested that

I be allowed to enter and see what I needed to see, and that he would stay at the gate. That didn't sound very good to me. I argued that I needed a guide to the mountain and that the book I was writing required this kind of fieldwork. I shamelessly dropped the names of scholars and scientists who had assisted the project, including several at the University of Arizona. I doubt that Ratje found such arguments convincing, but he may have recognized in my pleading the blueprint of the squawk I would make if he stuck to his guns. He may have concluded that we'd cause less grief if he let us in than if he kept us out. So he sternly adjured us to "surrender" the key and walkie-talkie to university police at the earliest opportunity, and he unlocked the gate.

Just inside stood a cluster of signs declaring multiple prohibitions: no weapons, no trespassing, etc. One sign said, *"Kode' Godntsfh,* Respect This Place," and acknowledged: "This area is considered sacred by the White Mountain Apache and San Carlos Apache." Warshall said that the Apaches, who fought hard against construction of the observatory, had insisted on the sign. It was virtually the only concession they won.

We radioed to the summit that we were coming—in order to forestall anyone coming down. The one-lane road ascended steeply, switchbacking through burn scars and shrub fields of recent vintage. It reentered forest as we arrived at the buildings housing the observatory's three telescopes. We parked next to the hut for the university policeman, whom we would meet later, together with his snarling sidekick, a K-9 corps German shepherd. A man in a pressed shirt and necktie came out of one of the telescope buildings and asked us what we were doing. We told him and he scowled silently at us while we organized our daypacks; then wordlessly he went back in. There were no windows in the portion of the building he'd come out of. Not knowing how he'd observed us, we were hesitant to linger or even to speak until we were deep in the woods.

THE BIOPOLITICAL HISTORY of Mt. Graham began in earnest in 1984 when the University of Arizona, as lead partner of an astrophysical consortium, proposed to build a series of telescopes along the crest of the mountain, each in its own building, together with necessary support structures, roads, and other infrastructure that would require clear-cutting a substantial portion of the mountain's spruce-fir forest. Because the Pinaleños Mountains are part of Coronado National Forest, the project required a Special Use Permit from the Forest Service. The agency approved the project and was set to issue the permit, when pressure from environmental groups forced it to backtrack and agree to prepare an environmental impact statement (EIS).[5]

The red squirrel, meanwhile, had embarked on a biopolitical journey of its own. Presumed abundant in the 1880s, its population fell as logging, hunting, road construction, and other disturbances encroached upon its world. By the 1960s some authorities believed the squirrel to be extinct.[6] A number of sightings in the early 1970s, however, drew new attention to the creature and initiated a series of declarations at both the state and federal level identifying it as a rare species. Ultimately, and much to the frustration of advocates for the Mount Graham International Observatory, the U.S. Fish and Wildlife Service formally listed the squirrel as an endangered species in 1987. Somewhere between 200 and 300 were then believed to exist.[7]

The showdown was set: big science versus a small mammal at the brink of extinction. And the science was very big. The consortium led by the University of Arizona included prestigious universities in Germany and Italy, plus an honor roll of similar institutions in the United States. It even included the Vatican, which had made a point of being on the right side of astronomy ever since it realized, centuries late, that its persecution of Galileo in the 1600s had been, to put it mildly, misguided. (Advocates of Vatican astronomy sometimes also mentioned—to the embarrassment of their allies—the need to monitor the universe for signs of extraterrestrial activity, lest life in distant corners of the galaxy prove to be susceptible of catechization.) The gem of the installation would be the $60 million Large Binocular Telescope (LBT), utilizing the two largest mirrors on Earth and, thanks to new technology allowing thousands of continuous and instantaneous adjustments for atmospheric conditions, achieving a degree of image resolution ten times greater than that of the Hubble spacecraft.[8]

Because the project would take place in habitat critical to an endangered species, the U.S. Fish and Wildlife Service was required to write a Biological Opinion (BO) on the likely impact of the observatory on the squirrel's survival. At the time, the Mt. Graham red squirrel was considered to be wholly dependent on spruce-fir forest, and not many biologists, inside or outside the Service, were sanguine about the destruction of even a portion of so limited a habitat. Multiple versions of the BO, addressing multiple locations and numbers of telescopes, were drafted and debated. Multiple alternatives were duly proposed for analysis in the EIS. Meanwhile, the astrophysical brain trust at the University of Arizona grew increasingly nervous. Its consortium would not hold together indefinitely. Large investments were at stake, and time was of the essence. They feared that the objections of environmental groups might metamorphose into lawsuits, which in turn might bring the creep of administrative progress to a halt. In addition, the San Carlos Apaches and the White Mountain Apaches, two separate tribal entities who were relative latecomers to the controversy, had asserted their interest in the mountain. Since time immemorial, they said, the mountain had been a sacred place, among the holiest in the life of their tribes. They resisted divulging the reasons for that sacredness because to do so would have been a violation of it.[9] Nevertheless, they were adamant that Big-Footed Mountain, a source of healing and spiritual strength and the object of many a pilgrimage, ought not to be disturbed. Ola Cassadore Davis, chair of the newly formed Apache Survival Coalition, emerged as a spokesperson for the Apache point of view. Because she was not a medicine person and thus not bound by a rigid code of silence, "she felt she was allowed to defend the mountain in public."[10]

In mid-1988, despite widespread objection, the Coronado National Forest approved an EIS that incorporated Fish and Wildlife's latest and most controversial BO (more on that later). The installation had been pared down to three telescopes that would be located, not on High Peak, the highest point on the mountain, but on Emerald Peak, about a mile away. There was also provision for up to four more telescopes to be added in the future. The stage was now set for the regents of the University of Arizona to make a fateful decision. They feared that the EIS and BO would be attacked in court. In an effort to forestall lawsuits and further delays of the project, the university hired the Washington lobbying firm of Patten, Boggs, and Blow to secure special legislation authorizing the project. By passing the legislation, Congress would say that it officially deemed the project to have met all federal environmental requirements. The signature of the president would make that determination law. The legislation would further direct the Forest Service to issue the needed Special Use Permit. Ultimately, the university would pay Patten, Boggs, and Blow over $1 million for these services. Its decision to seek a special act of Congress was expressly and baldly an end-run on both the Endangered Species Act and the National Environmental Policy Act, which mandates environmental impact statements for projects like the MGIO. It was also an effort to avoid implications of the American Indian Religious Freedom Act and the National Historic Preservation Act.

In October 1988, Arizona's senators, John McCain and Dennis DeConcini, attached the university's bill to the Arizona-Idaho Conservation Act, which, among other measures, created the San Pedro National Riparian Area, a trade-off for Mt. Graham. In the House, Reps. Morris Udall and Jim Kolbe did likewise. The rider received no committee review or discussion in either house. Congress ultimately approved the bill, and President Ronald Reagan signed it on November 18. The university now had what it wanted.

Of all the individual actions contributing to the drama, that of Morris Udall, a staunch defender of environmental laws, was perhaps most surprising. Udall was then suffering badly from Parkinson's disease. According to an aide, he lacked the personal vigor "to try to resolve things better."[11] Others commented that the only thing Morris Udall loved better than the environment was the University of Arizona. The university community knew this. A year prior to the university's legislative initiative, it established the Udall Center for Studies in Public Policy in his name.

THE GOLIATH OF THE MOUNTAIN: THE LARGE BINOCULAR TELESCOPE OF THE MT. GRAHAM INTERNA-TIONAL OBSERVAETORY, AUGUST 2010. *AUTHOR PHOTO.*

THE CONSEQUENCES OF the battle over Mt. Graham led in many directions. Those who opposed the observatory (and quite a few who supported it) were embittered that a prestigious state-supported university might use its power and influence to circumvent environmental laws. As a result, in the aftermath of the Arizona-Idaho Conservation Act, the polarization of environmental politics in the Southwest ratcheted upward by several costly notches.

The Apache tribes continued their efforts to protect their sacred mountain. They passed resolutions. They lobbied. A group of medicine men took the unprecedented step of affixing their names to a document asserting the mountain's sacredness. A delegation of tribal leaders even embarked for Rome, with Warshall as their guide, to plead their case to the Vatican. One religion should respect the holy places of another, they said. Surely the Roman Catholic Church, with so many sacred places of its own, would be sensitive to their plea. They called on the Papal Nunciator for Peace and Justice. The Apache elders made eloquent appeals to him describing the centrality of the mountain to Apache religion. They pointed out that at one time (in the early 1870s) the mountain had been part of the San Carlos reservation—until Mormon settlers, covetous of farmland in the Gila Valley, persuaded the government to "adjust" the reservation boundary. Dzil nchaa si'an—Big-Footed Mountain—belonged to them, if not legally, then morally and spiritually. And they belonged to it. It was their cathedral, their basilica.

The nunciator, speaking French (which for distant historical reasons is the language of the Vatican), expressed sympathy for their cause and admiration for their efforts. Unfortunately, he explained, he was powerless to help: *Ce n'est pas ma compétence.* "This is not my area." Given that each tribe is legally sovereign, he recommended that the Apache delegation pay a visit to the Nunciator for Sovereign Nations.

And so the delegation did. Unfortunately, the Nunciator for Sovereign Nations, after expressing sympathy for their cause and admiration for their efforts, told them he was powerless to help. "*Ce n'est pas ma compétence,*" he said, and sent them to another. Who also listened patiently, expressed his sympathy, and said the same thing. With each visit the Apaches' speeches grew shorter and more direct, but the correct *compétence*, in spite of visit after visit, audience after audience, could not be located. Finally, the exhausted delegation left Rome empty-handed, taking only memories of an exquisitely polite runaround, which they realized had been honed over the course of centuries.

The Mt. Graham battle also reverberated in Congress. Two Fish and Wildlife biologists accused their regional director, Mike Spear, of rewriting the Biological Opinion to meet political ends. A congressional hearing was convened "to see if someone was tampering with biology."[12] Spear testified, sitting next to his accusers at the witness table, that he had indeed rewritten the draft BO to accommodate the observatory, and that he had done so with the support and advice of other biologists on his staff. Furthermore, it was his prerogative to do so because, irrespective of the contributions of others, his was the signature of record and he was therefore the document's effective author. The logic of his decision was straightforward: he judged that the squirrel and its habitat faced two threats—the observatory and human activities—and that the observatory was the lesser. "You will have either people on top of the mountain or you will have telescopes. By accepting telescopes, you get less people."

At first blush, the trade-off made sense: you can't have everything, so take the best deal you can get. But Spear's many critics were unmoved. They noted that comparable astrophysical sites were available on other mountains where environmental and cultural conflicts were absent or more easily mitigated. (The alternate sites were less attractive to UA because they were less convenient to Tucson.) As to limiting human impacts in squirrel habitat, the Forest Service had the power to impose access restrictions on sensitive lands as it saw fit. Summoning the gumption to ban dogs, hunting, and hiking from a mountaintop did not require the licensing of an astrophysical observatory. More significant than the merits of the trade-off, however, was the fact that Spear had calculated one at all. His job under the Endangered Species Act— the job of the Fish and Wildlife Service—was to make a scientific assessment regarding the squirrel's survival. He'd done much more than that.

Spear's Biological Opinion, which remained in force, authorized construction of three telescopes on 8.6 acres, while establishing a squirrel refugium of 1,750 acres on the rest of the mountaintop. The refugium included most of the mountain's spruce-fir forest, and all but essential administrative and scientific activities would be barred from it. The BO also called for removal of fourteen Forest Service–sanctioned summer homes and a church summer camp in squirrel habitat, as well as closure and reforesting of certain forest roads.

Spear, a Naval Academy graduate and former nuclear submariner, was uncommonly direct for a senior federal bureaucrat. He admitted that his decision in the BO was not strictly scientific. It was shaped by his sense of pragmatism. The result, he believed, was that "the opinion we wrote was more protective of the squirrel than the status quo." Spear went further: he acknowledged that if the competing use of the mountaintop had been something trivial—a new campground or a few more summer homes—he would have dismissed it out of hand. He took seriously the MGIO proposal because it was a major undertaking that promised important benefits. His own interpretation was that his job was "not just to sit back and say 'no,' but to see if I can make things work."

His admission that nonbiological considerations influenced his decision drew heavy fire from observatory opponents and from defenders of the ESA in general. It also occasioned further investigation by the General Accounting Office (since renamed the Government Accountability Office). The GAO criticized his decision on multiple grounds and concluded that "weighing the risk of a species' extinction with the benefits of a project is a policy decision and should be left to a high-level Endangered Species Committee"—the so-called God Squad, which is specifically authorized by the Endangered Species Act to act in such cases—or to Congress itself.[13]

Spear asserts that the GAO's censure did not ruffle him. "From a purely legal point of view, they are probably right . . . , but GAO's job is relatively easy. They're not wearing your shoes." More than twenty years later, he says, "I felt strongly I handled it right," and he emphasizes that he took no action against the employees who spoke out against him (a decision that redounded to his credit within the agency). Spear went on to represent the Clinton administration on a succession of heated western issues, nearly all of them pitting endangered species against powerful economic and political forces.

Spear's was a singular assessment of the Endangered Species Act. As the GAO pointed out, the role of the Fish and Wildlife Service was not to broker a deal, but to determine what actions were necessary to prevent the extinction of a species. The question was not whether the red squirrel would be marginally better off as a result of a BO that established a refugium and accommodated the observatory, but whether the prescriptions in the BO were sufficient to ensure the squirrel's survival. The elimination of a Bible camp and summer homes from squirrel habitat on which Spear placed considerable importance was postponed by the Arizona–Idaho Conservation Act and as of 2010 had still not been effected, but no one I talked to, except Spear, thought it mattered. In fact, the BO as a whole seems an oddly dated document today. Too many things have changed since 1988, the climate being one of them.

EVEN WITH PASSAGE of the Arizona–Idaho Conservation Act, the University of Arizona was not finished soliciting special treatment from Congress. In 1989, before construction of MGIO facilities had far advanced, the squirrel population crashed below the level contemplated by the Fish and Wildlife Service's Biological Opinion. Opponents of the observatory argued that the increased peril of extinction should trigger fresh "consultation" between the Fish and Wildlife Service and the Forest Service. Such a reexamination would likely lead to a new BO, which might ban development of the MGIO altogether. When the agencies resisted, a host of entities sued, and with a precedent-bending twist named the squirrel itself as lead plaintiff. The squirrel's colitigants included the Sierra Club Legal Defense Fund, the National Audubon Society, local Audubon chapters, the National Wildlife Federation and its Arizona chapter, Defenders of Wildlife, the Mt. Graham Coalition, and the Apache Survival Coalition. Court battles continued for the next three years (and some issues continue to be litigated more than twenty years later). Sit-ins and other protests broke out on the UA campus, on the mountain, and on the campuses of some of the UA's current and prospective consortium partners. Several hundred scientists, including many astronomers, signed petitions opposing construction of the observatory. By 1994, only the UA, the Vatican, Italy's Arcetri Observatory, and the Max Planck Institute of Germany remained in the project. The universities of Chicago, Pittsburgh, and Texas, along with Cal Tech, Harvard, Michigan State, and the Carnegie Institute, among others, had pulled out.[14] It was going to be very hard to finance the Large Binocular Telescope.

Then it emerged that the site specified for the LBT in the Biological Opinion and the Forest Service's EIS would render the installation too vulnerable to wind disturbance and damage. The UA needed to move its telescope 1,300 feet to the east. The Forest Service announced that the move was permissible. No dice, said the project's critics: the move would violate the MGIO's permit. They sued and won an injunction barring all construction at the site. The UA turned again to the Arizona congressional delegation.

Congressman Jim Kolbe attached a rider to the Omnibus Appropriations Bill of 1996 that decreed the new site to be consistent with the Arizona–Idaho Conservation Act and therefore compliant with applicable laws and permits.

The Appropriations Bill was itself controversial. Repeated standoffs between a Republican-controlled Congress and the Clinton administration caused the government to shut down for want of funds—or to approach the verge of shutting down—not once but several times. The Appropriations Bill represented a compromise that would restore stability to government operations. It was long overdue and badly needed, but the administration objected to scores of unrelated riders attached to it, including the new Mt. Graham provision, and it threatened a veto. In eleventh-hour negotiations, the administration succeeded in removing over half of the objectionable provisions. The Mt. Graham rider was not one of them. President Clinton signed the bill on April 25, 1996, with Kolbe's rider intact, and the path was now clear for completion of the MGIO.[15]

PUTTING ASIDE THE question of whether structures of any kind should stand atop Mt. Graham, let alone Mike Spear's highly personal assessment that an observatory was less damaging to the squirrel's prospects than other human activities, the Biological Opinion of the Fish and Wildlife Service, together with the Mt. Graham Red Squirrel Recovery Plan that followed it in 1993, got two important things wrong. The first was the notion that the red squirrel was wholly dependent on spruce-fir habitat. Subsequent surveys have shown that Mt. Graham red squirrels, including the lactating female Warshall and I saw on Grant Hill, make heavy use of mixed conifer forests. Whether this has always been the case or is a learned response to habitat change remains an open question, but there can be no debate that the squirrel's adaptability is fortuitous, given what has happened to the protected area that was intended to ensure its survival. The refugium for the Mt. Graham Red Squirrel succumbed to the second miscalculation: neither the BO nor the 1993 recovery plan gave serious consideration to climate change.

The usual disclaimer applies: it is impossible to demonstrate that climate change specifically *caused* the impacts lately visited on the refugium, but the changes to ecosystems on the top of the mountain precisely match the impacts one would expect climate change to produce. Those impacts ensue from the two plagues that rising temperatures bring to forests: insects and fire.

PETER WARSHALL IS waxing poetic as we pick our way through the forest that borders the observatory. He calls it "the cathedral forest of the West" and "the most beautiful old-growth forest—definitely—in the whole Southwest." But I must have missed something. I am trailing behind him dodging deadfall. Fully half the standing trees are defunct, their bark fallen away and their branches broken. An equal number lie jumbled on the ground, piled and canted across each other. We are not walking so much as climbing through the woods, edging around tangles and clambering over logs that bristle with the splintered stubs of former limbs.

"So, Peter," I say, "When you are talking about the beauty of the forest, is that in the past tense?"

He laughs. "Oh, absolutely." Sunlight pours down through the trees and spangles the forest floor. Warshall gestures at the barricades of logs that surround us. "If you resurrected all these trees and made them alive, you'd have an old-growth forest with a open understory and about 85 percent canopy cover. So it was very cool, very shaded." It was also producing an abundance of cones each year, "so it was supporting the squirrels really well."

I am halfway over a barricade. "We're climbing through this. You used to walk?"

"Right. And bear trails were very obvious, but now bears have obviously had to switch their locations."

In 1996, a Geometrid moth (*Nepytia janetae* Rindge) began defoliating spruce and fir trees at the highest elevations of Mt. Graham. The weakened trees were soon assaulted by a second wave of insect predators, in this case bark beetles. One species of beetle (*Dendroctonus rufipennis* Kirby) preyed on spruce trees; the other (*Dryocoetes confusus* Swaine) feasted on corkbark fir. The bark beetles killed the majority of the trees they infested, leaving behind large swaths of forest that were a paradise for woodpeckers but a death trap for squirrels. Even where a few cone-producing trees remained (and where a squirrel, out of loyalty to place, might try to hang on), the newly opened canopy made adequate refrigeration of middens impossible. The increased warmth, sunlight, and dryness also inhibited growth of mushrooms, an important alternate food for squirrels, and the physical opening of forest structure, with the loss of branch-to-branch aerial escape routes, exposed the squirrels more than ever to predation by goshawks and other raptors.[16]

PETER WARSHALL IN THE FORMER "CATHEDRAL FOREST OF THE WEST," AUGUST 2010.
*AUTHOR PHOTO.*

The irruption of insects—a Hitchcockian nightmare from a tree's point of view—was largely a function of climate. A series of unusually warm winters allowed the bugs to survive to spring in large numbers, and then to resume reproduction early and continue through the unusually long growing season that followed. And then came another mild winter, and another long hot summer. A decade-long drought, meanwhile, had rendered the trees dehydrated and ill-equipped to resist the invaders, and so the insects feasted easily and reproduced with exponential fecundity.[17] This was precisely the scenario that drove the beetle infestation that killed piñon and ponderosa pine across millions of acres in 2002 and 2003. The difference on Mt. Graham was that the infestation lasted longer and involved more kinds of predators. It was a brief and splendid Golden Age for being a tree-eating bug.

Populations of the defoliating moth finally collapsed in 1999, but bark beetle outbreaks persisted through 2002. To make matters worse, a nonnative insect, the spruce aphid (*Elatobium abietinum*), made its appearance in 1999 and set about devouring spruce needles, further hastening the decline and death of trees across the mountaintop. Squirrel population in spruce-fir habitat plummeted.[18]

The insect infestations were only a beginning. In 1996, the human-caused Clark Peak Fire roared up the higher slopes of the mountain and into the spruce-fir zone, burning about 7,400 acres of squirrel habitat at various levels of intensity. According to tree-ring studies, the last time fire had penetrated the spruce-fir forests of Mt. Graham was more than three centuries earlier— in 1685.[19] Obviously, the mountain was in the throes of sweeping change.

Worse was yet to come. In June 2004, lightning strikes four days apart sparked two separate fires: the Gibson, which puttered along demurely for almost a week and then blew up into a monster, and the Nuttall, which was fierce from the outset. Ultimately, the two fires burned together (with help from firefighters, who intentionally burned out fuels between them). As had been the case two years earlier with the Rodeo-Chediski Fire, the Gibson and Nuttall Fires became known as a single blaze, the Nuttall Complex Fire.

Tom Swetnam and his colleagues at the UA's Laboratory of Tree-Ring Research have reconstructed a detailed fire history for the Pinaleños Mountains. It shows the expected pattern: frequent low-intensity fires in the pre-settlement era, which abruptly cease when domestic livestock grazing commences. But one anomaly stands out. The return frequency of pre-settlement fire in the mixed conifer zone on Mt. Graham was unusually short, comparable to that of ponderosa pine forests in other locations. Swetnam surmises this may have been due to the high proportion of Southwestern white pines among the Douglas-firs on Mt. Graham. The long needles of the white pines may have upped the flammability of the mixed conifer community. In any event, the unusually frequent fires almost certainly produced an open and grassy mixed conifer zone, which would have insulated the higher spruce-fir forests from the effects of crown fire. Because of the open landscape and its modest store of fuels, a fire moving upslope on Mt. Graham, says Swetnam, "couldn't get up a head of steam."[20]

That was then. With the advent of Forest Service management, new habits of fire suppression and logging—more of the expected pattern—removed the mountain's (and the squirrel's) natural defense. Where once the mixed conifer zone was airy and open, it became choked with trees. And in dry times, as was the case in 2004, those trees were far too numerous to get by on the moisture available. They were primed to feed a fire, and they did. The Gibson and Nuttall Fires stormed into squirrel habitat with an overabundance of steam.

The result was catastrophic. More than a third of standing spruce and fir were already dead, thanks to the previous work of the insects. The top of the mountain presented a landscape of vertical stovewood, which soon became a furnace. The Nuttall Complex Fire burned about 29,000 acres, including 9,400 acres of red squirrel habitat, most of it severely. Over large areas, every tree was killed, if not also consumed. What once had been mountain forest became mountain bald.

The fire was exceedingly dangerous to the men and women who fought it, forcing the deployment of fire shelters in one instance and emergency safety tactics in another.[21] The backfires set by firefighters to protect the observatory destroyed a great deal of squirrel habitat, thus substantiating the view, held by many, that the observatory remained a threat to the squirrel. Other observers, like Swetnam, are not so quick to point a finger. Swetnam acknowledges that fire-fighting tactics are now causing some of the worst impacts of wildfires, but adds, "How can you second guess burnout decisions with the incredible behavior of these fires and their risk of fatalities?" He says, "The general strategy with the powerful fires we have now is to set large backfires that leave big burnout holes and stand back."[22] One thing certain about the response to the fire is that once the flames were big and charging up the mountain, all thoughts of the squirrel and its dwindling habitat were forgotten. Neither the Forest Service nor anyone else showed an inclination to prioritize the squirrel's needs, or had a plan on hand with which to do so.

THE NUMBERS ON habitat acreage tell a sorry tale. In 1993, before the visitations of fire and insects, researchers counted 13,257 acres of potential habitat. That number dropped to 12,197 in 2003, registering losses from the Clark Peak Fire (the most intensely burned areas) and insect plagues, although effects of the latter were still manifesting. Following the Nuttall Complex Fire (and after shockingly long delays in assessing habitat loss), satellite imagery from June 2008 indicated only 6,427 acres of remaining potential habitat in both spruce-fir and mixed conifer forests. By this reckoning, more than half of the squirrel habitat available in 1993 is now gone. It can no longer be considered habitat.[23]

Comparatively speaking, that's the good news. The bad news is that the numbers do not reflect the decline of the remaining habitat, much of which fire and insects have rendered inhospitable to squirrels. Forest that looks intact to Landsat may yet be a desert or a death trap for a creature that needs a substantially closed canopy and plenty of live, cone-producing trees. Moreover, the heavy impacts of fire and insects on Mt. Graham have fragmented the remaining forest so much that squirrels, as they range across the mountain (up to a kilometer or more) seeking mates or new habitat, are increasingly exposed to raptors, foxes, and other predators.

THE GOSHAWK THAT flew over Warshall and me, cutting fast, tight slaloms against a robin's-egg sky, made a splendid sight. It skimmed across the tree-carcass stubble of the Nuttall burn and dissolved into a speck above High Peak. Like the Mt. Graham red squirrel, it is a protected species, although at a less dire level. The same is true of another squirrel predator, the Mexican spotted owl. All three species, their populations beleaguered, have suffered from shrinking habitat, the squirrel in its lone mountain redoubt, the owl and the hawk across the entirety of the region's montane forests, where much of the old-growth timber on which they depend has been cut for timber or destroyed by fire. It may be ironic that one rare species—or two, in this case— might further endanger another by preying on it, but this kind of irony will only grow more common as climate change and the relentless pressure of development continue to shrink the habitats of species around the globe.

Not every creature will be a loser. Elk, coyotes, and other animals that are large, mobile, adaptive, and opportunistic will do fine. Species with wings or long legs will travel to new habitats, and as long as their food and behavioral requirements aren't too specialized and their reproductive habits make them good colonizers, they will prosper. The story will be different, however, for many creatures that do not meet those specifications, including many birds, together with legions of short-legged hoppers and skitterers, not to mention turtles and snakes. It's hard to pack up and move to the next mountain range if you have to carry your house on your back or if you travel on your belly. If you are a legless invertebrate, like the endemic Pinaleño talussnail (*Sonorella grahamensis*) that inhabits rockslides on the northeast face of Mt. Graham, you will have to hitch a ride on someone else. Good luck.

Size and mobility are no guarantee of success, however. Grizzly bears in the greater Yellowstone ecosystem face an uncertain future, because pine beetles, on the same arc of increase as those that ravaged the spruce and fir of the Pinaleños, have decimated Yellowstone's whitebark pines, and the energy-rich seeds of those pines, robbed from red squirrel middens and the caches of other animals, constitute a major source of nourishment for the bears.[24] Without it, grizzlies gravitate ever more toward livestock, garbage, and human settlements, a recipe for conflict that is usually fatal to the bears.

Most vulnerable to a warming climate will be species that become stranded on isolated landforms or some other kind of habitat island. The Mt. Graham red squirrel, of course, falls in this category, but it is not alone. The Goat Peak pika (*Ochotona princeps nigrescens*), a squeaky rodent that looks like a rabbit without ears, dwells among the high-elevation rockslides of New Mexico's Jemez Mountains and may face a similar fate. Sometimes the dangers that beset such animals are subtle. John Koprowski, a UA professor who has studied squirrels on Mt. Graham since 2000, points out that warming temperatures have rendered some long-established squirrel middens unviable.[25] They don't stay cold enough to preserve their contents, even though the forest around them appears healthy. And without a viable midden, a red squirrel won't survive the winter. A failure of refrigeration can imperil a squirrel as much as insects or fire.

Another subtle threat of climate change involves what scientists call *synchrony*. The interactions that sustain an ecosystem can be imagined as an intricate dance with a vast number of participants. For the minuet to be successful, each dancer has to make the right step at the right time. The migrating bird or butterfly, when it arrives at its destination, must find food (in the form, say, of a hatch of insects or newly bloomed flowers), or concealment (trees in full leaf), or some other resource at the correct stage of development. If the cue that prompts the initiation of the migration falls out of synchrony with the readiness of resources along the way—if the bird arrives in Utah but no insects have hatched—not just individuals but the reproductive success of an entire population may be compromised.

Climate change is not a ramp. Its increments will not necessarily be linear and smooth. More often than not, its effects will be choppy and chaotic. If the cycles of algae and shrimp, or a plant and its pollinator, fall out of phase, if the cues for mating, migration, molting, feeding young, or some other essential function "decouple" from the timing of phenomena vital to the activity's success, a cascade of negative effects can ensue. Under such pressure, systems can and will collapse.[26]

One response is to say that this kind of thing has happened over and over again through the ages of evolutionary change. Indeed it has. The diversity of the world has emerged from endless cycles of creative destruction. The difference now is the speed of the change, which threatens to outstrip the adaptive capacity of many creatures, as well as the systems in which they are enmeshed.

A warming climate in the Southwest will also be a drying climate, even if precipitation stays steady (but don't bet against a decline). This means amphibians and fish will be especially vulnerable to habitat loss. They will suffer not just from quantity problems such as lowered pond depth and diminished streamflow, but also from changes in quality. Fish like trout, including cutthroats, the region's hard-hit natives, need cold water. Rising water temperatures, which reduce the productivity of invertebrates on which most fish feed, can shut down a fishery as surely as the ash flow from a forest fire. Often, quality takes a nose-dive when quantity declines, because there is less clean water to dilute pollutants. On the Rio Grande, where Albuquerque's wastewater system is the river's sixth largest New Mexico tributary, evidence is mounting that pollution from discarded and excreted prescription drugs (birth control pills, anti-depressants), endocrine disruptors (from pesticides and herbicides), and possibly other chemicals is producing sexual aberrations in downstream fish, including the endangered Rio Grande silvery minnow.[27]

MOST AUTHORITIES EMPHASIZE that *connectivity* will be the key to successful wildlife management as the climate changes. Habitat areas need to link to corridors that allow wildlife to migrate northward or upslope (hopefully without running out of room, like the Mt. Graham red squirrel). Such connectivity needs to be factored into land-use planning, highway design, and many other areas of human activity. But connectivity is not the whole story. Equally important to the preservation of biodiversity in an era of climate change is what you might call "societal promise-keeping," the business of sticking to agreements.

Let's say a couple of agencies and some NGOs cobble together a deal to guarantee a few acre-feet of water for a wetland somewhere. The immediate goal is to keep alive a pupfish or turtle or an endemic sedge, none of which will profit from any kind of wildlife corridor. Such creatures have to stay where they are. And protecting the fish or the turtle or the sedge is not the only issue. More generally, the wetland's defenders acknowledge that oases of all kinds—rivers, creeks, marshes, seeps—are the key to preserving the biota of a place, for the water they provide supports whole suites of creatures and their myriad ecological relationships. At certain times, for instance, the fox that dens on the dry mesa several miles away might depend on the wetland as utterly as the pupfish. So the agencies and the NGOs negotiate a deal, based on a stack of hydrological and biological reports and a year or more of meetings, stipulating that a certain amount of water will be bought or leased annually for the wetland (and one of the agencies will provide the money) or that specific water rights will be reserved and defended. It is a win for everyone, reported with gusto in press releases and on multiple websites, and everybody feels good.

Ten years pass. The people who negotiated the deal have retired or taken other jobs. The land is drought-stricken. Funds are tight. The cost of water has shot up. A new generation of domestic and commercial wells in the watershed (who let them get there?) are pulling down the water table at a great rate. Delivering water to the wetland will require extraordinary effort from all parties to the agreement. But every organization is fighting urgent battles on other fronts. Their harried workers, fewer than before because of budget cutbacks, are stretched thin. Money, time, energy—nobody has enough. The courts are unfriendly. No recourse there. And so the wetland dries. Most of the creatures that depend on it die in place. The rare species that started everything blinks out, and the land bleeds a little more in its death from a thousand cuts.

An adage of environmentalism holds that "all victories are temporary, but defeats are permanent." Too often, the adage proves true. Keeping a promise is always harder than making it, as the history of the Mt. Graham red squirrel attests. The promise implicit in listing a species as endangered is that the government, as society's representative, will take all practical steps to assure the species' survival. The general outline of those steps is spelled out in the applicable environmental laws. Putting aside the University of Arizona's dishonorable circumvention of the ESA and NEPA, and putting aside as well the controversial decisions made by Mike Spear and others, a key step in the protection of a species is the development of a "recovery plan," which is a recipe of actions intended to assure the species' survival.

Government biologists finally completed a recovery plan for the Mt. Graham red squirrel in 1993, almost five years after the Arizona–Idaho Conservation Act set construction of the MGIO in motion. The plan was slow in coming, but there it was. Then the insects arrived. Then the Clark Peak Fire in 1996. By 2002, it was clear that the squirrel refugium, the cornerstone of the existing recovery plan, no longer provided much refuge. In agency-speak, "it was determined that the status of the red squirrel and the threats it faced had changed," and therefore the recovery plan would have to be revised.[28] Two years later, came the devastating Nuttall Complex Fire, which should have spurred intense activity. If it did, the motion was all Brownian. At the close of 2010, the revision of the seventeen-year-old, profoundly off-the-mark recovery plan had still not been accomplished.

It is easy to blame the delay on foot-dragging by federal biologists and others who are charged with revising the plan. But the issue is not so simple. On the whole, the agency professionals are highly motivated and qualified, albeit overworked and underfunded. The problem is how things get done— or not. People change jobs; priorities get dropped or shifted. The Recovery Team, collective authors of the plan, includes dozens of people, and getting even half of them in one place at the same time is a cat-herding challenge of the highest order. Then, just when you might think things are coming together, the Fish and Wildlife Service or some other vital agency gets sued, and it's all hands on deck to produce a detailed report by some court-ordered deadline. The plan for the squirrel gets shunted to the side again.

Steady progress is extremely difficult under such circumstances. Case in point: in 2010, a few days after explaining to me that she hoped to complete the Recovery Plan Revision by year's end ("It would be such a big relief!"), the hard-working biologist serving as principal author was detailed to the Gulf Coast for four weeks to help deal with consequences of British Petroleum's Deepwater Horizon oil spill.[29] The red squirrel recovery plan would have to wait, again.

In the end, however, the recovery plan may prove to be less important than its name implies. The real theater of engagement, as far as the squirrel is concerned, has shifted from the Fish and Wildlife Service to the Forest Service, which is pushing forward what it calls the Pinaleño Ecosystem Restoration Project, an intended overhaul of Mt. Graham's forests that goes by the unfortunate acronym PERP.

In the aftermath of the Nuttall Complex Fire, PERP initially took shape as a fuels reduction program, an effort to thin the forests of the mountain in order to prevent future recurrence of crown fires. As planning advanced, however, the Forest Service broadened its circle of consultation, and PERP became more ambitious. It metamorphosed from a narrow focus on fuels to a broad embrace of ecological restoration.[30] A lot of people and a good deal of pressure from conservation groups helped nudge it in that direction, and one of the important influences was John Koprowski, a professor of ecology at the University of Arizona. Since 2000, Koprowski has headed the Mt. Graham Red Squirrel Monitoring Program, which is funded by the university as a condition of the MGIO's Special Use Permit. His impact on PERP derives not only from personal involvement but also from the impressive list of publications (more than fifty at last count) that he and his students and collaborators have produced over the past decade. It is probably fair to say that they have enlarged scientific understanding of the red squirrel by an order of magnitude.[31]

Enriched by this new learning, as well as by contributions from such other sources as the Arizona Game and Fish Department, PERP's planners made a series of changes, both large and small. They enlarged the area of nondisturbance required for each squirrel and its midden, and they intensified requirements for buffer zones and squirrel surveys. They also set specific goals for leaving large logs on the ground and generally tried to bend their fire and insect mitigation work toward the creation of squirrel-friendly habitat. This included being conscious that dense stands of mixed conifer favor the red squirrel, while more open stands favor the Abert's. With the demise of so much of the mountain's spruce and fir, the mixed conifer now holds the key to the squirrel's future. In those stands, the red squirrel directly competes for food and perhaps for territory with its larger, introduced cousin. The presence of the Abert's squirrel represents yet another threat to what Koprowski terms the red squirrel's "precarious" hold on existence.

PERP will require a decade and millions of dollars to accomplish its goals. In theory, it will buttress the red squirrel's survival by thinning a little here, a little there, year by year, while avoiding excessive disturbance to the squirrel or diminution of its remaining habitat. *In theory.* As with ambitious burn programs and other large-scale land treatments, PERP will test the ability of the Forest Service to carry out a complex program over an extended period of time. Sometimes the agency succeeds in doing what it sets out to do; sometimes its plans break down. And sometimes the treatments it executes on the ground do more harm than good.

Mt. Graham has already been the subject of more than a century of intense manipulation. Grazing, road construction, logging, fire suppression, summer homes, hunting and recreation, and at least one wildlife introduction have altered the mountain in profound ways. Absent those changes, the Mt. Graham red squirrel would stand a much better chance of breasting the challenges of climate change. Now, more manipulations are proposed to mitigate the manipulations of the past. These will be best-laid plans, based on present knowledge, which is never complete, and carried out by fallible individuals and even more fallible organizations. Meanwhile, climate and contingency, as ever, will have the last word. A drought, a fire, or a disease or infestation afflicting Douglas-fir, the squirrel's last stronghold, may send PERP and all related plans spiraling into irrelevance. It is no wonder that the Fish and Wildlife Service has initiated planning for a captive breeding program for the squirrel.[32]

Despite the daunting odds, PERP may nevertheless represent the red squirrel's best chance for prolonging its 10,000-year stay on Mt. Graham—and the planet. For PERP to succeed, the Forest Service and its congressional funders will have to show a degree of persistence and resoluteness of purpose that neither body is famous for. Society will have to keep a very complicated set of promises.

The outcome will be telling, and not only for Mt. Graham and the red squirrel. Koprowski soberly points out that "Mt. Graham is a relatively simple, isolated system" in both social and ecological terms. Notwithstanding the turmoil and contention described in the foregoing pages, it is a single mountaintop. There are a finite number of interested parties, and the terrain is well-bounded and encompasses at most twenty or so square miles. "But we still face enormous challenges," he says. If achieving consensus in a place so small can remain elusive over the course of decades, imagine the difficulty of securing long-distance connectivity for an animal with large range requirements or otherwise contending with the wildlife impacts of climate change at the scale of whole watersheds or multiple mountain ranges. Koprowski reminds us that, if exemplary work cannot be accomplished on Mt. Graham, the prospects elsewhere are less than good.

The present state of affairs offers yet more irony. Tendrils of the squirrel's fate are now tied to PERP, and PERP substantially draws on ten years of research by Koprowski and his shop. In an ideal world, the ESA, by itself, would have caused a comparably intense research program to spring into being, but it didn't. The research Koprowski and his team have accomplished would not have taken place without the MGIO intruding on the scene, together with the funding from UA that followed. Lest one conclude that the MGIO, in the end, proved beneficial to the squirrel, let it remembered that the requirement for monitoring and research was an outcome not of the observatory, but of the fight over the observatory. It was a small but perhaps hugely consequential concession to the many groups and individuals who had rallied to the other side of the conflict, campaigning not just against the MGIO *in that location*, but for the mountain, for the ESA and NEPA, and for the survival of Apache culture.

IN 1993, CONSTRUCTION of the Vatican telescope was under way. Workmen had cut down several acres of trees, and heavy equipment dug a great dark hole where the building to house the telescope would stand. The wounded earth gaped open. The mountain had been gored and torn.

That's how Warshall remembers the cold day when a small group of Native Americans came to the mountaintop to pray for Mt. Graham. Sioux, Cheyenne, Hopi, and native Hawaiian, among others, were represented. Ola Cassadore Davis, leader of the Apache Survival Coalition, was the dynamic center of the group. The other Apaches who came to Mt. Graham that day elected to stay behind, lower down on the mountain, explaining that "their prayers wouldn't go up if the land had been hurt this bad."[33]

Warshall was there, too. He led the group to a rock outcrop in the forest, perhaps a quarter mile from the construction site, and Gary Holy Bull, a Sioux medicine man from the Pine Ridge Reservation, led them in a modest ceremony. They made a pile of twigs about six inches high. They stood in a circle around the twigs and lit them. They threw sage on the fire for purification. They prayed in their various languages for the health of the mountain. Warshall remembers water drops falling on moccasins and steaming from the heat of the fire. It took him a moment to realize the drops of water were tears. It is a sentimental detail, but unforgettable.

Equally memorable was Gary Holy Bull's advice after the ceremony was over. They were walking downslope, toward the vehicles. Holy Bull asked, "Does this mountain have brothers or sisters?" Warshall was taken aback; he'd never thought in those terms. Brothers and sisters? Well, maybe, yes. There was Mt. Baldy, and also Mt. Turnbull, and the San Francisco Peaks. There were other Sky Islands.

AERIAL VIEW OF MT. GRAHAM SUMMIT AND MT. GRAHAM INTERNATIONAL OBSERVATORY, MARCH 2005. THE DOMINANT STRUCTURE AT CENTER RIGHT IS THE LARGE BINOCULAR TELESCOPE (LBT) OPERATED BY THE UNIVERSITY OF ARIZONA. DIRECTLY BELOW IT IS THE HEINRICH HERTZ SUBMILLIMETER TELESCOPE, AND BELOW AND TO THE RIGHT, OBSCURED BY TREES, IS THE VATICAN ADVANCED TECHNOLOGY TELESCOPE. VIRTUALLY ALL OF THE FOREST IN THIS VIEW SHOWS THE EFFECT OF EITHER INSECT-DRIVEN DIE-OFF OR CROWN FIRE (PRINCIPALLY THE NUTTALL COMPLEX FIRE OF 2004). *PHOTO COURTESY OF WILLIAM CROSBY.*

Holy Bull said, "You'd better go pray on those other mountains, because this mountain is too weak to defend itself."

Warshall believes some of the group later did exactly that, but if so, the prayers failed to fend off further construction, or insects, or fire. It is unscientific to say that Mt. Graham has become weak, but anyone familiar with conditions on top of the mountain would have to agree that the metaphor is apt. And meanwhile, a few hundred red squirrels, even with the heart of their habitat "whacked" (as Koprowski put it), somehow contrive to harvest cones and stash them and make it through the winter and go roving about for a few days a year looking for mates. How long they will be able to continue no one knows.

I asked Koprowski how he felt personally about the creature he has studied so closely. I wondered what the squirrel *meant* to him. He had a ready and immediate answer. "I have an incredible level of admiration—I don't want to make it sound strange," he said, "but this animal has persisted through so much, and they are still able to persist today in a situation that is very far from ideal."

The same might be said for a lot of humans today. And it will be said even more in years to come, as populations contend with rising seas, devastating storms, and shifting agricultural and ecological zones. Mt. Graham's final irony may be that the environmental predicament of humankind differs from that of the squirrel more in scale than in kind. We share a very important circumstance with *Tamiasciurus hudsonicus grahamensis*. The squirrel has its mountain; we have Earth. And that's all. Our small blue planet is a habitat island, too.

# 11

## HAWIKKU: WELCOME TO THE ANTHROPOCENE

HISTORY COMMENCED FOR a large portion of North America on July 7, 1540.

One may quibble over the dimension of territory affected, but the date is certain: it was duly written down, which is what distinguishes history, narrowly defined, from other ways of recording human experience. History is based on a documentary record; it begins when the documents start to pile up.

On that July day nearly five centuries ago, a small army of Spaniards and Mexican Indians gathered before a stone village in what is now New Mexico and informed the residents thereof that henceforward they owed obedience to someone called the Pope and to the Catholic Sovereigns of Castile, in whom the Pope had entrusted authority over the bodies, souls, and lands of every person living on the islands and continent of the Mar Océano. This included the people of the stone village. No doubt these ideas sounded strange to their intended hearers, if indeed they were translated with the remotest accuracy.

The intruders' puzzling message was shouted out by someone who appeared to stare at a cloth, or a material similarly thin, flat, and flexible. It was a sheet of thick paper, nearly the size of one of the natives' bison-hide shields.[1] It bore many small marks, but no one from the village was close enough to observe this. Not that it mattered.

The document asserting Spanish sovereignty over the village and its people was called the *Requerimiento*. Its purpose was to justify, in legal terms, the bloodshed and robbery that were soon to follow. An official of the Spanish host duly recited the entire document, which stipulated, on behalf of his commander, Captain-General Francisco Vásquez de Coronado, that if the natives failed to obey the commands of the *Requerimiento*, "I will make war against you everywhere and in every way I can. And I will subject you to the yoke and obedience of the Church and His Majesty. I will take your wives and children, and I will make them slaves.... I will take your property. I will do all the harm and damage to you that I can.... I declare that the deaths and injuries that occur as a result of this would be your fault and not His Majesty's, nor ours, nor that of these *caballeros* who have come with me."[2]

Certainly, communication of the *Requerimiento* was inexact, but not for lack of emphasis. In conformance with royal instruction, the drama of its reading was to be repeated three times, with notaries present. The interpreter who echoed the shouts of the first reader probably faked a good deal of his translation, since words in whatever Indian language he thought he was using would not have existed for quite a few concepts embedded in the Spanish message. Nevertheless, the natives did not need words to comprehend the intentions of the army before them. The intruders' armor, lances, and swords, to say nothing of their demeanor, spoke eloquently of their purpose. (It may have taken a little longer for the natives to register the implications of other Spanish technologies—arquebuses, crossbows, and even horses—but understanding, when it came, undoubtedly came fast.)

A few men from the stone village, which was called Hawikku, together with others from nearby communities, had probably tracked the strangers for at least several days previous. They knew that the intruders had followed the Little Colorado River to a point close to its confluence with the Zuni River. It is possible that some of them went all the way to the confluence, a place of great sacredness (known to non-Zunis as Zuni Heaven), which in the sixteenth century was still an expanse of marvelous wetlands. Some sources suggest that an element of Coronado's force may have interrupted native pilgrims traveling to or from the site. The encounter would have occurred close to the summer solstice, the time of such pilgrimages. It is also possible that the Spaniards angled away from the confluence, "straightening their route" as Coronado had ordered an advance detachment, in order to contact the Zuni River farther upstream. From there, they ascended the Zuni, which was then a substantial and perennial creek. Their trajectory put them on a collision course for Hawikku.

As the natives tracked the invading army, their intelligence may have been good enough to know that the intruders were starving. They had been trekking northward for seventy-six days since starting out from Culiacán on the west coast of Mexico. They traveled light, with little but the food that each man could carry or place on his horse, if he had one. Those stores had long since been consumed. The native scouts may have known that several days earlier one Spaniard, two Moors, and an unrecorded number of Indian allies had died from eating a poisonous plant, on account of their desperate hunger.

Prudently, Hawikku's women, children, and elders, except those who commanded the pueblo's defense, had been evacuated to safe refuges, probably most of them to Dowa Yalanne, or Corn Mountain, the majestic mesa that towers above modern Zuni. Only men of fighting age, at most 200 or 300, remained to face the strangers. For their part, the intruders included seventy-five soldiers mounted on gaunt horses and thirty Spanish footmen, carrying among them perhaps a dozen arquebuses (the terrifyingly loud, smoke-belching precursors of the musket), and several hundred "indios amigos," the unsung enablers of Coronado's conquest, who had been conscripted from tribes far to the south. The Spanish throng also included assorted friars, lay brothers, and camp followers, who might better have been called campmakers, for this was the task of the servants and the unknown number of women who accompanied the march. The force arrayed before Hawikku was only the advance guard of Coronado's expedition. The more numerous main body was even then moving northward, a month or so behind.

It is not recorded whether the *Requerimiento* received its full, triple-repetition recital before Hawikku on that fell July day. At some point the natives let loose their answer in the form of a volley of arrows, which did no serious harm, except that one pinned together the skirts of a friar, a comical touch on an uncomical day. The Spaniards then spurred their horses and charged, perhaps to the battle cry of "Santiago!," which had accompanied many a bloodletting in the Old World. They swiftly drove most of the defenders back within the walls of the village. Others fled to the hills. Once the village was surrounded, Coronado directed his crossbowmen and arquebusiers to pick off the defenders atop the walls at what seemed the principal entry, but "in short order the strings of the crossbowmen's weapons broke, and the arquebusiers accomplished nothing because they were so weak and debilitated that they could hardly stay on their feet."[3]

As a result, when Coronado and others stormed the walls, they were pelted by "an infinity of large stones," which the defenders rained down on them. Coronado got the worst of it. His gilded armor and the plume on his helmet attracted the defenders' energies. He was knocked to the ground, shot in the foot with an arrow, wounded twice in the face, and battered by stones on his arms and legs. The rock that knocked him down a second time nearly finished him, but his life was saved by his "excellent helmet" (now dented) and his comrade-in-arms, López de Cárdenas, who threw his body upon his fallen captain to shield him. Other "companions" thronged to Coronado and carried him from the field of battle.

In his subsequent report to the viceroy, Coronado detailed his own wounds and those that befell a handful of fellow Spaniards, none of which was severe. Of injuries suffered by his Indian allies he says nothing. As to the defenders of Hawikku, "Some Indians died, and more would have, if I had allowed them to be followed." Coronado records that, once the surviving natives had vanished and his men possessed the village, they attained their immediate and urgent tactical objective, which was food. In the storerooms of Hawikku they found "as great an abundance of corn as [we] were seeking [in] our necessity." They also availed themselves of beans, turkeys, and, as another chronicler put it, "the best and whitest salt I have seen in my whole life."[4] Centuries later, Zuni Salt Lake still produces splendid salt.

SOMEWHERE IN THE ruins of Hawikku lies the stone that conked Coronado and nearly killed him. Presumably it is buried in the mounds of rubble that lie fathoms deep on the south face of the low mesa where the village stood. Time has chinked those incalculable tons of sandstone with a quarry's worth of windblown soil, the remains of earthen roofs and plasters, the decayed wood of lintels and roofbeams, bushels of potsherds, and the random, decomposed remains of tumbleweeds and packrats. The ruins, in complete collapse, descend the face of the mesa obedient to the angle of repose. They are all but indistinguishable from the mesa itself.

I am standing at the top of the ruins with Edward Wemytewa, who gestures toward the plain where first contact between native flesh and Spanish steel took place. "There is blood in that sand," he says.

The plain stretches away to the south and west, yielding to scrub-tree ridges on the horizon. Coronado's starving army would have arrived under a flutter of tattered banners and a plume of dust rising from a thousand shuffling feet. It would have arrayed before the pueblo in full view of the angry,

RUINS OF HAWIKKU, MARCH 2010, LOOKING WEST-SOUTHWEST. IN 1540, CORONADO AND HIS ARMY PROBABLY CAME INTO VIEW SOMEWHERE NEAR THE EXTREME LEFT OF THE IMAGE. *AUTHOR PHOTO.*

nervous, outnumbered defenders. The Hawikku elders responsible for defense would have stood on a rooftop, surveying the same view of the plain that now spreads before Edward and me. The sun blazes. We shield our eyes from its low-angled light. A raven glides by, inspecting us. We can hear the rip of air against its wings. I find myself wondering if the history of Spanish America and the native Southwest might have turned out differently had Coronado's rock been ten pounds heavier or thrown with fatal force. A intriguing prospect, but doubtful. The history he had started was not about to stop. In 1540, the momentum of Spain's imperial impulse, like the momentum of contemporary climate change today, had far to run. If Coronado had died a martyr to King and Church, other ambitious knights would have taken up his crusade of conquest. Undoubtedly, their efforts, like his, would have ended in disappointment. Coronado was looking for wealth. His chief hope was to discover populations large and prosperous enough that their payment of tribute (in corn, labor, hides, precious metals, or other commodities) would render colonization profitable. He never found the cities he dreamed of, but in time, colonization limped forward anyway.

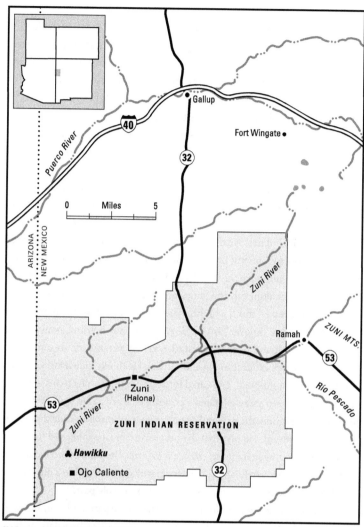

**MAP 10:** ZUNI AND SURROUNDINGS.

THE WARM APRIL day is now cooling, and the late sun flings our shadows to infinity. We walk carefully in the spring soil, which is as soft as a bruise. Edward is not happy with the condition of the ruins. Truck tracks coming from the mesa behind Hawikku have rutted the approach, and they reach nearly to the mounded remains of the room blocks and to the bowls of former kivas. No effort has been made to protect the site, let alone interpret it. The prevailing view within the community, says Edward, "is that the place is full of bad memories. They say, 'Why do you want to bring that up and remind people?'" At Zuni, the events of 1540 still hurt.

Edward is no stranger to the intricacies of politics within his community. At the time of our visit to Hawikku, he was a former member of the Zuni Tribal Council, soon to be elected to the council again. He has also represented the tribe on a number of delicate matters. When an Arizona utility sought to open a coal mine close to Zuni Salt Lake, the campaign to block it and thereby protect the groundwater on which the lake depends began in Edward's kitchen. It took years of effort, but ultimately the campaign secured the lake's protection.[5]

Edward is dressed this day in baggy jeans and a thick wine-colored shirt that he has just put on over his tee. He wears his gray hair in a ponytail and sports a wispy salt-and-pepper moustache. When he takes off his sunglasses, the look in his eyes is questioning and earnest. He thinks Hawikku should be a tribal park or monument, and cared for accordingly. Parallel to his life as an activist, Edward is a storyteller and a performance artist. He works from the premise that to ignore the past and its unhappy emotions is to cede them control over the present. He says that the best way to escape inherited trauma is to face it head-on, even if the experience is painful. This kind of reconciliation is the goal of his storytelling. He believes Hawikku is integral to Zuni history and that it should be part of the construction of a usable past.

Zuni does not lack for reminders of how harsh the past has been, and continues to be. Innumerable encroachments by neighboring populations have pared away access to extensive areas of traditional land use. The sacred springs at Ojo Caliente are in decline. The Zuni River, which bisects the reservation, is more a gutter than a river nowadays. Every time he glances at it, Edward is reminded of how much the tribe has lost. Notwithstanding that the Zuni River sustained native communities for more than a thousand years, it is now, like the Gila, bereft of water most of the time. Mormon settlers came to its upper watershed in the late 1800s and dammed a tributary to impound water for irrigation. The result was Ramah Lake, a scenic gem and an agricultural reservoir whose water never gets to Zuni. Additional impoundments, multiple dam failures on other tributaries, and heavy-handed logging and grazing in the Zuni Mountains have combined to entrench and dry up the river, lowering the water table of the lands through which it flowed, where the Zunis made their farms. Today, 80 percent of the tribe's former agricultural lands won't grow corn or any other useful crop.[6] The proliferation of household wells among the real estate developments of the upper watershed has sealed the river's fate. "If you want to see what happened to the Zuni River," says Edward, "just look at the State Engineer's map." It records the locations for which well permits have been issued, and the map for the Zuni watershed, upstream of the reservation, is stippled with hundreds of pinpoints, each one tapping the underground life of the river, draining it away, gallon by gallon.

AT THE EDGE of the sun-washed plain where Coronado marched, the past seems less hidden than in other places. In the aridlands, dehydration assists preservation, and especially in places like Zuni, where an emphatic human presence is measured not in centuries but in millennia, the bones of earlier eras are less decayed than merely weathered, and they lie wherever you look. Hawikku is like a whale that has merged with a mesa, the curves of its giant ribs echoing the curves of the land. Its presence is brooding, not dead. In such a place, the future can seem more than usually visible, too. The Zuni River has already endured what lies ahead for many aridland watercourses. Its downcut and water-starved channel presents a picture of what violent storms and a warming, aridifying climate may do to scores of other streams. Even more, the idea of history restarting, or a new history beginning, as happened here in 1540, has resonance for the present. It is as though we stand today before our own stone village, and in the distance we see a rising plume of dust, signaling the approach of a new kind of future.

In 1807, a group of scientists interested in ascertaining the age of Earth and pursuing other provocative questions formed the London Geological Society. It was the world's first national organization dedicated to the earth sciences. Two centuries later, in 2008, the Society's Stratigraphy Commission, a respected voice on the classification of geologic time, delivered a portentous report to the main body. The impact of human-induced change on the planet, said the commission, "has so intensified to make our present interval comparable to major global perturbations of the geological past." As a result, "a new geological epoch, worthy of formalisation, may indeed have commenced."

The members of the commission were not merely alluding to the age-old human capacity to alter ecosystems. Rather, they sought to draw attention to something new under the sun: a cumulative and determinative human effect on core planetary processes, primarily climate. They said that the Holocene epoch, the period since the end of the last ice age and the incubator of human civilization, was now at an end. A new geologic epoch had begun. The name the commission suggested for the new epoch was already in currency in certain circles: *Anthropocene*—the human or human-shaped epoch. The commission postulated that one might mark its onset at about the time the Society was formed, at the start of the Industrial Revolution. In the life of the planet, as well as the history of humankind, the Anthropocene constitutes a new beginning, the dawning of a new drama, and it is happening now—in the "long now" of the geologic present.[7]

IT MAY BE a new drama, but the dramatics are familiar. The essence of tragedy in Shakespeare's time was considered to be an essential blindness, an inability of the tragic figure to recognize the flaw in himself that causes his fall. Other classical traditions may place the root of tragedy in the stars or immutable fate, but the Bard lays it squarely, and for the most part curably, in human laps. Hamlet could have pulled himself together. Lear might have appraised the motives of his daughters, and himself, more shrewdly. Macbeth might have ignored his wife, and Othello trusted his. The problems that exist at the level of these invented characters exist equally at the level of society. Whom to believe? Whether and when to act? How to put aside doing what one wants and commence doing what one must?

Notwithstanding a large cast of senatorial ideologues, right-wing bloviators, and modern-day Iagos lobbying for Big Coal and Big Oil, the protagonists in this drama are the rest of us, our collectivity, the commonweal, including the creatures and communities of nonhuman nature.[8]

Climate change denial is reckoned by many to be a particularly contemporary form of blindness, and it has spawned a cottage industry of psychological investigation. The American Psychological Association, in an extensive report, says the problem may be less a matter of reasoning than of cognition.[9]

It seems that the human brain and nervous system are well adapted to respond to immediate threats like a cobra in the grass or a menacing fellow with a club coming over the hill. But when it comes to longer-term problems that engage reason alone while leaving the emotions and adrenal glands at rest, humans are not much better than the legendary frog in the gradually warming pot.

The list of mental obstacles to action on climate change identified by the APA report reads like a catalogue of biblical afflictions: Ignorance, Uncertainty, Mistrust and Reactance (i.e., mulishness), Denial, Habit, Impotence ("What can I do?"), Tokenism, Conflicting Goals, and Belief in Divine Intervention.[10] This is but a partial list; the categories of paralysis are nearly inexhaustible. Other researchers have probed people's habitual dislike for delayed benefits, or the notion that most of us possess "a finite pool of worry"—that we can fret about only so much at a time and that new calamities like a plunging stock market or a loved one's illness will push old concerns out of sight and out of mind.[11] In the end, however, all of these defenses against change are probably different aspects of the same phenomenon, which no one has pegged more concisely or wryly than the economist John Kenneth Galbraith. He is credited with saying, "Faced with the choice between changing one's mind and proving that there is no need to do so, almost everyone gets busy on the proof."

Such behavior was hardly unknown in Coronado's day. In 1510, the Spanish Court got busy with proving that nothing was morally amiss in the New World. Members of the court wanted to allay their own fears and the charges of their critics that Spain's *conquistadores* were abusing the natives. The result was the *Requerimiento*, which Castile's foremost jurist drew up in order to clarify the Holy Authority that justified Spain's conquests and to tidy the ethical side issues of murder, enslavement, and theft. With the *Requerimiento* in place, the entire Court—the royal family, their friends, ministers, and advisers—felt better, and revenues were unaffected.

FOR MANY PEOPLE, climate science is a jumble of numbers, but there are really only a few that the informed citizen need keep an eye on. The current reading of the Keeling Curve (discussed in Chapter 1) is one of them. It also makes sense to know that the average annual temperature in the United States has risen roughly 2°F during the past half century, and that there is already enough momentum in the climate system to push it significantly higher. If present trends of carbon pollution hold steady, the increase could amount to

8°F for parts of the country (including parts of the Southwest) by the end of the century. And that's the *average*. Extreme weather, which will become more common, will likely spike higher than past highs, and perhaps lower than past lows. Phoenix's brutal old high of 122°F, set in 1990, may one day be remembered with nostalgia.

There are two broad avenues for addressing the problem of climate change: mitigation and adaptation. Mitigation involves reducing the amount of carbon pollution being injected into the atmosphere and embarking on the difficult journey back to 350 ppm. The most reasonable strategy toward this end is for government to place a price on carbon, essentially to tax it at the point in the production cycle at which it becomes a pollutant. The majority of the revenues thus generated might be redistributed as dividends to the consumers to whom the cost of the tax is passed. Competitive pressures within industry would do the rest.[12] Were the United States to take such action, other industrial nations, including China, which recently passed the United States as the globe's leading carbon polluter, would likely follow suit. Absent U.S. leadership, however, prospects for concerted global action are close to nil. Perhaps only catastrophe will change the current intransigence.

Which leaves adaptation. Given that the climate is changing, and the lands and waters with it, where should we position our boat to enter the coming rapids? What should be our angle and velocity at the point of entry?

The curious thing about adapting to climate change in the North American Southwest (and indeed in most places) is that the adaptations commanding highest priority are tasks that have needed doing for a long time, irrespective of climate. They are the difficult, postponed chores that never went away. None should come as a surprise; all involve the pursuit of resilience. Near the top of anybody's list, I should think, would be the achievement of water security, the rehabilitation of forests, and the task of devising a responsible program for dealing with displaced and work-starved populations. A word on each, in reverse order:

When the conditions that sustain human occupation of a place change, people move. This works in both directions, push and pull. The advent of air conditioning helped "pull" people to the hot deserts of the aridlands. Now something like the reverse may be taking place. Increasing heat and drought will push people away, especially from northern Mexico, where the resources to counteract such changes are limited and where the livelihoods of large numbers of people are slipping below the margin of subsistence. The experience of Sand Canyon Pueblo and other communities of the ancient world reminds us that responding to population flux has been a crucial challenge for human society since time immemorial. That the task has never been easy is perhaps the best excuse for the recent performance of the United States in this arena. Demagoguery too often carries the day; resilience, in the form of public consensus on a comprehensive set of policies, seems unachievable amid the distortions of the current (mostly shouted) public conversation.

Perhaps a little further into the future, beset by interlinked shortages of both water and energy, and thus deprived of power to mollify the climate, many of those now shouting loudest about newcomers (together with their more tolerant neighbors) may themselves be forced to pull up stakes and become newcomers elsewhere. If so, their shouts will surely strike a different note. Ultimately no community is immune to climatic change, even when it is as well defended as Phoenix. Large, scaled-up systems can fail for the same reasons as small ones. As the Hohokam demonstrated, if they cannot be adapted to new circumstances, the people who rely on them will have to move.

Forest management takes place in less contested terrain. Substantial agreement exists on two key points: first, that managers need to reduce the density of forest stands and the build-up of fuels in key ecosystems, and second, that public funds will never be adequate to do all that needs doing. Therefore, triage is necessary. A responsive strategy would take advantage of the mosaic of burns already present in the landscape, as Tom Swetnam advises, and sanely prioritize the protection of human communities and singular natural areas, including places like Mount Graham. All of this is possible, yet difficult, and will require an abundance of insight, flexibility, and constancy. Meanwhile, the continued penetration of fire-prone wildlands by subdivisions that place people and property in harm's way needs to be arrested. This particular form of "playing with fire" persists as an item of urgent but unfinished business in hundreds of jurisdictions, as well as within the sleepy precincts of the insurance industry, which underwrites the risks.

Water security tops all other issues for difficulty and importance. It is the key to stable communities and viable agriculture. It is also vital to ecological protection, because none of the region's surviving rivers and springs is safe from raiding, if not hijacking, if human communities are thirsty enough.

The story of the West is essentially a story about water, and its lack. The central issue of the region, identified in modern times by John Wesley Powell and eloquently echoed by Wallace Stegner, among others, is formed by the tension between "the aridity that breeds sparseness and the denial of that condition, which leads to overdevelopment."[13] Previous generations quested after water security in the arid West, and at times largely attained it. The colossal dams, reservoirs, and aqueducts that constitute the plumbing of the Colorado and other western rivers created an abundance that, to the popular eye, seemed inexhaustible. But continuous economic and demographic growth has a way of turning abundance into scarcity, and the hydraulic cornucopia of the late twentieth century now shrinks in the twenty-first. The immense bright wall of the high dam, its smooth arc holding back a liquid prairie, is the icon of the booming, postwar West. But the icon soon to replace it will likely be the mudflat, the dead pool, and the ominous bone-white bathtub ring—the calcic band that marks where the water used to be.

Eventually, under any set of circumstances, the continuous growth of western America would push the curve of water need to the limit of what the rivers and aquifers can provide. Climate change only accelerates the day of reckoning. It doesn't wait for a curve to approach the limit; it moves the limit toward and even past the curve, hastening the final mad scramble to reconcile a finite and declining resource with unbounded appetite.

A first order of business would be to resolve the water-related tasks that have awaited the people of the region since long before the prospect of climate change began to loom. An appreciation of the megadroughts of medieval times, by itself, should prompt preparation for water shortages more severe than any known from the modern era. Clearly, the region's water budget must be brought into balance, which means, foremost, eliminating the current overdraft of 1.2 to 1.3 million acre-feet in annual water consumption in the Lower Basin. The deficit exists independent of climate change, and even if the flow of the Colorado River held to its mean from now until kingdom come, the excess of demand would ultimately bankrupt the system.

Quite beyond the overdrafts—and the inherited injustices and outsized expectations embedded in current water use—and still independent of climate change, there remains an underlying problem in the economy of water. It ensures futility in the pursuit of water security for as long as the current model of continuous economic growth prevails. The problem is that water conservation doesn't do what people think it does. As currently practiced, it doesn't relieve long-term shortage; it can actually make things worse.

When people take shorter showers, install low-flow appliances, xeriscape the yard, and do all the things that a decent sense of social responsibility dictates should be done, they tend to think that they have contributed to their own and their community's water security. They haven't. Except temporarily, the water they save isn't reserved to augment the resilience of the system that provides it. In reality, the saved water they've "produced" now becomes available for consumption by the next strip mall or housing development down the road. It fuels growth.

In the bad old days when water use was lavish, drought was relatively easy to handle. You stopped sprinkling the lawn and washing the car. The city let the medians and ball fields turn brown; other uses tightened up, and current demand dropped like a stone.

But when each person and entity, public and private, conserves aggressively with drip irrigation and low-flow everything, demand for water "hardens." As mentioned earlier, the uses that remain are then essential; you can't turn them off; sometimes you can barely pare them back. Conservation enables a community with fixed water resources to continue growing, but the more it grows on the strength of conservation, the more demand hardens and the more vulnerable the community becomes to drought. When dry times inevitably come, its system has less flex than ever, less room for adaptation.

One response might be to limit growth, but no community has managed that trick without dire secondary effects. Consider Bolinas, California. Because of limited water resources, Bolinas put a cap on the number of water meters its utility would support. To no one's surprise, the cost of acquiring a water meter in Bolinas shot up. And up. In the spring of 2010, a Bolinas water meter changed hands for a cool $300,000.[14] Outside of similarly small, self-contained communities, even a small spike in the cost of access to water would be deemed politically unacceptable, if not a violation of human rights.

Pushed to explain their long-term strategy, many water managers invoke plans for "augmentation"—desalination, interbasin transfers, tapping new aquifers, or rain-making—which might keep the water-supply hamster wheel spinning for another generation, at considerable fiscal and environmental cost. But no such strategy will stop scarcity from arising out of abundance, and none offers more than temporary security against the shrinkage of supply that climate change is likely to visit on the region.

Optimists say that conservation at least buys time by putting off the day of reckoning. True enough. The question all rate-payers should ask their utilities is, "How are we using the time we've bought?"

An insightful body of analysis holds that sudden catastrophes, like earthquakes, fires, and great storms, bring people together. They pitch in, cooperate, and ignore the economic and social divisions that previously held them apart. Survivors of the 1906 San Francisco earthquake, for instance, reported an almost utopian sense of community in the shantytowns that sprang up to shelter the homeless rich and the homeless poor alike.[15] But drought is different. It is gradual and drawn out. You don't know you are in it until it is already well begun, and you never know when it will end. An earthquake shudders and is over; a fire blazes and dies; a storm finally passes. But a drought creeps on. It is not a thing; it is the absence of things—rain and snow—that might cancel it. Drought doesn't dissolve differences in the shock of thunderbolt change; it gives people plenty of time to erect defenses, pick sides, and meditate on the defects of their neighbors.[16] Drought divides people, a fact that should remind us that solving the conundrum of water, growth, and hardened demand is work best done in the present, before the curve of rising need and the downshifting line of limits slam together.

ZUNI WAS SHAPED by climate. The survival of its people through the megadroughts of the twelfth and thirteenth centuries, their repeated decisions over time to aggregate into larger communities or to spread out, to abandon the old site here by the spring and start anew over there by the river, to succor this group of starving immigrants but to turn away that one—all of these myriad choices, spread over generations, mirror those over which the people of Sand Canyon Pueblo and thousands of other ancient communities also deliberated. Taken together, they add up to a dance with the facts of place. Climate is not the only such fact, but it is central and implacable. It calls the tune.

Modern society, empowered by technology and global in its reach, might seem exempt from place-based quadrilles and minuets. Indeed, the notion of place-based restrictions—in a world where blueberries arrive daily from Chile, coffee comes from Kenya, and the mineral components of a smart phone might hail from six of the seven continents—seems almost quaint.

Still, the dance is there, the steps new-learned, inconceivably long, and impossibly intricate. Place has not gone away; it has been magnified to the scale of the planet.

One of its incontestable facts is the dynamism of its seas and its atmosphere, the agents of its fluid life. Another is the collective influence humankind has had on that dynamism, as a result of our continuing alteration of the chemistry of the air. We are no longer just dancing with the facts; we are changing them, and the new facts will force us to step more lively than ever.

In 2008, a team of scientists (including two who have been quoted in these pages) published an article that examined the engineering concept of "stationarity." They found it wanting. Stationarity holds that rivers and streams operate within discernable and predictable limits: if you design a culvert or a levee for a "hundred-year" flood, you postulate that the level of protection you are providing will be adequate for 99 out of 100 years. In appropriately circumspect language, the authors disagreed with so simple a formulation. They said that, as a result of climate change, all bets were off. Stationarity was dead. Predictions of the future would henceforth depend on new kinds of "probabilistic models." The environment, they said, would not simply reenact what it had done in the past; it would press into new territory. Their message echoed the standard disclaimer of the investment prospectus: "past performance is no guarantee of future results."[17]

Essentially, they said that the dance of the future will be a new dance, but no one yet knows how it will go. Nevertheless, a dry-eyed handicapper might say that some trends are evident and that prospects do not look good. At least three big factors cast gloomy shadows forward. Already mentioned is the failure of governments, especially that of the United States, to take meaningful action to limit greenhouse gas emissions. The underlying cause of this political paralysis, which is the intransigence of industries that are the main source of carbon pollution, comes as no surprise. This factor sets the stage for two more. First is the certainty that the global appetite for energy will increase steeply, not decrease, in the years ahead, driven particularly by consumption in nations heretofore called "undeveloped" or "developing." (China and India, of course, lead the group, which also includes many smaller countries.) Prospects for curbing atmospheric pollution are therefore dim at best. Second is the probability that continued perturbation of world climate will eventually produce nonlinear shifts in key processes—ocean currents, monsoon systems, and the like—generating dire consequences for vulnerable populations and the global community at large. On the one hand, the problem is ancient and Malthusian: more people consuming more resources in a finite world. On the other, it is new: climate stability has itself become finite, and the prospect of its "expiration" shortens the time available for world civilization to make the adaptations that would allow it to inhabit the planet in a manner that can be sustained.

The hope for reaching a new and sustainable equilibrium has centered on the idea that, before things fall apart, the developing world will undergo the "demographic transition" that the industrial, developed world has largely already experienced. This is the process by which rising affluence leads to falling birth rates and a host of other changes, including attainment of the means and, where it has been lacking, the desire to implement environmental protection. Even in the best readings of this debatable scenario, however, climate change intrudes a severe complication: the planet may simply be unable to absorb the effects of so much industrial growth.[18] Its systems may break down before the transition is complete. Taken together, the fateful combination of present inaction, rising energy and resource consumption, and climatic

EDWARD WEMYTEWA, MARCH 2010, HAWIKKU. *AUTHOR PHOTO.*

vulnerability make it difficult to envision a safe landing for humankind. Still, we know where to begin if we would soften the crash.

When we were at Hawikku, Edward Wemytewa translated a term from the Zuni language to describe the people who inhabited the stone village when Coronado appeared before it. He called them "fiber people," meaning that they were exceptionally and involuntarily lean, their bodies all sinew and gristle, devoid of fat. They lived too close to famine to store much surplus in their bodies. What surplus they managed to amass from their meticulously tended farms, they sealed in jars and granaries in the inner rooms of their village. Their sense of time was certainly different from ours, but their sense of the future was acute: they laid up reserves against the hard days they knew would inevitably come. They prepared.

The idea of fiber lends itself to metaphor: it will take fiber to contend with the changes of the future and not least to endure the uncertainty that will attend them. Ultimately, the best answers to the climate change predicament in the North American Southwest lead back to mundane matters: we need to get on with what we should have been doing all along, including limiting greenhouse gases. We need to take care of unfinished business on the border, in our forests, and in water management. It wouldn't hurt to love the desert, too—there will be so much more of it—and to protect the rivers and to give the diversity of nature our serious respect. No silver bullet will make the coming decades of the Anthropocene more tolerable. There is only the age-old duty to extend kindness to other beings, to work together and with discipline on common challenges, and to learn to live in the marvelous aridlands without further spoiling them. It is an old calling and a great one. We have already had a lot of practice. We should be better at it. We can be.

## ACKNOWLEDGMENTS

This book started out, years ago, as a general environmental history of the Southwest, but environmental change intervened. The dynamics of the past never lost their fascination, but the galloping dynamics of the present overtook them, and I resolved to follow the action.

Research was a joy. Nothing could be better than taking to the road, learning new places, and talking to the smart, dedicated people who know the most about them. I am deeply indebted to everyone I interviewed, some of whom also tolerated follow-up emails and phone calls, or reviewed and commented on draft chapters, or sent along articles or charts or photographs, or—far beyond the call of duty—lent me a place to stay or to work. Some of them did all of the above and more besides. It is hard to imagine how the book could have come about without the help of Tom Swetnam, Ed Fredrickson, Christa Sadler, Craig Allen, Peter Warshall, Luther Propst, and Dinah Bear. I would also hasten to add Edward Wemytewa, Mark Varien, Dave Breshears, Brad Udall, Jonathan Overpeck, Richard Seager, Chris Milly, Ricky Lightfoot, Eric Blinman, Julio Betancourt, Jeff Dean, Kathy Jacobs, Connie Woodhouse, Mark Muro, Patricia Mulroy, Gregory Hobbs, Brian Hurd, Bill Odle, Dan Millis, Gene and Sue Lefebvre, Daniel Nelson, Tom McCann, Keith Basso, Grady Gammage Jr., John Bedell, Mike Spear, Marit Alanen, and John Koprowski. More help came from Tim Barnett, David Pierce, Tom Wolf, Peter and DeeDee Decker, Esteban Lopez, Claire Zugmeyer, Henry Adams, James Brooks, Bob Barrett, Rodrigo Sierra Corona, Wendy Glenn, Don Ostler, Diana Hadley, Curt Meine, Susan Flader, Conci Bokum, Kris Havstad, Henry Diaz, Patrick McCarthy, Terry Sullivan, Wayne Cornelius, Melissa Savage, Steve Pyne, John Allman, Steve Harris, Sky Crosby, Ken Raffa, Ralph Keeling, John Weisheit, Richard White,

Richard and Shirley Flint, and the late and much missed David Weber. Rita Sudman and the Water Education Foundation extended great kindness, and Kent Ellett and the staff of the Safford Ranger Station, Coronado National Forest, afforded access to Mt. Graham. Although these helpful people saved me from many errors, any that remain are mine and mine alone.

I was able to embark on this journey thanks to the support of the John Simon Guggenheim Memorial Foundation, which provided the opportunity of a lifetime. Vital additional assistance came from the General Service Foundation, the McCune Charitable Foundation, and the Thaw Charitable Trust. My heartfelt thanks to Lani Shaw, Owen Lopez, Gene Thaw, Sherry Thompson, and Wendy Lewis, as well as to Arturo Sandoval, who sheltered the project at the Center of Southwest Culture, and to Phil Smith and Rob Elliott, who together made possible a sojourn in the Grand Canyon.

Others essential to this enterprise include my close colleague and friend Tony O'Brien and our students at the now sadly defunct College of Santa Fe, who helped me to understand what the effort to document real-world situations properly entailed.

Special thanks also go to Yvonne Bond, who transcribed hundreds of pages of interviews with great timeliness and accuracy, and to Deborah Reade, mapmaker extraordinaire, whose patience is as great as her artistic skill. Besides giving welcome encouragement, Jack Loeffler and Don Usner provided desperately needed assistance in audio recording and photography, respectively. More encouragement came from Marty Peale, Joan Halifax, Sally Denton, Sandy Blakeslee, Carl Moore, Sara Dant, Dan Flores, Ted and Betsy Rogers, Jonathan Cobb, and Phil Smith, even when they did not know they were providing it. The office of Encourager-in-Chief, however, belongs to Don Lamm, to whom I am profoundly indebted. His belief in the project, coupled with the support of his colleague Christy Fletcher, led to its finding a home with Oxford University Press, where Tisse Takagi and Tim Bent skillfully guided its publication.

The encouragement of family was vital, and so was their tolerance. Thanks to Anne and Katie and David for being in my corner, to Laura and Ginny for ferrying me through terrain where I was not competent to navigate, and to Joanna for everything.

# NOTES

## INTRODUCTION

1. As quoted by the anthropologist Edward T. Hall, who tells this story in his memoir *West of the Thirties* (New York: Doubleday, 1994), 156.
2. Jonathan Overpeck and Bradley Udall, "Dry times ahead," *Science* 328 (June 25, 2010): 1642–1643.
3. Ibid. The Overpeck and Udall article offers a handy synthesis of current research.
4. For example, see James A. Parks, Jeffrey S. Dean, and Julio L. Betancourt, "Tree rings, drought, and the Pueblo abandonment of south-central New Mexico in the 1670s," in David E. Doyel and Jeffrey S. Dean, *Environmental change and human adaptation in the ancient American Southwest* (Salt Lake City: University of Utah Press, 2006), 214–227.
5. Richard A. Kerr, "Climate change hot spots mapped across the United States," *Science* 321 (August 15, 2008): 909.
6. Mark Muro, et al., "Mountain Megas: America's newest metropolitan places and a federal partnership to help them prosper," The Brookings Institution Metropolitan Policy Program, Washington, DC, 2008, 4.
7. Jonathan Overpeck, personal communication, October 10, 2008, Tucson, Arizona.

## CHAPTER 1

1. Hadley was fundamentally right that tropical convection powers the trades. Having descended onto the aridlands, the dry air of the Hadley cell flows back toward the Equator to complete its circuit. On an unspinning planet, its path would follow a north–south axis, but the rotation of the Earth deflects the returning air to produce northeasterly trade winds in the Northern Hemisphere and southeasterlies in the Southern.
2. Subtropical areas with oceans to the west experience the opposite dynamic and tend to be extremely dry: Baja and southern California, the coasts of Peru and northern Chile, Namibia, western Australia, and so forth.
3. See, for instance, Jian Lu, Gabriel Vecchi, and Joellen Russell, "Expansion of the Hadley cell under global warming," *Geophysical Research Letters* 34 (2007): L06805.

4. Richard Seager, et al., "Model projections of an imminent transition to a more arid climate in southwestern North America," *Science* 316 (May 25, 2007): 1181–1184.

5. Siegfried Schubert, et al., "On the cause of the 1930s Dust Bowl," *Science* 303 (March 19, 2004): 1855–1859.

6. Thomas Painter, et al., "Response of Colorado River runoff to dust radiative forcing in snow," *Proceedings of the National Academy of Sciences* 107/40 (October 5, 2010): 17125–17130; Yun Qian, et al., "Effects of soot-induced snow albedo change on snowpack and hydrological cycle in western United States based on weather research and forecasting chemistry and regional climate simulations," *Journal of Geophysical Research* 114 (2009): D03108.

7. Raymond S. Bradley, Frank T. Keimig, and Henry F. Diaz, "Projected temperature changes along the American Corillera and the planned GCOS Network," *Geophysical Research Letters* 31 (2004): L16210, doi: 10.1029/2004GL020229. Also Henry F. Diaz, personal communication, January 24, 2009.

8. Stephanie A. McAfee and Joellen L. Russell, "Northern Annular Mode impact on spring climate in the western United States," *Geophysical Research Letters* 35 (2008): L17701, doi: 10.10299/2008GL034828.

9. Overpeck points out that another force inhibiting the El Niño storm track results from destruction of stratospheric ozone, which peaked in the 1980s, and augmented temperature differentials between upper and lower portions of the atmosphere, especially at the poles. This seems to be intensifying the circumpolar winds, which in turn pull the jet streams poleward.

10. P. C. D. Milly, K. A. Dunne, and A. V. Vecchia, "Global pattern of trends in streamflow and water availability in a changing climate," *Nature* 438 (November 17, 2005): 347–350.

11. DeBuys, *Salt dreams*, 97.

CHAPTER 2

1. Jane Poynter, *The human experiment: Two years and twenty minutes inside Biosphere II* (New York: Basic Books, 2006), 225, 245.

2. USDA Forest Service, "Initial attack progression of Pina, Rodeo and Chediski Fires," Rodeo-Chediski Fire official website, no date, www.wmat.us/narrative.html.

3. USDA Forest Service, "Forest insect and disease conditions in the Southwestern Region, 2003," Publication PR-R3-04-02, Albuquerque, New Mexico: US Forest Service, Southwestern Region, 2004, 20, www.fs.fed.us/r3/publications/documents/fidc2003.pdf.

4. As quoted, Craig D. Allen and David D. Breshears, unpublished ms. in author's possession, 2002.

5. D. D. Breshears, "Drought-induced vegetation mortality and associated ecosystem responses: Examples from semiarid woodlands and forests," 2007, 89–95 in Appendix D. "Understanding multiple environmental stresses: Report of a workshop board on atmospheric sciences and climate," Board on Atmospheric Sciences and Climate, National Research Council, National Academies Press, Washington, DC.

6. Jim Robbins, "Bark beetles kill millions of acres of trees in West," *New York Times*, November 18, 2008.

7. David Dunn and James P. Crutchfield, "Insects, trees, and climate: The bioacoustic ecology of deforestation and entomogenic climate change," Santa Fe Institute Working Paper 06-12-055, December 11, 2006.

8. Carbon starvation can also take a second form in which the tree's phloem, the tissue through which sugars move, breaks down, preventing delivery of resources where they are needed, "for example from leaves to roots.... This is sort of like someone eating but being unable to digest their food because of some toxin in their system." Henry Adams, personal communication, October 8, 2010.

9. Henry D. Adams, et al., "Temperature sensitivity of drought-induced tree mortality portends increased regional die-off under global-change-type drought," *Proceedings of the National Academy of Sciences* 106/17 (2009): 7063–7066.

10. Jason P. Field, et al., "On the ratio of wind- to water-driven transport: Conserving soil under global-change-type extreme events," *Journal of Soil and Water Conservation* 66/2 (March–April 2011): 51A–56A.doi:10.2489/jswc.66.2.51A. See also Jason P. Field, et al., "The ecology of dust," *Frontiers in Ecology and the Environment* 8/8 (2010): 423–430, doi:10.1890/090050.

11. Thomas H. Painter, et al., "Response of Colorado River runoff to dust radiative forcing in snow."

12. Benjamin I. Cook, Ron L. Miller, and Richard Seager, "Amplification of the North American 'Dust Bowl' drought through human-induced land degradation," *Proceeding of National Academy of Sciences* 106/13 (2009): 4997–5001; doi:10.1073/pnas/0810200106.

13. See, for instance, Aldo Leopold, "A biotic view of land" (1939), in Susan L. Flader and J. Baird Callicot, eds., *The River of the Mother of God and other essays* (Madison: University of Wisconsin Press, 1991).

14. Donald Worster, *Nature's Economy* (San Francisco: Sierra Club Books, 1977), Chapter 11.

15. C. S. Holling, "What barriers? What bridges?" in Lance H. Gunderson, C. S. Holling, and Stephen S. Light, eds., *Barriers and bridges to the renewal of ecosystems and institutions* (New York: Columbia University Press, 1995).

16. David N. Cole, et al., "Naturalness and beyond: Protected area stewardship in an era of global environmental change," *George Wright Forum* 25/1 (2008): 36–56.

17. Craig D. Allen, et al., "A global overview of drought- and heat-induced tree mortality reveals emerging climate change risks for forests," *Forest Ecology and Management* 259 (2010): 660–684, doi:10.1016/j.foreco.2009.09.001.

18. David D. Breshears, et al., "Vegetation synchronously leans upslope as climate warms," *Proceedings of the National Academy of Sciences* 105/33 (August 19, 2008): 11591–11592.

19. Allen was not alone in observing the oak's drought strategy. See J. M. Fair and David D. Breshears, "Drought stress and fluctuating asymmetry in *Quercus undulata* leaves: Confounding effects of absolute and relative amounts of stress," *Journal of Arid Environments* 62 (2005): 235–249.

CHAPTER 3

1. The Village Ecodynamics Project (VEP) produced a mid-1200s population estimate of 19,200 for its study area of 1,817 square kilometers in southwest Colorado. Sand Canyon Pueblo is located in the south central part of the study area. The VEP's

methodology and primary findings are reported respectively in Scott G. Ortman, Mark D. Varien, and T. Lee Gripp, "Empirical Bayesian methods for archaeological survey data: An application for the Mesa Verde region," and Mark D. Varien, et al., "Historical ecology in the Mesa Verde region: Results from the Village Ecodynamics Project," both articles in *American Antiquity* 72/2 (2007): 241–272 and 273–299 respectively. For a digest of the VEP and related study of Sand Canyon Pueblo, see Timothy A. Kohler, et al., "Mesa Verde migrations," *American Scientist* 96 (2008): 146–153.

2. The basic source on Sand Canyon Pueblo is Kristin Kuckelman, ed., *The archaeology of Sand Canyon Pueblo: Intensive excavations at a late-thirteenth-century village in southwestern Colorado* (2007), http://www.crowcanyon.org/sandcanyon. Discussion of push versus pull may be found in Chapter 9, "Summary and Conclusions," paragraph 70. See also William D. Lipe, ed., *The Sand Canyon Archaeological Project: A progress report*, Crow Canyon Archaeological Center Occasional Paper No. 2, 1992.

3. Scott Ortman, "Evidence of a Mesa Verde homeland for the Tewa Pueblos," in Timothy A. Kohler, Mark D. Varien, and Auron Wright, eds., *Time of peril, time of change: Explaining thirteenth-century Pueblo migration* (Tucson: University of Arizona Press, 2010), 233–269.

4. Harry Walters and Hugh Rogers explore the translation in "Anasazi and 'Anaasázi: Two worlds, two cultures," *Kiva* 66/3 (2001): 317–326. For 'anaa they give "those who live beside us but not among us," and for *sazi*, "ancestors of greater than five generations old, ones whose bodies have returned to the earth and are now scattered about."

5. A description of the McElmo Dome carbon dioxide field may be found in Scott H. Stevens, C. E. Fox, and L. S. Melzer, *McElmo Dome and St. Johns natural $CO_2$ deposits: Analogs for geologic sequestration*, Greenhouse Gas Control Technologies, CSIRO Minerals, East Melbourne, Australia, 2001. The gas, when injected into petroleum-bearing formations, reduces the viscosity of the oil, enhancing recovery. The process results in the $CO_2$ ultimately being re-sequestered underground, and is considered a model for potential $CO_2$ sequestration efforts.

6. Scott Ortman and Bruce Bradley, "Sand Canyon Pueblo: The container in the center," in Mark D. Varien and Richard H. Wilhusen, eds., *Seeking the center place: Archaeology and ancient communities in the Mesa Verde region* (Salt Lake City: University of Utah Press, 2002).

7. Varien, et al., "Historical ecology in the Mesa Verde region," 281.

8. Jeff Dean, personal communication, October 8, 2008. Eric Blinman, personal communication, February 16, 2009. Also, Eric Blinman, "2000 years of cultural adaptation to climate change in the southwestern United States," Ambio Special Report No. 14, Royal Swedish Academy of Sciences, November 2008.

9. In addition to Kristin Kuckelman, ed., *The archaeology of Sand Canyon Pueblo*, cited earlier, see also Kristin A. Kuckelman, Ricky R. Lightfoot, and Debra L. Martin, "The bioarchaeology and taphonomy of violence at Castle Rock and Sand Canyon Pueblos, southwestern Colorado," *American Antiquity* 67/3 (2002): 486–513.

10. Mark D. Varien, "The depopulation of the northern San Juan region: A historical perspective," in Kohler, Varien, and Wright, eds., *Time of peril, time of change*.

11. For detailed information on Castle Rock Pueblo, see Kristin A. Kuckelman, ed., *The archaeology of Castle Rock Pueblo: A thirteenth-century village in southwestern Colorado*, (2000), http://www.crowcanyon.org/castlerock. See also Kuckelman, et al., "The bioarchaeology and taphonomy of violence," cited above.

12. Kristin A. Kuckelman, "Thirteenth-century warfare in the central Mesa Verde region," in Varien and Wilhusen, eds., *Seeking the center place.*

13. Kohler, et al., "Mesa Verde migrations."

14. Jeffrey S. Dean and Carla R. van West, "Environment-behavior relationships in southwestern Colorado," in Varien and Wilhusen, eds., *Seeking the center place.*

15. The literature on Chaco can fill a library. Among the most accessible sources: John Kantner, *Ancient Puebloan Southwest* (New York: Cambridge University Press, 2004), and David G. Noble, ed., *In search of Chaco: New approaches to an archaeological enigma* (Santa Fe: School of American Research Press, 2004).

16. Jeffrey S. Dean, personal communication, October 8, 2008; Eric Blinman, personal communication, February 16, 2009; Eric Blinman, "2000 years of cultural adaptation to climate change in the southwestern United States"; K. L. Petersen, "Climate and the Dolores River Anasazi," University of Utah Anthropological Papers, 11/3, Salt Lake City: 1988; Matthew W. Salzer, "Temperature variability and the Northern Anasazi: Possible implications for regional abandonment," *Kiva* 65 (2005): 295–318.

17. Jeffrey S. Dean and Gary S. Funkhouser, "Dendroclimatic reconstructions for the Southern Colorado Plateau" (1995), in *Climate change in the Four Corners and adjacent regions: Implications for environmental restoration and land-use planning,* W. J. Waugh, ed., Conf-9409325, U.S. Dept. of Energy, Grand Junction Projects Office, Grand Junction, Colorado, 85–104.

18. Blinman interview and article.

19. Jeffrey S. Dean, personal communication, March 11, 2009.

20. Fray Atanasio Dominguez, *Missions of New Mexico, 1776* (Albuquerque: University of New Mexico Press, 1975), 213.

21. One observer of the "longue durée" in the Southwest says, "I don't really think there are that many lessons that we can learn from the collapse of the Anasazi. The argument on the part of anthropologists...is that there's a lot to be gained from looking at the similarities between us and them. And I actually think there's a hell of a lot more to gain from focusing on the differences" (Julio Betancourt, personal communication, October 7, 2008). Aside from the obvious differences in technological capacity, Betancourt points out that the ancient Puebloans relied mainly on summer rain to produce the crops on which they subsisted, whereas the main components of contemporary urban society depend on snowpack and the rivers and reservoirs that are fed mainly by winter storms. Point taken. Agriculture in the region around Chaco Canyon primarily depended on summer rain. But in certain instances, winter moisture was also crucial to the ancient world; without it, the maize and beans of the Sand Canyon people would not have germinated in the loess soils of the McElmo Dome. In any event, the most severe, society-changing droughts of the ancient world appear to have entailed acute shortages of winter moisture, as well as failure of summer rains.

22. Ortman and Bradley, "Sand Canyon Pueblo: Container in the center," 62–65.

23. Bruce A. Bradley, "Pitchers to mugs: Chacoan revival at Sand Canyon Pueblo," *Kiva* 61/3 (1996): 241–255.

24. From Scott Ortman, "Evidence of a Mesa Verde homeland for the Tewa Pueblos": "These findings do not necessarily indicate that the entire population of the Mesa Verde region moved to the northern Rio Grande during the thirteenth century.

Indeed, it seems far more likely that the Mesa Verde region population dispersed broadly, and that groups of immigrants joined existing communities across the Pueblo world. What these data do suggest, however, is that the many thousands of people who migrated to the Tewa and Galisteo basins in the thirteenth century colonized a frontier landscape in such a way that the language, gene pool, and historical identity of the immigrants was preserved."

CHAPTER 4

1. Vernon O. Bailey, "Improvement of public range," 1908, unpublished field reports in Record Group 7176, Box 72, Folder 15, Archives and Special Collections of the Smithsonian Institution, Washington, DC.

2. E. A. Goldman, "Big Hatchet Mts. and Hatchet Ranch, Grant [now Hidalgo] Co.," July 19–24, 1908, unpublished field reports in Record Group 7176, Box 74, Folder 1, Archives and Special Collections of the Smithsonian Institution, Washington, DC.

3. Lee C. Buffington and Carlton H. Herbel, "Vegetational changes on a semidesert grassland range from 1858 to 1963," *Ecological Monographs* 35 (1965): 139–164.

4. The ecological transformation of southwestern rangelands has spawned a vast literature. Two summary articles with extensive bibliographies are O. W. Van Auken, "Shrub invasions of North American semiarid grasslands," *Annual Review of Ecology and Systematics* 31 (2000): 197–215; and Herbert D. Grover and H. Brad Musick, "Shrubland encroachment in southern New Mexico, USA: An analysis of desertification processes in the American Southwest," *Climatic Change* 17 (1990): 305–330.

5. R. S. Campbell, "Vegetative succession in the *Prosopis* sand dunes of southern New Mexico," *Ecology* 10/4 (October 1929): 392–398.

6. Walter G. Whitford and Brandon T. Bestelmeyer, "Chihuahuan desert fauna: Effects on ecosystem properties and processes." Also, Jeffrey E. Herrick, Kris M. Havstad, and Albert Rango, "Remediation research in the Jornada Basin: Past and future," both sources in Kris Havstad, Laura F. Huenneke, and William H. Schlesinger, eds., *Structure and function of a Chihuahua desert ecosystem: The Jornada Basin long-term ecological research site* (Oxford: Oxford University Press, 2006), 248, 292.

7. Rurik List, et al., "The Janos Biosphere Reserve, northern Mexico," *International Journal of Wilderness* 16/2 (August 2010): 35–41.

8. Rurik List, et al., "Historic distribution and challenges to bison recovery in the northern Chihuahuan Desert," *Conservation Biology* 21/6 (2007): 1487–1494.

9. Krista Schlyer, "The lost herd of Janos-Hidalgo," *Wildlife Conservation* (January–February 2009): 47–55.

10. Ed L. Fredrickson, et al., "Mesquite recruitment in the Chihuahuan Desert: Historic and prehistoric patterns with long-term impacts," *Journal of Arid Environments* 65 (2006): 285–295.

11. Aldo Leopold, "Conservationist in Mexico," in Susan L. Flader and J. Baird Callicot, eds., *The river of the Mother of God and other essays* (Madison: University of Wisconsin Press, 1991), 239.

12. Curt Meine, *Aldo Leopold: His life and work* (Madison: University of Wisconsin Press, 1988), 94.

13. Ibid., 181. See also "'Piute forestry' vs. forest fire prevention" in Flader and Callicott, *The river of the Mother of God.*

14. Susan Flader, interview with Jack Loeffler, reproduced in "Aldo Leopold in the Southwest" (radio program), a production of Lore of the Land, Santa Fe, 2009.

15. Leopold, "Conservationist in Mexico," 241, 240.

16. "1936 Mexico Trip," located under "Diaries and Journals, 1917–1945," digitally archived at http://digital.library.wisc.edu/1711.dl/AldoLeopold.

17. Leopold, "Conservationist in Mexico," 242.

18. Fred Wilbur Powell, *The railroads of Mexico* (Boston: Stratford, 1921), 158. See also Eli Bartra, "Engendering clay: Las Ceramistas of Mata Ortiz," in Eli Bartra, ed., *Crafting gender* (Durham, NC: Duke University Press, 2003); and "Ferrocarril Noroeste de México: An inventory of the records at the Benson Latin American Collection," http://www.lib.utexas.edu/taro/utlac/00020/00020-P.html.

19. Juan Mata Ortiz was second in command in the 1880 battle in which Victorio and seventy-seven other Apaches were killed and over sixty taken prisoner.

20. Richard Schwartzlose, "Mormon settlements in Mexico" (1952), http://thecardonfamilies. org/Documents/MormonSettlementsInMexico. See also: Brandon Morgan, "From brutal ally to humble believer: Mormon colonists' image of Pancho Villa," *New Mixico Review* 85/2 (2010): 109–129.

21. The Río Gavilán flows to the Bavispe, which feeds the Yaqui, which finds the Gulf of California.

22. Harlo Johnson is among the oral history sources acknowledged in Schwartzlose's "Mormon settlements."

23. Leopold, "Conservationist in Mexico," 240.

24. Schwartzlose, "Mormon settlement," 24. See also Shelly Bowen Hatfield, *Chasing shadows: Indians along the United States–Mexico Border 1876–1911* (Albuquerque: University of New Mexico Press, 1998), 114 ff. Neil Goodwin reports, without documentation, that the 1900 confrontation occurred on November 16, and that the "battle" resulted in the deaths of twelve Apaches and "a number of Mormons" (*The Apache diaries*, 246).

25. Goodwin, *The Apache diaries*, 55–57 and passim.

26. David J. Weber, *New Spain's far northern frontier* (Albuquerque: University of New Mexico Press, 1979), 60.

27. Friedrich Katz, *The life and times of Pancho Villa* (Stanford: Stanford University Press, 1998), 19 and passim.

28. Aldo Leopold, "Song of the Gavilan," in *A Sand County almanac and sketches here and there* (New York: Oxford University Press, 1949, 1968), 151.

29. Leopold, "Wilderness," in *A Sand County Almanac*, 197; Meine, *Aldo Leopold*, 380.

30. Schwartzlose, "Mormon Settlement," 83.

31. William Fleming and William Forbes, "Following in Leopold's footsteps: Revisiting and restoring the Rio Gavilan watershed," *Ecological Restoration* 24/1 (March 2006): 25–31. Notably, however, at the time of this writing the Río Gavilán still supported a population of otters.

32. Leopold, "Conservationist in Mexico," 240.

33. Susan Flader, "Evolution of a land ethic," in Thomas Tanner, ed., *Aldo Leopold: The man and his legacy* (Ankeny, IA: Soil Conservation Society of America, 1987), 16–17.
34. Meine, *Aldo Leopold*, 527–528.
35. Luna Leopold, *A view of the river* (Cambridge, MA: Harvard University Press, 1994), 19.
36. Ibid., 9.

CHAPTER 5

1. Lester Snow, comment in plenary session, Colorado River Project Symposium (a project of the Water Education Foundation, www.watereducation.org), The Bishop's Lodge, Santa Fe, New Mexico, September 17, 2009. Transcript (see p. 79) available in "Colorado River Project: Symposium Proceedings, September 16–18, 2009," Water Education Foundation, Sacramento, CA, 2010.
2. Ibid., 71–72, 78.
3. Mark Reisner, *Cadillac desert* (New York: Viking Pengiun, 1986), 125.
4. Officially, there is also "Lee Ferry" about a mile downstream from "Lees Ferry." The former is the exact dividing point between the two basins, and the latter is the location of the all-important U.S. Geological Survey river gauge. The Pariah River enters the Colorado between the two, which means deliveries to the Lower Basin are the sum of measurements at both the Pariah gauge and the Lees Ferry gauge. For simplicity's sake, *Lees Ferry* will be used here to refer to the general area.
5. Stewart Udall, *The forgotten founders: Rethinking the history of the Old West* (Washington: Island Press, 2002), 63–76.
6. A sympathetic appreciation of Udall's legacy may be found in Charles Wilkinson, *Fire on the Plateau*, 216 ff. See also Charles Coate, "'The biggest water fight in American history': Stewart Udall and the Central Arizona Project," *Journal of the Southwest* 37/1 (spring 1995): 79–101.
7. Brad Udall, personal communication, September 17, 2009, Bishop's Lodge, Santa Fe, New Mexico. Comments by Udall in this chapter are also derived from his panel presentation and answers to questions the following day during the Colorado River Project Symposium of the Water Education Foundation (www.watereducation.org), September 16–18, 2009.
8. Tim P. Barnett and David W. Pierce, "When will Lake Mead go dry?" *Water Resources Research* 44 (2008): W03201, doi:10.1029/2007WR006704.
9. The figure of 25 percent reflects "actual" flow at Lees Ferry—or, rather, what would have been actual flow if Glen Canyon Dam weren't in the way. This figure is lower than a reconstruction of "native flow" because it includes depletions from upstream diversions. The year's "yield" was closer to 40 percent of average native flow, which ignores those depletions. At the risk of stating the obvious, any discussion of the Colorado River requires clarity about the pedigree of the numbers being used.
10. Tim P. Barnett, personal communication, November 30, 2009. Barnett and Pierce also used a statistical methodology that estimated the river's native flow better than the models used by the Bureau of Reclamation. (Because there are many diversions upstream of Lees Ferry, estimates of the natural flow at that location are necessarily reconstructions.)

11. *Las Vegas Review-Journal*, February 14, 2008.
12. California, Arizona, Nevada, and (a very small interest) New Mexico.
13. "Water use in Santa Fe: A survey of residential and commercial water use in the Santa Fe urban area," Planning Division, Planning and Land Use Department, City of Santa Fe, New Mexico, February 2001.
14. Charles J. Meyers and Richard L. Noble, "The Colorado River: The treaty with Mexico," *Stanford Law Review* 19 (1966–1967), 405–406 and passim. A similar strategic concern lay behind the 1973 agreement by which the United States agreed to meet a salinity standard for the river water it delivered to Mexico. At that time OPEC had shut off the flow of Middle Eastern oil to supporters of Israel, and the United States wanted to guarantee access to newly discovered petroleum reserves in Mexican waters off the Yucatán.
15. James Lawrence Powell, *Dead pool: Lake Powell, global warming, and the future of water in the West* (Berkeley: University of California Press, 2008), 69. See also wwa.colorado.edu/treeflow/lees/compact.html. Note: In 1963 the Supreme Court ruled that the 1.0-maf tributary flow of the Gila River should not count against the 2.8 maf allocated to Arizona under the Colorado River Compact and subsequent agreements. This raises to roughly 18.0 maf the total estimate of consumable river flow underpinning the "Law of the River."
16. Charles W. Stockton and Gordon C. Jacoby Jr., "Long-term surface-water supply and streamflow trends in the Upper Colorado River Basin based on tree-ring analysis," *Lake Powell Research Project Bulletin* 18 (1976), Institute of Geophysics and Planetary Physics, University of California, Los Angeles.
17. David M. Meko, et al., "Medieval drought in the Upper Colorado River Basin," *Geophysical Research Letters* 34 (2007): L10705, doi:10.1029/2007GL029988.
18. This figure includes the mysterious contribution of "unmeasured returns"—about 0.25 maf/year consisting of groundwater flows reentering the river or, possibly, measurement errors. David Pierce, personal communication, November 30, 2010.
19. Personal communication, Don Ostler, Upper Colorado River Commission, December 4, 2009; Estevan Lopez, Director, New Mexico Interstate Stream Commission, December 11, 2009; Brad Udall, November 9, 2010. Also see U.S. Bureau of Reclamation, *Hydrologic determination 2007, water availability from Navajo Reservoir and the Upper Colorado River Basin for use in New Mexico*, dated April 2007, www.usbr.gov/envdocs/eis/navgallup/FEIS/vol1/attach-N.pdf.
20. Personal communication, Justice Gregory Hobbs, Colorado Supreme Court, December 3, 2009.
21. See Balaji Rajagopalan, et al., "Water supply risk on the Colorado River: Can management mitigate?" *Water Resources Research* 45 (2009): W08201, doi:10.1029/2008WR007652.
22. Tim P. Barnett and David W. Pierce, "Sustainable water deliveries from the Colorado River in a changing climate," *Proceedings of the National Academy of Sciences*, 106 (2009): 7334–7338; published online before print April 20, 2009, doi:10.1073/pnas.0812762106.
23. Patricia Mulroy, personal communication, September 18, 2009, Bishop's Lodge, Santa Fe, New Mexico. Comments by Mulroy in this chapter are also derived from her panel presentation and answers to questions in plenary session (of the same date) during the Colorado River Project Symposium of the Water Education Foundation (www.watereducation.org), September 16–18, 2009.

24. Mark Muro, Robert E. Lang, and Andrea Sarzynski, "Mountain Megas," 35. Also, Emily Green and Michael Weissenstein, in note 26 below.
25. When I met with Mulroy, the name of the character based on Mulholland in Roman Polanski's *Chinatown* somehow seemed important, but I could not bring it to mind. Later I looked it up: the water engineer who is drowned at the outset of the movie is Hollis Mulwray; his wife, the film's tragic heroine, played by Faye Dunaway, is Evelyn Mulwray, a name close enough to Mulroy to make Patricia sound like a relation.
26. Jon Christensen, "Las Vegas wheels and deals for Colorado River water," *High Country News*, February 21, 1994. For a portrait of Mulroy and a profile of her career, see Emily Green, "The chosen one," *Las Vegas Sun*, June 8, 2008; and Michael Weissenstein, "The water empress of Vegas," *High Country News*, April 9, 2001.
27. Matt Jenkins, "Vegas forges ahead on pipeline plan," *High Country News*, October 5, 2009, 16.
28. "Lake Mead is considered at 100 percent capacity at 1,219.6 feet. However, there is exclusive flood control space above 1,219.6 feet to elevation 1,229.0 feet." Personal communication, October 26, 2009, Water Operations Control Center, Boulder Canyon Operations Office, Bureau of Reclamation.
29. Jenkins, "Vegas forges ahead," *High Country News*, October 5, 2009.
30. http://www.usbr.gov/lc/hooverdam/faqs/powerfaq.html.
31. Robert Glennon, *Unquenchable: America's water crisis and what to do about it* (Washington: Island Press, 2009), 59.
32. Joan F. Kenny, et al., "Estimated use of water in the United States in 2005," U.S. Geological Survey Circular 1344, 2009.
33. Stacy Tellinghuisen, presentation, September 17, 2009, Santa Fe, New Mexico, part of the Colorado River Project Symposium of the Water Education Foundation (www.watereducation.org), September 16–18, 2009.
34. Mike Hightower, presentation, September 17, 2009, Santa Fe, New Mexico, part of the Colorado River Project Symposium of the Water Education Foundation (www.watereducation.org), September 16–18, 2009.
35. Mike Hightower and Suzanne A. Pierce, "The energy challenge," *Nature* 452/20 (March 2008): 285–286.
36. As quoted by Jon Christensen, "Las Vegas wheels and deals for Colorado River water," *High Country News*, February 21, 1994.
37. Brad Udall, personal communication, September 17, 2009. Transmission losses include water lost to instream evaporation, evapotranspiration, infiltration, and other "inefficiencies." Inflows from rivers and springs between Lees Ferry and Hoover Dam cover some of these losses, too.
38. Barnett and Pierce, "Sustainable water deliveries," "Discussion and conclusions."
39. Muro, et al., "Mountain Megas," 21.
40. El Paso, Texas, which taps a saline aquifer and mixes desalted water with freshwater drawn from the Rio Grande and other well fields, is a rare example of successful desalination at an inland location.

41. California Water Plan Update 2009, www.waterplan.water.ca.gov/CWPU2009/1208prd/vol1/1-4_CAWaterToday. New Mexico Office of the State Engineer, www.ose.state.nm.us/publications_technical_reports_wateruse_basino5.html. Arizona Department of Water Resources, www.azwater.gov/AzDWR/StatewidePlanning/WaterAtlas/documents/statewide_demand_web.pdf.

42. In 2010, Lyndon LaRouche and his followers advocated revival of NAWAPA. The gist of their message is that "NAWAPA will overturn a century-long policy of deliberate underdevelopment imposed by President Teddy Roosevelt and perpetuated under the evil of environmentalism." See www.larouchepac.com/nawapa.

43. As Tom McCann points out in Chapter 7, such a scheme might also help the Lower Basin reduce the impact of additional future Upper Basin diversions. For current diversions, see Lori E. Apodaca, Verlin C. Stephens, and Nancy E. Driver, "What affects water quality in the Upper Colorado River Basin?" USGS Factsheet, http://pubs.usgs.gov/fs/fs-109-196/pdf/fs109-96.pdf.

44. Barnett and Pierce, "Sustainable Water Deliveries," "Discussion and Conclusions."

CHAPTER 6

1. Thomas E. Sheridan, *Arizona: A history* (Tucson: University of Arizona Press, 1995), 224.

2. John U. Carlson and Alan Boles Jr., "Contrary views of the law of the Colorado River: An examination of rivalries between the Upper and Lower Basin states," *Rocky Mountain Mineral Law Institute* 32 (1986): 63–70 See also Charles J. Meyers, "The Colorado River," *Stanford Law Review* 19 (November 1966): 37 ff.

3. Meyers, "The Colorado River." See the detailed analysis of *Arizona v. California* at 36–75.

4. Sheridan, *Arizona*, 341–343.

5. "Record of decision, Colorado River interim guidelines for Lower Basin shortages," December 2007, http://www.usbr.gov/lc/region/programs/strategies/Recordof Decision. pdf.

6. "Final EIS—Colorado River Basin shortage guidelines and coordinated operations for Lake Powell and Lake Mead, Appendix E: Colorado River entitlement priority systems within Arizona, California, and Nevada," November 2007, http://www.usbr.gov/lc/region/programs/strategies/FEIS/AppE.pdf.

7. "A Udall warning," *Twenty-First Annual Report of the Arizona Interstate Stream Commission*, July 1, 1967, to June 30, 1968, 27.

8. Executive summary, "Nuclear power and water desalting plants for southwest United States and northwest Mexico: A preliminary assessment conducted by a joint United States–Mexico–International Atomic Energy Agency study team, September 1968," copy provided courtesy of Thomas W. McCann, Assistant General Manager, Central Arizona Project. Discussions of bi-national seawater desalination soon took a backseat to negotiations over the salinity of Colorado River water the United States was delivering to Mexico. The issue of river water salinity was eventually settled by adoption of

Minute 242 to the International Water Treaty, but the prospective seawater desalination project was not revived.

9. Public Law 90-537, 90th Congress, S. 1004, September 30, 1968, http://www.usbr.
   gov/lc/region/pao/pdfiles/crbproj.pdf. See also Meyers, "Colorado River," 73.

10. The CAP's "firm" water right is considered to be 1.49 maf, of which, after transmission
    losses, 1.415 maf is available for long-term contracting. In good years, more than 1.49
    maf may be delivered to CAP. Source: Tom McCann, Assistant General Manager,
    CAP, personal communication, November 28, 2010.

11. Tim Barnett, personal communication, December 1, 2010.

12. See, generally, William deBuys, ed., *Seeing things whole: The essential John Wesley Powell*
    (Washington: Island Press, 2001).

## CHAPTER 7

1. The National Bureau of Economic Research marks the onset of the recession at
   December 2007 and its "trough," the point at which economic contraction ceases and
   recovery begins, at June 2009 (www.nber.org/cycles/sept2010.html). The technical
   end of a recession is not the same as a return to pre-recession prosperity; it only marks
   a reversal in the direction of change from negative to positive. If, as in the case of the
   Great Recession, the negative falloff was precipitous and prolonged (the worst since
   the 1930s), a long period of sustained growth is required before the economy reattains
   the vigor it possessed prior to its decline.

2. See, for instance, Christopher B. Leinberger, "The next slum," *The Atlantic*, March
   2008, http://www.theatlantic.com/doc/200803/subprime.

3. Grady Gammage Jr., "Megapolitan: Arizona's Sun Corridor," July 2009 update, http://
   morrisoninstitute.asu.edu/morrison-update/megapolitan-arizonas-sun-corridor.
   (The body of the report was issued May 2008.).

4. Gammage adds: "Some of those of us who live here worry about [the Sun Corridor
   becoming so hot and dry that it's not a habitable place], but I don't think people in
   Indiana worry about that particularly." Grady Gammage Jr., personal communication,
   March 8, 2010. Additional quotes from Gammage in this chapter derive from this
   interview.

5. Gammage, "Megapolitan," 23; Muro, et al., "Mountain Megas," 4.

6. Gammage, "Megapolitan," 27.

7. Mark Muro, personal communication, November 24, 2008.

8. Muro, et al., "Mountain Megas," 19.

9. Ibid., passim.

10. Gammage, "Megapolitan," 24, 41.

11. J. Brett Hill, et al., "Prehistoric demography in the Southwest," *American Antiquity*
    64/4 (2004): 689–716. See also Suzanne K. Fish and Paul R. Fish, eds., *The
    Hohokam millennium* (Santa Fe: School for Advanced Research, 2008).

12. Shaun McKinnon, "Valley has worst dust pollution in U.S.," *Arizona Republic*, April
    28, 2010.

13. As quoted, Sheridan, *Arizona*, 241. Census information is from Michael F. Logan, *Desert cities: The environmental history of Phoenix and Tucson* (Pittsburgh: University of Pittsburgh Press, 2006).

14. "EPA report ranks Arizona as high water user," http://www.ag.arizona.edu/azwater/awr/246c9d60-7f00-0101-014c-28dbodbbeefo.html.

15. The closest attempt to sketching a comprehensive water future for Arizona is Grady Gammage Jr., et al., "Superstition Vistas: Water matters," Morrison Institute for Public Policy, Arizona State University, July 2005, available at www.morrisoninstitute.org. As of early 2010, Gammage and others were seeking funding for a statewide study.

16. "Arizona Water Institute latest budget casualty," *Arizona Daily Star*, posted February 5, 2009, http://azdailysun.com/news/local/article_c07cfc0c-35d7-57d5-b7bc-a6dfa2bbf3b2.html.

17. Most analyses presume that the Upper Basin will absorb declines in flow consequent to climate change. Certainly it is in a better position than the Lower Basin to absorb them, but the liability for those losses raise an important question about the compact: which provision should take precedence—the obligation of the Upper Basin to deliver 750 maf every ten years, or the principle of equal division of waters between Upper and Lower Basins? If courts decide in favor of the latter and if the river declines as much as predicted, the consequences would devastate the Lower Basin.

18. Brad Udall, personal communication, November 9, 2010.

19. Thomas W. McCann, personal communication, March 8, 2010.

20. William deBuys, *Salt dreams: Land and water in low-down California* (Albuquerque: University of New Mexico Press, 1999), 148–149. The Bureau of Reclamation website has abundant information on the Yuma Desalting Plant (www.usbr.gov/lc/yuma/facilities/ydp). See also www.usbr.gov/newrom/newsrelease/detail.cfm?RecordID=30721.

21. National Research Council, *Colorado River Basin water management* (Washington: National Academies Press, 2007).

22. Larry R. Dozier and Thomas W. McCann, "Restoration of CAP's priority to Colorado River water," August 2004, courtesy Thomas W. McCann.

23. Anthony Brazel, et al., "Mitigating urban heat island effects with water- and energy-sensitive urban designs," paper presented at the 89th American Meteorological Society Annual Meeting, Phoenix, Arizona, January 15, 2009, extended abstract, http://ams.confex.com/ams/89annual/techprogram/paper_145231.htm.

24. Anthony Brazel, et al., "Determinants of changes in the regional urban heat island in metropolitan Phoenix (Arizona, USA) between 1990 and 2004," *Climate Research* 22/2 (2007): 171–182, doi:103354/cro33171. Subhrajit Guhathakurta and Patricia Gober, "The impact of the Phoenix urban heat island on residential water use," *Journal of the American Planning Association* 73/3 (summer 2007): 317–329.

25. Sharon L. Harlan, et al., "Neighborhood microclimates and vulnerability to heat stress," *Social Science and Medicine* 63 (2006): 2847–2863.

26. http://azforeclosurereliefnow.com/blog/2009/12/06/arizona-foreclosures-fall-in-august-its-economy-improves-slightly/.

27. Mark Muro, interview, and "Mountain Megas," passim.

28. Gammage, "Megapolitan," 40.

CHAPTER 8

1. Randal C. Archibold, "U.S. falters in screening Border Patrol near Mexico," *New York Times*, March 11, 2010.
2. Shuaizhang Feng, Alan B. Krueger, and Michael Oppenheimer, "Linkages among climate change, crop yields, and Mexico–U.S. cross-border migration," *Proceedings of the National Academy of Sciences* 107/32 (August 10, 2010): 14257–14262.
3. The policy is fully discussed in Joséph Nevins, *Operation Gatekeeper: The Rise of the illegal alien and the making of the U.S.–Mexico boundary* (New York: Routledge, 2002).
4. Maria Jimenez, "Humanitarian crisis: Migrant deaths at the U.S.–Mexico border," October 1, 2009, a report for the American Civil Liberties Union and Mexico's National Commission of Human Rights, www.aclu.org/immigrants-rights/humanitarian-crisis-migrant-deaths-us-mexico-border.
5. www.defense.gov/NEWS/casualty.pdf.
6. Following my visit to the aid station, I was given essentially the same information about birth control pills and rape by several other sources. Tim Vanderpool also writes about it in "Price of admission: Along the border, sexual assault has become routine," *Tucson Weekly*, June 5, 2008.
7. In 2008, "nearly 359,000 aliens were removed from the United States—the sixth consecutive record high." Department of Homeland Security, Office of Immigration Statistics, Annual Report, "Immigration Enforcement Actions: 2008," July 2009, www.dhs.gov/xlibrary/assets/statistics/publications/enforcement_ar_08.pdf.
8. Alejandro Nadal, "The environmental and social impacts of economic liberalization on corn production in Mexico," a report for Oxfam GB and WWF International, September 2000, http://ase.tufts.edu/gdae/Pubs/rp/NadalOxfamWWFMaizeMexico2000.pdf. See also Eric Perramond, *Political ecologies of cattle ranching in northern Mexico* (Tucson: University of Arizona Press, 2010), 164–165; and Richard Seager, et al., "Mexican drought: An observational, modeling and tree ring study of variability and climate change," *Atmosfera* 22/1 (2009): 1–31.
9. www.state.gov/secretary/rm/2009a/11/13/1666.htm and www.state.gov/secretary/rm/2009a/11/13/131724.htm.
10. U.S. Government Accountability Office, "Secure border initiative fence construction costs," GAO-09-244R, January 20, 2009.
11. Until January 2011, work also continued—at staggering cost—on SBInet, the virtual Border Wall. SBInet (SBI stands for Secure Border Initiative) was a high-tech chain of towers and remote sensors, sited in Arizona, which proponents hoped might one day extend the entire length of the border, provided of course that Congress ponied up its estimated total cost of $8 billion. SBInet's motion detectors, microphones, and night-vision cameras were advertised to observe, feel, and listen to everything that moved along the border. But they never worked correctly and their costs spiraled out of control, causing the Obama administration to cancel the program. Now SBInet's idle sentry towers soar uselessly above the desert.
12. See, for instance, Aaron D. Flesch, et al., "Potential effects of the United States–Mexico border fence on wildlife," *Conservation Biology* 24/1 (2010): 171–181.

13. Joanna Lydgate, "Assembly-line justice: A review of Operation Streamline," policy brief, Warren Institute on Race, Ethnicity and Diversity, University of California, Berkeley School of Law, January 2010, table p. 6 and note 22 at p. 5; Department of Homeland Security, Office of Immigration Statistics, Annual Report, "Immigration Enforcement Actions: 2008," July 2009, www.dhs.gov/xlibrary/assets/statistics/publications/enforcement_ar_08.pdf.

14. Lydgate, "Assembly-line justice."

15. Total apprehensions in the Southwest were 1,072,018 in 2006, almost even with 2005's 1,189,018.

16. Wayne A. Cornelius, David Fitzgerald, Pedro Lewin Fischer, and Leah Muse-Orlinoff, eds., "Mexican migration and the U.S. economic crisis: A transnational perspective," presentation to the Bishop's Forum, St. Paul's Cathedral, San Diego, November 22, 2009. See also Wayne A. Cornelius, et al., *Mexican migration and the U.S. economic crisis: A transnational perspective* (San Diego: Center for Comparative Immigration Studies, 2010).

17. Jimenez, "Humanitarian crisis."

18. "Study: Legalizing pot won't hinder Mexican cartels," *New York Times*, October 12, 2010. James C. McKinley Jr., "U.S. is arms bazaar for Mexican cartels," *New York Times*, February 26, 2009.

19. U.S. Government Accountability Office, "Firearms trafficking: U.S. efforts to combat arms trafficking to Mexico face planning and coordination challenges," Report No>: GAO-09-709, June 2009.

20. Warner Glenn, Malpai Borderlands Group Newsletter, November 2010.

21. Beau Kilmer, et al., "Reducing drug trafficking revenues and violence in Mexico: Would legalizing marijuana in California help?" Rand Corporation Occasional Paper, 2010. http://www.rand.org/pubs/occasional_papers/2010/RAND_OP325.pdf.

22. No More Deaths, "Crossing the line: Human rights abuses of migrants in short-term custody on the Arizona/Sonora border," September 2008, www.nomoredeaths.org/index.php/Abuse-Report/.

23. The bill takes acronym invention beyond jabberwocky. Its full name is Rearing and Empowering America for Longevity against acts of International Destruction.

24. Two petitions for *certiorari* were filed with the Supreme Court challenging the constitutionality of this provision: *County of El Paso v. Napolitano*, 129 S. Ct. 2789 (2009), and *Defenders of Wildlife v. Chertoff*, 128 S. Ct. 2962 (2008). The Supreme Court considered the 2009 petition for eight weeks before denying it (http://www.supremecourt.gov/Search.aspx?FileName=/docketfiles/08-751.htm). The Court accepts only a fraction of the thousands of cases filed before it annually, and seasoned observers noted that the unusually long period of consideration betrayed interest in hearing the relevant arguments. Because of the short (sixty-day) period in which cases must be filed after invocation of the waiver, no further challenges to existing waivers can be brought, but future invocation of the waiver authority will likely result in additional applications to the Supreme Court.

25. By mid-2011, the Obama administration had neither invoked the waiver authority nor disavowed its use.

26. At least, the Congressional Research Service was unable to identify any other provision comparable in sweep to this one. See Memorandum from Stephen R. Viña and Todd Tatelman, Am. Law Division, Cong. Research Serv., "Sec. 102 of HR 418, Waiver of Laws Necessary for Improvement of Barriers at Borders," February 9, 2005, http://graphics8.nytimes.com/packages/pdf/national/20080408_CRS_report.pdf.

27. See, for instance, Raekha Prasad, "India builds a 2,500-mile barrier to rival the Great Wall of China," *Times* (London), December 28, 2005, http://www.timesonline.co.uk/tol/news/world/asia/article782933.ece.

CHAPTER 9

1. John Bedell, personal communication, April 1, 2010. Bedell was supervisor of the Apache–Sitgreaves National Forests (note plural), two forests administered jointly.

2. "Apache–Sitgreaves National Forests, Rodeo–Chediski fire effects summary report," August 2002. Copy obtained courtesy of Victoria Lowe, Black Mesa Ranger Station (Overgaard, AZ), Sitgreaves NF, April 16, 2010. Information on the behavior of the Rodeo-Chediski Fire is drawn primarily from this report and the recollections of John Bedell.

3. Summary data on firefighting efforts and acreages burned per day derive from miscellaneous fact sheets obtained courtesy of Victoria Lowe, Black Mesa Ranger Station (Overgaard, AZ), Sitgreaves NF, April 16, 2010.

4. Lisa Dale and unnamed coauthors, "The true cost of wildfire in the western U.S.," Western Forestry Leadership Coalition, Lakewood, Colorado, April 2009, updated 2010, www.wflccenter.org/news_pdf/324_pdf.pdf.

5. Burn severity was classed as "high" on 147,560 acres and "moderate" on 99,606 acres, representing 32 percent and 22 percent, respectively, of 462,375 acres in the burned area. (Other accounts give a total acreage of 467,066 acres.) Even in "moderately burned" areas, virtually all trees in all age classes were killed. The difference between the two categories has mainly to do with watershed function and vulnerability to erosion. (Fact sheet obtained courtesy of Victoria Lowe, Black Mesa Ranger Station, Sitgreaves NF, April 16, 2010.)

6. "Hayman fire starter testifies she was emotionally 'a mess,'" *Rocky Mountain News*, September 12, 2008, http://www.rockymountainnews.com/news/2008/sep/12/hayman-fire-starter-testifies-she-was-emotionally-/. Terry Barton served six years for starting the Hayman Fire. Leonard Gregg's ten-year term for igniting the Rodeo Fire will end in 2014. Gregg, a Cibecue resident, was a seasonal firefighter; he told investigators he was motivated by the hope of being hired to help put the fire out. Prosecutors shocked many local residents when they opted not to charge Valinda Jo Elliot, who unleashed the Chediski Fire. She suffered no legal penalty for lighting her ill-advised signal fire.

7. Dale, et al., "True cost of wildfire." Also, "Hayman Fire case study: Summary," USDA Forest Service General Technical Report RWRS-GTR-114, 2003.

8. Dale, et al., "True cost of wildfire."

9. The prescribed burn's purpose was to "reduce hazard fuels in the burn unit while allowing fire to be restored as a keystone natural process.... This will reduce the threat

of unwanted fires moving onto nonpark lands." (Cerro Grande prescribed fire investigative report, May 18, 2000, Appendix 3, electronic copy in author's possession.) "Nonpark lands" in this context means the adjoining National Forest. Dr. Craig Allen of the USGS, who is based at Bandelier National Monument, the park unit where the prescribed fire began, explains that the "worst-case scenario" that the fire was intended to prevent was an ignition along a highway bisecting park land that prevailing winds would drive into heavy fuels on adjoining National Forest and thence to the city and laboratory (Craig Allen, personal communication, December 12, 2010).

10. "Valles Fuels Management Project: Environmental assessment," Santa Fe National Forest, USDA Forest Service, spring 2000, unpublished manuscript in author's possession. See also William deBuys, "Burned," *Conservation in Practice* 5/4 (fall 2004): 12–19.

11. For an overview of the subject, see Thomas W. Swetnam and Christopher H. Baisan, "Tree-ring reconstructions of fire and climate history in the Sierra Nevada and southwestern United States," in Thomas T. Veblen, William L. Baker, Gloria Montenegro, and Thomas W. Swetnam, eds., *Fire and climatic change in temperate ecosystems of the western Americas* (Berlin: Springer, 2002).

12. John H. Dieterich and Thomas W. Swetnam, "Dendrochronology of a fire-scarred ponderosa pine," *Forest Science* 30/1 (1984): 238–247.

13. Melissa Savage and Thomas W. Swetnam, "Early 19th-century fire decline following sheep pasturing in a Navajo ponderosa pine forest," *Ecology* 71/6 (1990): 2374–2378.

14. Julio Betancourt, personal communication, Tucson, Arizona, October 7, 2008.

15. Thomas W. Swetnam and Julio L. Betancourt, "Fire–Southern Oscillation relations in the southwestern United States," *Science* 249 (August 31, 1990): 1017–1020.

16. Anthony L. Westerling, H. G. Hidalgo, D. R. Cayan, and Thomas W. Swetnam, "Warming and earlier spring increase western U.S. forest wildfire activity," *Science* 313 (August 18, 2006): 940–943.

17. Forest carbon budgets are a subject of pressing interest and research. See April Reese, "Interior West forests on verge of becoming net carbon emitter," *Land Letter*, November 11, 2010, www.eenews.net/Landletter/2010/11/11/1/.

18. "The age of megafires," 60 *Minutes*, October 21, 2007, www.cbsnews.com/stories/2007/10/18/60minutes/main3380176.shtml?tag=contentMain;contentBody.

19. A. Park Williams, et al., "Forest responses to increasing aridity and warmth in the southwestern United States," *Proceedings of the National Academy of Sciences* 107/50 (December 14, 2010): 21289–21294.

20. Craig D. Allen, et al., "A global overview of drought and heat-induced tree mortality reveals emerging climate change risks for forests," *Forest Ecology and Management* 259 (2010): 660–684, doi:10.1016/j.foreco.2009.09.001.

21. Bradford Torrey, ed., *The writings of Henry David Thoreau: Journal, 1850–September 15, 1851* (Boston: Houghton Mifflin, 1906), 21–25. See also John Pipkin, "How a forest fire may have pushed Thoreau to Walden Pond," *Boston Globe*, April 12, 2009. Pipkin pursues this theme in his novel *Woodsburner* (New York: Nan A. Talese, 2009). Tom Swetnam points out that Thoreau aggravated his final, fatal illness by spending too much time "in the cold and snow at Walden Pond counting tree rings on stump tops."

22. Mark Twain, *Roughing It*, Chapter 23.

23. William deBuys, ed., *Seeing things whole: The essential John Wesley Powell* (Washington: Island Press), 289.

24. Ibid., 290–291.

25. Timothy Egan, *The big burn: Teddy Roosevelt and the fire that saved America* (New York: Houghton Mifflin Harcourt), 101.

26. Pulaski was blinded in one eye and all but crippled with burns. Despite the pleas of many friends, the government never compensated him for his injuries. He died embittered, not just for himself, but for many others who rendered heroic service and on whom the government disgracefully turned its back.

27. Stephen J. Pyne, *Fire in America* (Princeton, NJ: Princeton University Press, 1982), 102. Aldo Leopold, "'Piute fire' vs. forest fire prevention," *Southwestern Magazine*, 1920, reprinted in Flader and Callicot, eds., *The river of the Mother of God.*

28. Aldo Leopold, "Grass, brush, timber, and fire in Southern Arizona," *Journal of Forestry* (October 1924); "Conservationist in Mexico," *American Forests*, 1937, both articles reprinted in Flader and Callicot, eds., *The river of the Mother of God.* See also Curt Meine, *Aldo Leopold: His life and work*, 216–218.

29. Pyne, *Fire in America*, 114–119.

30. A splendid source on the mind-set of the early conservation movement is Samuel P. Hayes, *Conservation and the gospel of efficiency* (Cambridge, MA: Harvard University Press, 1959).

31. "Wildland fires: Forest Service and BLM need better information and a systematic approach for assessing the risks of environmental effects," GAO-04-705, June 24, 2004, 62–63.

32. In her 2005 M.S. thesis at Northern Arizona University, Barb Strom clearly showed that thinned forest stands fared much better than unthinned stands in the Rodeo-Chediski Fire. "Pre-fire treatment effects and post-fire forest dynamics on the Rodeo-Chediski burn area, Arizona" is available on her website, barbstrom.net.

33. Melissa Savage and Joy Nystrom Mast, "How resilient are southwestern ponderosa pine forests after crown fires?" *Canadian Journal of Forest Research* 35 (2005): 967–977.

34. These variations are called Milankovich cycles, after the Serbian engineer who identified them early in the twentieth century.

35. Keith H. Basso, *Wisdom sits in places* (Albuquerque: University of New Mexico Press, 1996), xv.

36. Keith Basso, personal communication, March 5, 2010.

CHAPTER 10

1. Peter Warshall, "Southwestern sky island ecosystems," in E. T. LaRoe, et al., eds., *Our living resources: A report to the nation on the distribution, abundance, and health of U.S. plants, animals, and ecosystems* (Washington: U.S. Department of the Interior, National Biological Service, 1995), 318–322; and "The Madrean Sky Island Archipelago: A planetary overview," in L. F. DeBano, et al., technical coordinators, *Biodiversity and management of the Madrean Archipelago: The sky islands of southwestern United States and northwestern Mexico* (USDA Forest Service, General Technical Report RM-GTR-264, 1995), 6–18.

2. Peter Warshall, personal communication, August 9, 2010, and December 29, 2010. Lawrence L. C. Jones, "Natural history and management of the Pinaleño Mountains with emphasis on Mt. Graham red squirrel habitats," in H. Reed Sanderson and John L. Koprowski, *The last refuge of the Mt. Graham red squirrel* (Tucson: University of Arizona Press, 2009).

3. In addition to Peter Warshall, John Koprowski also provided information on squirrel biology (personal communication, September 15, 2010). Another source is Mount Graham Red Squirrel Recovery Team, "Recovery plan for the Mount Graham red squirrel," draft of January 19, 2009, prepared for Region 2 U.S. Fish and Wildlife Service, Albuquerque, New Mexico, 27.

4. Robert Sullivan and Terry Yates, "Population genetics and conservation biology of relict populations of red squirrels," in C. A. Istock and R. S. Hoffmann, eds., *Storm over a mountain island* (Tucson: University of Arizona Press, 1994), 193–208.

5. Peter Warshall, "The biopolitics of the Mt. Graham red squirrel," *Conservation Biology* 8 (1994): 977–988.

6. W. L. Minckley, "Possible extirpation of the spruce squirrel from the Pinaleño (Graham) Mountains, south-central Arizona," *Journal of the Arizona Academy of Sciences* 5 (1968): 110.

7. Mount Graham Red Squirrel Recovery Team, "Recovery Plan," draft of January 16, 2009, 6.

8. John R. Ratje, "Exploring the larger environment: Astrophysical explorations from Mt. Graham," in H. Reed Sanderson and John L. Koprowski, eds., *The last refuge of the Mt. Graham red squirrel.*

9. Keith Basso, personal communication, August 24, 2010.

10. Peter Warshall, personal communication, December 29, 2010.

11. As quoted by Donald W. Carson and James W. Johnson, *Mo: The life and times of Morris K. Udall* (Tucson: University of Arizona Press, 2001).

12. Mike Spear, personal communication, August 26, 2010. Quotations elsewhere in this chapter that are attributed to Mike Spear also derive from this interview.

13. James Duffus III, director, Natural Resources Management Issues, Resources, Community, and Economic Development Division, General Accounting Office T-RCED-90-92, "Views on Fish and Wildlife Service's biological opinion addressing Mt. Graham Astrophysical Facility," June 26, 1990.

14. Warshall, "Biopolitics."

15. John Ratje, "Astrophysical explorations"; Warshall, "Biopolitics"; Dinah Bear, former General Counsel, Council on Environmental Quality, personal communication, August 23, 2010.

16. John L. Koprowski, Marit I. Alanen, and Ann M. Lynch, "Nowhere to run and nowhere to hide: Response of endemic Mt. Graham red squirrels to catastrophic forest damage," *Biological Conservation* 126 (2005): 491–498. Claire A. Zugmeyer and John L. Koprowski explore the concept of "ecological traps" in "Habitat selection is unaltered after severe insect infestation: Concern for forest-dependent species," *Journal of Mammalogy* 90/1 (2009): 175–182, and in "Severely insect-damaged forest: A temporary trap for red squirrels," *Forest Ecology and Management* 257 (2009): 464–470.

17. U.S. Fish and Wildlife Service, "Mount Graham red squirrel (*Tamiasciurus hudsonicus grahamensis*), 5-year review: Summary and evaluation," www.fws.gov/southwest/es/arizona/Documents/SpeciesDocs/MGRS/Mount%20Graham%20Red%20Squirrel%205-Year%20Review.pdf.

18. Koprowski, et al., "Nowhere to run and nowhere to hide."

19. Tom Swetnam, personal communication, August 23, 2010. Also, H. D. Grissino-Mayer, C. H. Baisan, and T. W. Swetnam, "Fire history in the Pinaleño Mountains of southeastern Arizona: Effects of human-related disturbances," in DeBano, et al., *Biodiversity and management of the Madrean Archipelago*, 1995. Acreage for Clark Peak Fire is taken from U.S. FWS, "5-year review," cited above in note 17.

20. Tom Swetnam, personal communication, August 23, 2010.

21. USDA Forest Service, Southwestern Region, "Nuttall Complex fire shelter deployment review," December 6, 2004, www.wildfirelessons.net/documents/Nuttall_Deployment_Review_Final_2004.pdf.

22. Robin Silver, Center for Biological Diversity, comments on January 16, 2009, draft MGRS Recovery Plan, in author's possession. Tom Swetnam, personal communication, August 23, 2010. In a formal consultation following the fire, the Fish and Wildlife Service concluded that "burn-out" areas would have burned in any case and that firefighting efforts, on balance, saved critical habitat. (Letter of Steven L. Spangle, Field Supervisor, USFWS, Tucson Office, to Jeanne Derby, Supervisor, Coronado National Forest, June 8, 2007, www.fws.gov/southwest/es/arizona/Documents/Biol_Opin/040299_NuttallComplexFire.pdf.)

23. Marit Alanen, lead biologist, Mt. Graham Red Squirrel Recovery Team, U.S. Fish and Wildlife Service, Tucson, Arizona, personal communication, September 9, 2010. Ms. Alanen indicated that these figures, based on work by James R. Hatten, would be included in a new and (she earnestly hoped) imminently final draft of the MGRS Recovery Plan. (See James R. Hatten, "Mapping and monitoring Mt. Graham red squirrel habitat with GIS and thematic mapper imagery," in Sanderson and Koprowski, *The Last Refuge*.)

24. Laura Koteen, "Climate change, whitebark pine, and grizzly bears in the Greater Yellowstone Ecosystem," in Schneider and Root, *Wildlife responses to climate change*.

25. John Koprowski, personal communication, September 15, 2010.

26. The Nature Conservancy has issued a number of reports on the likely effects of climate change on the biodiversity of the Southwest. The one I found most helpful was Carolyn Enquist and Dave Gori, "Implications of recent climate change for conservation priorities in New Mexico," April 2008, courtesy of Patrick McCarthy, TNC, Santa Fe, New Mexico. See also Schneider and Root, *Wildlife responses to climate change*.

27. Laura Paskus, "Rise of the she-fish?" *Santa Fe Reporter*, August 25–30, 2010.

28. Mount Graham Red Squirrel Recovery Team, "Recovery Plan," draft of January 16, 2009, 2.

29. Marit Alanen, personal communication, September 9, 2010, and intermittently thereafter.

30. "Restoration" is a term that gives some scientific semanticists conniptions. Restoration to what? they ask. How do you define "pre-settlement conditions"? How do you know what those conditions were? Etc., etc. Often the best alternative is to substitute the

less evocative word "rehabilitation" and get on with work, but the Forest Service nevertheless adheres to "restoration" as the goal of PERP.

31. See http://ag.arizona.edu/research/redsquirrel/research.html.

32. http://www.fws.gov/southwest/es/arizona/mount.htm.

33. Peter Warshall, personal communication, August 10, 2010.

## CHAPTER 11

1. Richard and Shirley Flint, preeminent scholars of the Coronado expedition, kindly confirmed that paper, not parchment, was the material of choice for official Spanish communications and record-keeping in the sixteenth century. Personal communication, August 4, 2010.

2. Richard Flint and Shirley Cushing Flint, *Documents of the Coronado Expedition, 1539–1542* (Dallas: Southern Methodist University Press, 2005), 616 ff.

3. Letter of Vásquez de Coronado to Viceroy Mendoza, August 3, 1540, in Flint and Flint, *Coronado Expedition*, 257.

4. Ibid., 257, 293.

5. "Zuni Salt Lake and Sanctuary Zone protected," press release of Zuni Salt Lake Coalition, August 7, 2003, www.minesandcommunities.org/article.php?a=1536.

6. E. Richard Hart, "The Zuni Land Conservation Act of 1980," in E. Richard Hart, editor, ed., *Zuni and the courts: A struggle for sovereign land rights* (Lawrence: University Press of Kansas, 1995), 96.

7. Geological Society of London, "The Anthropocene Epoch: Today's context for governance and public policy," 2008, at http://www.geolsoc.org.uk/gsl/views/letters.

8. The message of the suited Iagos prowling Capitol Hill is that alteration of the nation's energy economy would lead to economic tragedy. To which the Nobel-laureate economist Paul Krugman replies: "It has always been funny, in a gallows humor sort of way, to watch conservatives who laud the limitless power and flexibility of markets turn around and insist that the economy would collapse if we were to put a price on carbon. All serious estimates suggest that we could phase in limits on greenhouse gas emissions with at most a small impact on the economy's growth rate." ("Who cooked the planet?" *New York Times*, July 25, 2010, http://www.nytimes.com/2010/07/26/opinion/26krugman.html?scp=1&sq=%22Who%20cooked%20the%20planet%22&st=cse.)

9. Janet Swim, et al., "Psychology and global climate change: Addressing a multifaceted phenomenon and set of challenges," report of the American Psychological Association Task Force on the Interface between Psychology and Global Climate Change, www.apa.org/science/about/publications/climate-change.aspx, originally posted August 2009, final version March 2010.

10. Ibid., 65–68. The idea that the APA's catalogue of denial mechanisms evokes a biblical list of sins or plagues is borrowed from Andrew Revkin's blog "Dot Earth" for the *New York Times* ("Is the climate problem in our heads?" August 5, 2009, dotearth.blogs.nytimes/2009/08/05/).

11. Jon Gertner, "Why isn't the brain green?" *New York Times*, April 19, 2009.

12. Yes, this is an oversimplification, but nevertheless essentially true. The literature on use of "Pigovian tax" to reduce a market's negative externalities is large and persuasive. "Cap and trade," an alternative strategy, requires less government intervention and better avoids the opprobrium of a *tax*, but is less effective in addressing the cost uncertainty of reducing carbon emissions. The two strategies may be used in concert. An extensively documented discussion of these issues may be found at http://en.wikipedia.org/wiki/Carbon_tax.

13. Philip L. Fradkin, *Wallace Stegner and the American West* (Berkeley: University of California Press, 2008), 167.

14. Fred A. Bernstein, "The price for building a home in this town: $300,000 water meter," *New York Times*, April 13, 2010.

15. Rebecca Solnit, *A Paradise built in Hell: The extraordinary communities that arise in disaster* (New York: Viking, 2009).

16. James Workman, *Heart of dryness: How the last Bushman can help us endure the coming age of permanent drought* (New York: Walker, 2009).

17. P. C. D. Milly, et al., "Stationarity is dead: Whither water management?" *Science* 319 (2008): 573–574.

18. James H. Brown, et al., "Energetic limits to economic growth," *BioScience* 61/1 (January 2011): 19–26.

# BIBLIOGRAPHY

Adams, Henry D., Maite Guardiola-Claramonte, Greg A. Barron-Gafford, Juan Camilo Villegas, David D. Breshears, Chris B. Zou, Peter A. Troch, and Travis E. Huxman. "Temperature sensitivity of drought-induced tree mortality portends increased regional die-off under global-change-type drought." *Proceedings of the National Academy of Sciences* 106/17 (2009): 7063–7066.

Allen, Craig D. "Cross-scale interactions among forest dieback, fire, and erosion in northern New Mexico landscapes." *Ecosystems* 10 (2007): 797–808.

Allen, Craig D., and David D. Breshears. "Drought, tree mortality, and landscape change in the southwestern United States: Historical dynamics, plant–water relations, and global change implications." Unpublished manuscript, 2002.

Allen, C. D., and D. D. Breshears. "Drought-induced shift of a forest/woodland ecotone: Rapid landscape response to climate variation." *Proceedings of the National Academy of Sciences* 95 (1998): 14839–14842.

Allen, Craig D., et al. "A global overview of drought and heat-induced tree mortality reveals emerging climate change risks for forests." *Forest Ecology and Management* 259 (2010): 660–684, doi:10.1016/j.foreco.2009.09.001.

Bahre, Conrad J., and Marlyn L. Shelton. "Historic vegetation change, mesquite increases, and climate in southeastern Arizona." *Journal of Biogeography* 20/5 (September 1993): 489–504.

Bahre, Conrad J., and Marlyn L. Shelton. "Rangeland destruction: Cattle and drought in southeastern Arizona at the turn of the century." *Journal of the Southwest* 38/1 (spring 1996): 1–22.

Bailey, Vernon Orlando. 1903–1908. Unpublished field reports in Record Group 7176, Box 72, Folders 7, 9, and 15. Archives and Special Collections of the Smithsonian Institution, Washington, DC.

Bailey, Vernon Orlando. 1931. *Mammals of New Mexico*, No. 53 in the series North American Fauna, U.S. Department of Agriculture, Bureau of Biological Survey. Subsequently republished as *Mammals of the Southwestern United States*. New York: Dover, 1971.

Barnett, Tim P., and David W. Pierce. "Sustainable water deliveries from the Colorado River in a changing climate." *Proceedings of the National Academy of Sciences* 106 (2009): 7334–7338; published online before print April 20, 2009, doi:10.1073/pnas.0812762106.

Barnett, Tim P., and David W. Pierce. "When will Lake Mead go dry?" *Water Resources Research* 44 (2008): W03201, doi:10.1029/2007WR006704.

Barnett, T. P., et al. "Human-induced changes in the hydrology of the western United States." *Science* 319 (2008): 1080–1083.

Bartra, Eli, editor. *Crafting gender*. Durham, NC: Duke University Press, 2003.

Basso, Keith H. *Wisdom sits in places*. Albuquerque: University of New Mexico Press, 1996.

Blinman, Eric. "2000 years of cultural adaptation to climate change in the southwestern United States." Ambio Special Report No. 14. Stockholm: Royal Swedish Academy of Sciences, November 2008.

Bradley, Bruce A. "Pitchers to mugs: Chacoan revival at Sand Canyon Pueblo." *Kiva* 61/3 (1996): 241–255.

Bradley, Raymond S., Frank T. Keimig, and Henry F. Diaz. "Projected temperature changes along the American cordillera and the planned GCOS network." *Geophysical Research Letters* 31 (2004): L16210, doi:10.1029/2004GL020229.

Brazel, Anthony, P. Gober, L. Seung-Jae, S. Grossman-Clarke, J. Zehnder, B. Hedquist, and E. Comparri. "Determinants of changes in the regional urban heat island in metropolitan Phoenix (Arizona, USA) between 1990 and 2004." *Climate Research* 22/2 (2007): 171–182, doi:103354/cr033171.

Breshears, David D. "Drought-induced vegetation mortality and associated ecosystem responses: Examples from semiarid woodlands and forests." In Appendix D: "Understanding multiple environmental stresses: Report of a workshop board on atmospheric sciences and climate," 89–95. Board on Atmospheric Sciences and Climate, National Research Council. Washington: National Academies Press, 2007.

Breshears, David D., Travis E. Huxman, Henry D. Adams, Chris B. Zou, and Jennifer E. Davison. "Vegetation synchronously leans upslope as climate warms." *Proceedings of the National Academy of Sciences* 105/33 (August 19, 2008): 11591–11592.

Breshears, David D., et al. "Regional vegetation die-off in response to global-change type drought." *Proceedings of the National Academy of Sciences* 102 (2005): 15144–15148.

Brown, David E., and Carlos A. López González. "Notes on the occurrences of jaguars in Arizona and New Mexico." *Southwestern Naturalist* 45/4 (December 2000): 537–542. www.jstor.org/stable/3672607.

Brown, James H., et al. "Energetic limits to economic growth." *BioScience* 61/1 (January 2011): 19–26.

Buffington, Lee C., and Carlton H. Herbel. "Vegetational changes on a semidesert grassland range from 1858 to 1963." *Ecological Monographs* 35 (1965): 139–164.

Campbell, R. S. "Vegetative succession in the *Prosopis* sand dunes of southern New Mexico." *Ecology* 10/4 (October 1929): 392–398.

Carlson, John U., and Alan Boles Jr. "Contrary views of the law of the Colorado River: An examination of rivalries between the Upper and Lower Basin states." *Rocky Mountain Mineral Law Institute* 32 (1986).

Carson, Donald W., and James W. Johnson. *Mo: The life and times of Morris K. Udall*. Tucson: University of Arizona Press, 2001.

Coate, Charles. "'The biggest water fight in American history': Stewart Udall and the Central Arizona Project." *Journal of the Southwest* 37/1 (spring 1995): 79–101.

Cole, David N., et al. "Naturalness and beyond: Protected area stewardship in an era of global environmental change." *George Wright Forum* 25/1 (2008): 36–56.

Cook, Benjamin I., Ron L. Miller, and Richard Seager. "Amplification of the North American 'Dust Bowl' drought through human-induced land degradation." *Proceedings of the National Academy of Sciences* 106/13 (2009): 4997–5001; doi:10.1073/pnas.0810200106.

Cornelius, Wayne A., et al. *Mexican migration and the U.S. economic crisis: A transnational perspective*. San Diego: Center for Comparative Immigration Studies, 2010.

Dale, Lisa, and unnamed coauthors. "The true cost of wildfire in the western U.S." Lakewood, CO: Western Forestry Leadership Coalition, 2009, updated 2010. www.wflccenter.org/news_pdf/324_pdf.pdf.

Dean, Jeffrey S. "Complexity theory and sociocultural change in the American Southwest." In Roderic J. McIntosh, Joseph A. Tainter, and Susan Keech McIntosh, editors, *The way the wind blows: Climate, history, and human action*. New York: Columbia University Press, 2000.

Dean, Jeffrey S. "Demography, environment, and subsistence stress." In Joseph A. Tainter and Bonnie Bagley Tainter, editors, *Evolving complexity and environmental risk in the prehistoric Southwest*. Santa Fe Institute Studies in the Sciences of Complexity Proceedings 24. Reading: Addison Wesley, 1996.

Dean, Jeffrey S., and Gary S. Funkhouser. "Dendroclimatic reconstructions for the Southern Colorado Plateau." In W. J. Waugh, editor, *Climate change in the Four Corners and adjacent regions: Implications for environmental restoration and land-use planning*. Conf-9409325. U.S. Dept. of Energy, Grand Junction Projects Office, Grand Junction, CO, 1995, 85–104.

deBuys, William. "Burned." *Conservation in Practice* 5/4 (fall 2004): 12–19.

deBuys, William. *Salt dreams: Land and Water in low-down California*. Albuquerque: University of New Mexico Press, 1999.

deBuys, William, editor. *Seeing things whole: The essential John Wesley Powell*. Washington: Island Press, 2001.

Dieterich, John H., and Thomas W. Swetnam. "Dendrochronology of a fire-scarred ponderosa pine." *Forest Science* 30/1 (1984): 238–247.

Diffenbaugh, N. S., F. Giorgi, and J. S. Pal. "Climate change hotspots in the United States." *Geophysical Research Letters* 35 (2008): L16709, doi:10.1029/2008GL035075.

Doyel, David E., and Jeffrey S. Dean. *Environmental change and human adaptation in the ancient American Southwest*. Salt Lake City: University of Utah Press, 2006.

Dunn, David, and James P. Crutchfield. "Insects, trees, and climate: The bioacoustic ecology of deforestation and entomogenic climate change." Santa Fe Institute Working Paper 06-12-055, December 11, 2006.

Egan, Timothy. *The big burn: Teddy Roosevelt and the fire that saved America*. New York: Houghton Mifflin Harcourt, 2009.

Fair, J. M., and David D. Breshears. "Drought stress and fluctuating asymmetry in *Quercus undulata* leaves: Confounding effects of absolute and relative amounts of stress." *Journal of Arid Environments* 62 (2005): 235–249.

Feng, Shuaizhang, Alan B. Krueger, and Michael Oppenheimer. "Linkages among climate change, crop yields, and Mexico–U.S. cross-border migration." *Proceedings of the National Academy of Sciences* 107/32 (August 10, 2010): 14257–14262.

Ffolliott, Peter F., Cody L. Stropki, and Daniel G. Neary. *Historical wildfire impacts on ponderosa pine tree overstories: An Arizona case study*. Res. Pap. RMRS-RP-75. Fort Collins,

CO: U.S. Department of Agriculture, Forest Service, Rocky Mountain Research Station, 2008.

Field, Jason P., Jayne Belnap, David D. Breshears, Jason C. Neff, Gregory S. Okin, Jeffrey J. Whicker, Thomas H. Painter, Sujith Ravi, Marith C. Reheis, and Richard L. Reynold. "The ecology of dust." *Frontiers in Ecology and the Environment* 8/8 (2010): 423–430, doi:10.1890/090050.

Field, Jason P., David D. Breshears, Jeffrey J. Whicker, and Chris B. Zou. "On the ratio of wind- to water-driven transport: Conserving soil under global-change-type extreme events." *Journal of Soil and Water Conservation* 66/2 (March–April 2011): 51A–56A. doi:10.2489/jswc.66.2.51A.

Fish, Suzanne K., and Paul R. Fish, editors. *The Hohokam millennium*. Santa Fe: School for Advanced Research, 2008.

Fleming, William, and William Forbes. "Following in Leopold's footsteps: Revisiting and restoring the Rio Gavilan watershed." *Ecological Restoration* 24/1 (March 2006): 25–31

Flesch, Aaron D., Clinton W. Epps, James W. Cain III, Matt Clark, Paul R. Krausman, and John R. Morgart. "Potential effects of the United States–Mexico border fence on wildlife." *Conservation Biology* 24/1 (2010): 171–181. www3.interscience.wiley.com/journal/122465326/abstract.

Flint, Richard, and Shirley Cushing Flint. *Documents of the Coronado Expedition, 1539–1542*. Dallas: Southern Methodist University Press, 2005.

Fradkin, Philip L. *Wallace Stegner and the American West*. Berkeley: University of California Press, 2008.

Fredrickson, Ed L., R. E. Estell, A. Laliberte, and D. M. Anderson. "Mesquite recruitment in the Chihuahuan Desert: Historic and prehistoric patterns with long-term impacts." *Journal of Arid Environments* 65 (2006): 285–295.

Gammage, Grady, Jr. "Megapolitan: Arizona's Sun Corridor." Tempe, Arizona: Morrison Institute for Public Policy, Arizona State University, May 2008, updated July 2009. http://morrisoninstitute.asu.edu/morrison-update/megapolitan-arizonas-sun-corridor.

Gammage, Grady, Jr., Bruce Hallin, Jim Holway, Terri Sue Rossi, and Rich Siegel. "Superstition Vistas: Water matters." Tempe, Arizona: Morrison Institute for Public Policy, Arizona State University, July 2005. www.morrisoninstitute.org.

Geological Society of London. "The Anthropocene Epoch: Today's context for governance and public policy." 2008. www.geolsoc.org.uk/gsl/views/letters.

Glennon, Robert. *Unquenchable: America's water crisis and what to do about it*. Washington: Island Press, 2009.

Goldman, E. A. "Big Hatchet Mts. and Hatchet Ranch, Grant [now Hidalgo] Co." July 19–24, 1908. Unpublished field report in Record Group 7176, Box 74, Folder 1. Archives and Special Collections of the Smithsonian Institution, Washington, DC.

Goodwin, Neil. *The Apache diaries*. Lincoln: University of Nebraska, 2000.

Graham, Russell T., technical editor. Hayman Fire case study. Gen. Tech. Rep. RMRS-GTR-114. Ogden, UT: U.S. Department of Agriculture, Forest Service, Rocky Mountain Research Station, 2003.

Grissino-Mayer, H. D., C. H. Baisan, and T. W. Swetnam. "Fire history in the Pinaleno Mountains of southeastern Arizona: Effects of human-related disturbances." In Leonard F. DeBano, Gerald J. Gottfried, Robert H. Hamre, Carleton B. Edminster, Peter F. Ffolliott, and Alfredo Ortega-Rubio, technical coordinators, *Biodiversity and management of the Madrean Archipelago: The sky islands of southwestern United States and northwestern Mexico.* RM-GTR-264. Fort Collins, CO: Rocky Mountain Forest and Range Experiment Station, 1995.

Grover, Herbert D., and Brad Musick. "Shrubland encroachment in southern New Mexico, U.S.A.: An analysis of desertification processes in the American Southwest." *Climatic Change* 17 (1990): 305–330.

Guhathakurta, Subhrajit, and Patricia Gober. "The impact of the Phoenix urban heat island on residential water use." *Journal of the American Planning Association* 73/3 (summer 2007): 317–329.

Gunderson, Lance H., C. S. Holling, and Stephen S. Light, editors. *Barriers and bridges to the renewal of ecosystems and institutions.* New York: Columbia University Press, 1995.

Hall, Edward T. *West of the thirties.* New York: Doubleday, 1994.

Hansen, James, Makiko Sato, Pushker Kharecha, David Beerling, Robert Berner, Valerie Masson-Delmotte, Mark Pagani, Maureen Raymo, Dana L. Royer, and James C. Zachos. "Target atmospheric $CO_2$: Where should humanity aim?" *Open Atmospheric Science Journal* 2 (2008): 217–231.

Harlan, Sharon L., Anthony J. Brazel, Lela Prashad, William L. Stefanov, and Larissa Larsen. "Neighborhood microclimates and vulnerability to heat stress." *Social Science and Medicine* 63 (2006): 2847–2863.

Hart, E. Richard, editor. *Zuni and the courts: A struggle for sovereign land rights.* Lawrence: University Press of Kansas, 1995.

Hatfield, Shelly Bowen. *Chasing shadows: Indians along the United States–Mexico border, 1876–1911.* Albuquerque: University of New Mexico Press, 1998.

Havstad, Kris, Laura F. Huenneke, and William H. Schlesinger, editors. *Structure and function of a Chihuahua Desert ecosystem: The Jornada Basin Long-Term Ecological Research Site.* New York: Oxford University Press, 2006.

Hayes, Samuel P. *Conservation and the gospel of efficiency.* Cambridge, MA: Harvard University Press, 1959.

Hightower, Mike, and Suzanne A. Pierce. "The energy challenge." *Nature* 452/20 (March 2008): 285–286.

Hill, J. Brett, Jeffery J. Clark, William H. Doelle, and Patrick D. Lyons. "Prehistoric demography in the Southwest: Migration, coalescence and Hohokam population decline." *American Antiquity* 69/4 (2004): 689–716.

Indermühle, A., et al. "Holocene carbon-cycle dynamics based on $CO_2$ trapped in ice at Taylor Dome, Antarctica." *Nature* 38 (March 11, 1999): 121–126.

Jimenez, Maria. "Humanitarian crisis: Migrant deaths at the U.S.–Mexico border." A report for the American Civil Liberties Union and Mexico's National Commission of Human Rights, October 1, 2009. www.aclu.org/immigrants-rights/humanitarian-crisis-migrant-deaths-us-mexico-border.

Jornada Experimental Range. "A laboratory without walls." General information booklet. Las Cruces, NM: USDA Agricultural Research Service, Jornada Experimental Range, 1999.

Kantner, John. *Ancient Puebloan Southwest.* New York: Cambridge University Press, 2004.

Katz, Friedrich. *The life and times of Pancho Villa.* Stanford, CA: Stanford University Press, 1998.

Kenny, Joan F., Nancy L. Barber, Susan S. Hutson, Kristin S. Linsey, John K. Lovelace, and Molly A. Maupin. "Estimated use of water in the United States in 2005." U.S. Geological Survey Circular 1344. 2009.

Kerr, Richard A. "Climate change hot spots mapped across the United States." *Science* 321 (August 15, 2008): 909.

Kilmer, Beau, Jonathan P. Caulkins, Brittany M. Bond, and Peter H. Reuter. "Reducing drug trafficking revenues and violence in Mexico: Would legalizing marijuana in California help?" RAND Corporation Occasional Paper 2010. www.rand.org/pubs/occasional_papers/2010/RAND_OP325.pdf.

Kohler, Timothy A., Mark D. Varien, and Aaron Wright, editors. *Time of peril, time of change: Explaining thirteenth-century Pueblo migration.* Tucson: University of Arizona Press, 2010.

Kohler, Timothy A., Mark D. Varien, Aaron M. Wright, and Kristin Kuckelman. "Mesa Verde migrations." *American Scientist* 96 (2008): 146–153.

Koprowski, John L., Marit I. Alanen, and Ann M. Lynch. "Nowhere to run and nowhere to hide: Response of endemic Mt. Graham red squirrels to catastrophic forest damage." *Biological Conservation* 126 (2005): 491–498.

Kuckelman, Kristin, editor. *The archaeology of Sand Canyon Pueblo: Intensive excavations at a late-thirteenth-century village in southwestern Colorado.* Cortez, CO: Crow Canyon Archaeological Center, 2007. Electronic documents and database at www.crowcanyon.org/sandcanyon.

Kuckelman, Kristin A., Ricky R. Lightfoot, and Debra L. Martin. "The bioarchaeology and taphonomy of violence at Castle Rock and Sand Canyon Pueblos, southwestern Colorado." *American Antiquity* 67/3 (2002): 486–513.

Lamar, Howard R. *The far Southwest, 1846–1912: A territorial history.* New York: Norton, 1966, 1970.

Leinberger, Christopher B. "The next slum." *The Atlantic,* March 2008. www.theatlantic.com/doc/200803/subprime.

Leopold, Aldo. *The river of the Mother of God and other essays.* Susan L. Flader and J. Baird Callicot, editors. Madison: University of Wisconsin Press, 1991.

Leopold, Aldo. *A Sand County almanac and sketches here and there.* New York: Oxford University Press, 1949, 1968.

Lipe, William D., editor. *The Sand Canyon Archaeological Project: A progress report.* Crow Canyon Archaeological Center Occasional Paper No. 2. Cortez, CO: Crow Canyon Archaeological Center, 1992.

List, Rurik, Gerardo Ceballos, Charles Curtin, Peter J. P. Gogan, Jesús Pacheco, and Joe Truett. "Historic distribution and challenges to bison recovery in the northern Chihuahuan Desert." *Conservation Biology* 21/6 (2007): 1487–1494, doi:10.1111/j.1523–1739.2007.00810.x.

List, Rurik, Jesús Pacheco, Eduardo Ponce, Rodrigo Sierra-Corona, and Gerardo Ceballos. "The Janos Biosphere Reserve, northern Mexico." *International Journal of Wilderness* 16/2 (August 2010): 35–41.

Logan, Michael F. *Desert cities: The environmental history of Phoenix and Tucson.* Pittsburgh: University of Pittsburgh Press, 2006.

Lu, Jian, Gabriel Vecchi, and Thomas Reichler. "Expansion of the Hadley cell under global warming." *Geophysical Research Letters* 34 (2007): L06805.

Lydgate, Joanna. "Assembly-line justice: A review of Operation Streamline." Policy brief, Warren Institute on Race, Ethnicity & Diversity, University of California, Berkeley School of Law, January 2010.

McAfee, Stephanie A., and Joellen L. Russell. "Northern Annular Mode impact on spring climate in the western United States." *Geophysical Research Letters* 35 (2008): L17701, doi:10.10299/2008GL034828.

Meine, Curt. *Aldo Leopold: His life and work.* Madison: University of Wisconsin Press, 1988.

Meko, David M., Connie A. Woodhouse, Christopher A. Baisan, Troy Knight, Jeffrey J. Lukas, Malcolm K. Hughes, and Matthew W. Salzer. "Medieval drought in the upper Colorado River Basin." *Geophysical Research Letters* 34 (2007): L10705, doi:10.1029/2007GL029988.

Meyers, Charles J. "The Colorado River." *Stanford Law Review* 19 (November 1966): 1–75.

Meyers, Charles J., and Richard L. Noble. "The Colorado River: The treaty with Mexico." *Stanford Law Review* 19 (1966–1967): 367–419.

Milly, P. C. D., Julio Betancourt, Malin Falkenmark, Robert M. Hirsch, Zbigniew W. Kundewicz, Dennis P. Lettenmaier, and Ronald J. Stouffer. "Stationarity is dead: Whither water management?" *Science* 319 (2008): 573–574.

Milly, P. C. D., K. A. Dunne, and A. V. Vecchia. "Global pattern of trends in streamflow and water availability in a changing climate." *Nature* 438 (November 17, 2005): 347–350.

Minckley, W. L. "Possible extirpation of the Spruce Squirrel from the Pinaleño (Graham) Mountains, south-central Arizona." *Journal of the Arizona Academy of Sciences* 5 (1968): 110.

Morgan, Brandon. "From brutal ally to humble believer: Mormon colonists' image of Pancho Villa." *New Mexico Historical Review* 85/2 (2010): 109–129.

Mount Graham Red Squirrel Recovery Team. "Recovery plan for the Mount Graham red squirrel." Draft of January 19, 2009. Prepared for Region 2 U. S. Fish and Wildlife Service, Albuquerque, NM.

Muro, Mark, Robert E. Lang, and Andrea Sarzynski. "Mountain Megas: America's newest metropolitan places and a federal partnership to help them prosper." Washington: The Brookings Institution Metropolitan Policy Program, 2008. www.brookings.edu/metro/intermountain_west.aspx.

Nadal, Alejandro. "The environmental and social impacts of economic liberalization on corn production in Mexico." A report for Oxfam GB and WWF International, September 2000. http://ase.tufts.edu/gdae/Pubs/rp/NadalOxfamWWFMaizeMexico2000.pdf.

National Research Council. "Colorado River Basin water management." Washington: National Academies Press, 2007.

Nevins, Joseph. *Operation Gatekeeper: The rise of the illegal alien and the making of the U.S.–Mexico boundary.* New York: Routledge, 2002.

Noble, David G., editor. *In search of Chaco: New approaches to an archaeological enigma.* Santa Fe: School of American Research Press, 2004.

Ortman, Scott G., Mark D. Varien, and T. Lee Gripp. "Empirical Bayesian methods for archaeological survey data: An application for the Mesa Verde region." *American Antiquity* 72/2 (2007): 241–272.

Overpeck, Jonathan, and Bradley Udall. "Dry times ahead." *Science* 328 (June 25, 2010): 1642–1643.

Painter, Thomas H., Jeffrey S. Deems, Jayne Belnap, Alan F. Hamlet, Christopher C. Landry, and Bradley Udall. "Response of Colorado River runoff to dust radiative forcing in snow." *Proceedings of the National Academy of Sciences* 107/40 (October 5, 2010): 17125–17130.

Perramond, Eric P. *Political ecologies of cattle ranching in northern Mexico.* Tucson: University of Arizona Press, 2010.

Petersen, K. L. "Climate and the Dolores River Anasazi." University of Utah Anthropological Papers, 11/3. Salt Lake City: University of Utah Press, 1988.

Powell, Fred Wilbur. *The railroads of Mexico.* Boston: Stratford, 1921.

Powell, James Lawrence. *Dead pool: Lake Powell, global warming, and the future of water in the West.* Berkeley: University of California Press, 2008.

Poynter, Jane. *The human experiment: Two years and twenty minutes inside Biosphere II.* New York: Basic Books, 2006.

Pyne, Stephen J. *Fire in America: A cultural history of wildland and rural fire.* Princeton, NJ: Princeton University Press, 1982.

Qian, Yun, William I. Gustafson Jr., L. Ruby Leung, and Steven J. Ghan. "Effects of soot-induced snow albedo change on snowpack and hydrological cycle in western United States based on weather research and forecasting chemistry and regional climate simulations." *Journal of Geophysical Research* 114 (2009): D03108, doi:10.1029/2008JD011039.

Raffa, Kenneth F., Brian H. Aukema, Barbara J. Bentz, Allan L. Carroll, Jeffrey A. Hicke, Monica G. Turner, and William H. Romme. "Cross-scale drivers of natural disturbances prone to anthropogenic amplification: The dynamics of bark beetle eruptions." *BioScience* 58/6 (June 2008): 501–517.

Rajagopalan, Balaji, K. Nowak, J. Prairie, M. Hoerling, B. Harding, J. Barsugli, A. Ray, and B. Udall. "Water supply risk on the Colorado River: Can management mitigate?" *Water Resources Research* 45 (2009): W08201, doi:10.1029/2008WR007652.

Rauscher, S. A., J. S. Pal, N. S. Diffenbaugh, and M. M. Benedetti. "Future changes in snowmelt-driven runoff timing over the western U.S." *Geophysical Research Letters* 35 (2008): L16703, doi:10.1029/2008GL034424.

Reisner, Mark. *Cadillac desert: The American West and its disappearing water.* New York: Viking Penguin, 1986.

Salzer, Matthew W. "Temperature variability and the Northern Anasazi: Possible implications for regional abandonment." *Kiva* 65 (2002): 295–318.

Sanderson, H. Reed, and John L. Koprowski. *The last refuge of the Mt. Graham red squirrel: Ecology of endangerment.* Tucson: University of Arizona Press, 2009.

Savage, Melissa, and Joy Nystrom Mast. "How resilient are southwestern ponderosa pine forests after crown fires?" *Canadian Journal of Forest Research* 35 (2005): 967–977.

Savage, Melissa, and Thomas W. Swetnam. "Early 19th-century fire decline following sheep pasturing in a Navajo ponderosa pine forest." *Ecology* 71/6 (1990): 2374–2378.

Schlyer, Krista. "The lost herd of Janos-Hidalgo." *Wildlife Conservation* (January–February 2009): 47–55.

Schneider, Stephen H., and Terry L. Root. *Wildlife responses to climate change: North American case studies.* Washington: Island Press, 2002.

Schubert, Siegfried, Max J. Suarez, Philip J. Pegion, Randal D. Koster, and Julio T. Bacmeister. "On the cause of the 1930s Dust Bowl." *Science* 303 (March 19, 2004): 1855–1859.

Schwartzlose, Richard. "Mormon settlements in Mexico." 1952. http://thecardonfamilies.org/Documents/MormonSettlementsInMexico.

Seager, R., M. F. Ting, M. Davis, M. A. Cane, N. Naik, J. Nakamura, C. Li, E. Cook, and D. W. Stahle. "Mexican drought: An observational, modeling and tree ring study of variability and climate change." *Atmosfera* 22/1 (2009): 1–31.

Seager, Richard, et al. "Model projections of an imminent transition to a more arid climate in southwestern North America." *Science* 316 (May 25, 2007): 1181–1184.

Sheridan, Thomas E. *Arizona: A history.* Tucson: University of Arizona Press, 1995.

Solnit, Rebecca. *A Paradise built in Hell: The extraordinary communities that arise in disaster.* New York: Viking, 2009.

Stockton, C. W., and G. C. Jacoby Jr. "Long-term surface-water supply and streamflow trends in the Upper Colorado River Basin based on tree-ring analysis." *Lake Powell Research Project Bulletin 18*, Institute of Geophysics and Planetary Physics, University of California, Los Angeles, 1976.

Sullivan, Robert, and Terry Yates. "Population genetics and conservation biology of relict populations of red squirrels." In C. A. Istock and R. S. Hoffmann, editors, *Storm over a mountain island*, 193–208. Tucson: University of Arizona Press, 1994.

Swetnam, Thomas W., Craig D. Allen, and Julio L. Betancourt. "Applied historical ecology: Using the past to manage for the future." *Ecological Applications* 9/4 (1999): 1189–1206.

Swetnam, Thomas W., and Christopher H. Baisan. "Tree-ring reconstructions of fire and climate history in the Sierra Nevada and southwestern United States." In Thomas T. Veblen, William L. Baker, Gloria Montenegro, and Thomas W. Swetnam, editors, *Fire and climatic change in temperate ecosystems of the western Americas.* New York: Springer, 2002.

Swetnam, Thomas W., and Julio L. Betancourt. "Fire–Southern Oscillation relations in the southwestern United States." *Science* 249 (August 31, 1990): 1017–1020.

Swim, Janet, Susan Clayton, Thomas Doherty, Robert Gifford, George Howard Joseph Reser, Paul Stern, and Elke Weber. "Psychology and global climate change: Addressing a multifaceted phenomenon and set of challenges." Report of the American Psychological Association Task Force on the Interface between Psychology and Global Climate Change, www.apa.org/science/about/publications/climate-change.aspx, originally posted August 2009, final version March 2010.

Tanner, Thomas, editor. *Aldo Leopold: The man and his legacy.* Ankeny, IA: Soil Conservation Society of America, 1987.

Torrey, Bradford, editor. *The writings of Henry David Thoreau: Journal, 1850–September 15, 1851.* Boston: Houghton Mifflin, 1906.

Twain, Mark. *Roughing it.* New York: New American Library, 1962.

Udall, Stewart. *The forgotten founders: Rethinking the history of the Old West.* Washington: Island Press, 2002.

USDA Forest Service. "Forest insect and disease conditions in the Southwestern Region, 2003." Publication PR-R3–04–02. Albuquerque: U.S. Forest Service, Southwestern Region, NM, 2004. www.fs.fed.us/r3/resources/health.

USDA Forest Service. "Initial attack progression of Pina, Rodeo and Chediski Fires." Rodeo-Chediski Fire official website. No date. www.wmat.us/narrative.html.

U.S. Government Accountability Office. "Firearms trafficking: U.S. efforts to combat arms trafficking to Mexico face planning and coordination challenges." Report No. GAO-09-709, June 2009.

U.S. Government Accountability Office. "Secure Border Initiative fence construction costs." Report No. GAO-09-244R. January 20, 2009.

Van Auken, O. W. "Shrub invasions of North American semiarid grasslands." *Annual Review of Ecology and Systematics* 31 (2000): 197–215.

Varien, Mark D., Scott G. Ortman, Timothy A. Kohler, Donna M. Glowacki, and C. David Johnson. "Historical ecology in the Mesa Verde region: Results from the Village Ecodynamics Project." *American Antiquity* 72/2 (2007): 273–299.

Varien, Mark D., and Richard H. Wilhusen, editors. *Seeking the center place: Archaeology and ancient communities in the Mesa Verde region.* Salt Lake City: University of Utah Press, 2002.

Walters, Harry, and Hugh Rogers. "Anasazi and 'Anaasází: Two worlds, two cultures." *Kiva* 66/3 (2001): 317–326.

Warshall, Peter. "The biopolitics of the Mount Graham red squirrel." *Conservation Biology* 8 (1994): 977–988.

Warshall, Peter. "The Madrean sky island archipelago: A planetary overview." In L. F. DeBano, P. F. Ffolliott, A. Ortega-Rubio, G. J. Gottfried, H. Robert, and C. B. Edminster, technical coordinators, *Biodiversity and management of the Madrean Archipelago: The sky islands of southwestern United States and northwestern Mexico,* 6–18. USDA Forest Service, General Technical Report RM-GTR-264, 1995.

Warshall, Peter. "Southwestern sky island ecosystems." In E. T. LaRoe, G. S. Farris, C. E. Puckett, P. D. Doran, and M. J. Mac, editors, *Our living resources: A report to the nation on the distribution, abundance, and health of U.S. plants, animals, and ecosystems,* 318–322. U.S. Department of the Interior, National Biological Service, Washington, DC, 1995.

Water Education Foundation. "Colorado River Project: Symposium Proceedings, Sept. 16–18, 2009." Sacramento, CA: Water Education Foundation, 2010.

Webb, Robert H. *Grand Canyon, a century of change: Rephotography of the 1889–1890 Stanton Expedition.* Tucson: University of Arizona Press, 1996.

Weber, David J. *New Spain's far northern frontier.* Albuquerque: University of New Mexico Press, 1979.

Westerling, Anthony L., H. G. Hidalgo, D. R. Cayan, and Thomas W. Swetnam. "Warming and earlier spring increase western U.S. forest wildfire activity." *Science* 313 (August 18, 2006): 940–943.

Wilkinson, Charles. *Fire on the plateau.* Washington: Island Press, 1999.

Williams, A. Park, Craig D. Allen, Constance I. Millar, Thomas W. Swetnam, Joel Michaelsen, Christopher J. Still, and Steven Leavitt. "Forest responses to increasing aridity and warmth in the southwestern United States." *Proceedings of the National Academy of Sciences* 107/50 (December 14, 2010): 21289–21294.

Woodhouse, Connie A., and Jonathan T. Overpeck. "2000 years of drought variability in the central United States." *Bulletin of the American Meteorological Society* 79 (1998): 2693–2714.

Wooton, E. O. *The range problem in New Mexico.* New Mexico Agricultural Experiment Station Bulletin 66, 1908.

Workman, James. *Heart of dryness: How the last Bushman can help us endure the coming age of permanent drought.* New York: Walker, 2009.

Worster, Donald. *Nature's economy: The roots of ecology.* San Francisco: Sierra Club Books, 1977.

Zugmeyer, Claire A., and John L. Koprowski. "Habitat selection is unaltered after severe insect infestation: Concern for forest-dependent species." *Journal of Mammalogy* 90/1 (2009): 175–182.

Zugmeyer, Claire A., and John L. Koprowski. "Severely insect-damaged forest: A temporary trap for red squirrels." *Forest Ecology and Management* 257 (2009): 464–470.

# INDEX

Illustrations indicated by letter *i*, maps by letter *m*, boxes by letter *b*.

CPSIA information can be obtained
at www.ICGtesting.com
Printed in the USA
BVHW01s0535071217
502187BV00005B/14/P